FUTURE ASPECTS OF TUMOR SUPPRESSOR GENE

Edited by **Yue Cheng**

Future Aspects of Tumor Suppressor Gene

http://dx.doi.org/10.5772/56471
Edited by Yue Cheng

Contributors

Kiyoshi Ohtani, Yue Cheng, Fani Papagiannouli, Bernard M. Mechler, Hitoshi Yagisawa, Nelly Etienne-Selloum, Israa Dandache, Tanveer Sharif, Cyril Auger, Valérie Schini-Kerth, Yu-Wen Zhang, George Vande Woude, Nobuko Mori, Damjan Glavac

Published by InTech

Janeza Trdine 9, 51000 Rijeka, Croatia

Copyright © 2013 InTech

Notice

Publishing Process Manager Danijela Duric
Technical Editor InTech DTP team
Cover InTech Design team

Additional hard copies can be obtained from orders@intechopen.com

Future Aspects of Tumor Suppressor Gene, Edited by Yue Cheng
p. cm.
ISBN 978-953-51-1063-7

For EU product safety concerns:
IN TECH d.o.o., Prolaz Marije Krucifikse Kozulić 3, 51000 Rijeka, Croatia,
info@intechopen.com or visit our website at intechopen.com.

Meet the editor

Dr. Yue Cheng is currently working in the University of Hong Kong as an active researcher in cancer biology, molecular genetics, and stem cell biology. He obtained his Ph.D. degree from the Hong Kong University of Science & Technology in cancer biology. He began his research career on TSGs in Dr. Eric Stanbridge's laboratory at the University of California at Irvine. For a long time, he has been interested in the study of nasopharyngeal carcinoma and made landmark achievements by studying tumor suppressive activities in this cancer. His seminal works provided functional evidence of a tumor suppressive region mapped on the chromosome 3p21.3, which was subsequently confirmed as a TSG cluster region in human genome. During his career at National Cancer Institute of USA, he started to be interested in the stem cell biology and mechanisms of relevant signaling regulation in these cells. His recent works focus on how Wnt pathway is a critical signaling in the control of TSG networks, epithelial-mesenchymal transition, and tumor development in human cells.

Contents

Preface

Losses of specific chromosomal regions are frequently reported in different human tumors, suggesting that these regions may contain important genes associated with tumor development. Cell fusion studies provided the first functional evidence for a class of negatively-acting tumor suppressor genes (TSGs) harbored on certain human chromosomes. Based on Knudson's "two-hit hypothesis", the first TSG, RB was identified. Since 1980s, many TSGs have been discovered by using different approaches. Accumulated knowledge indicates that TSGs not limited to tumor suppression play critical roles in various biological activities in human cells.

In 20132, InTech published a book called "Tumor Suppressor Genes", which covers the most important fields, from cell cycle control, signaling pathways, epigenetic regulation, and current challenges to therapeutic applications of known TSGs. Some well-studied TSGs, such as p53 and p16, and their regulatory mechanisms in tumor development are addressed in this book. However, TSG research is a fast growing area, and many novel approaches and findings have been discovered recently. Therefore, it is necessary to publish a new open access book that may provide future directions for TSG studies.

This book, "Future Aspects of Tumor Suppressor Genes", contains some important areas that were not mentioned in the previous book. The majority of known TSGs were identified from hereditary tumor syndromes. However, more than 90% of human tumors are sporadic cases, so it is always a challenge to identify tumor susceptibility loci in sporadic tumors. Using animal models, authors in this book investigated whether strain-specific allele loss was an important clue to identify tumor suppressors involved in tumor susceptibility, which should be interesting to many researchers. Other basic researches contain investigations of several TSG signaling pathways from different laboratories: START-GAP/DLC family proteins and their molecular pathways involved in the control of cell growth, E2F-mediated tumor suppressive mechanism associated with RB, p53, ARF, p27Kip1 and TAp73 transcription factors, and TSGs in the regulation of receptor tyrosine kinase signaling via a negative feedback loop. Understanding these signaling regulatory mechanisms may lead to findings of molecular targets for cancer therapy. In recent years, it has been well-accepted that microRNAs are an abundant class of endogenous small RNA molecules that can regulate tumor development. To reflect the trends of these novel researches, authors in this book present an extensive review for current knowledge of microRNAs that play in the control of tumor growth and therapeutic application.

This book also includes some other fascinating fields and emerging subjects in TSG studies. For example, the application of Drosophila as a special model for tumor suppression studies is addressed, and future directions used for the pharmacological screening and therapy strategies are also proposed. Natural compounds, such as polyphenols, interfere with the initiation and progression of cancer development via multiple TSG pathways. Recent evidence, demonstrating that these compounds are able to modulate various cellular activities, such as cell cycle arrest, anti-angiogenesis, and metastasis suppression, are summarized in the relevant chapter. Finally, the regulatory role of TSGs, such as p16, p53 and RB, in cell reprogramming, stemness transition process, and signaling networks of these genes during these cellular processes are extensively reviewed, which indicates that TSGs are actively involved in many aspects of stem cell biology and regenerative medicine.

I would like take this opportunity to express my gratitude to all authors and InTech staff for their contributions in this publication project, and I hope that this book will be helpful for students, researchers and clinicians.

Yue Cheng, PhD
Department of Clinical Oncology
The University of Hong Kong

Chapter 1

Strain-Specific Allele Loss: An Important Clue to Tumor Suppressors Involved in Tumor Susceptibility

Nobuko Mori and Yoshiki Okada

Additional information is available at the end of the chapter

http://dx.doi.org/10.5772/55159

1. Introduction

Development of tumors is controlled by multiple genes such as cellular oncogenes and tumor suppressors activated or inactivated by somatic mutations and/or epigenetic mechanisms. Tumor development is also controlled by heritable factors as well as environmental factors, i. e., diet, oxidative stress and sustained inflammation, as reviewed by a large number of recent reports [1-12]. Both heritable and environmental factors are important targets for clinical controls and prevention of cancers.

Heritable factors underlying cancer risks have been identified in familial cancer-prone pedigrees. In the pedigree members, tumors develop in a Mendelian dominant inheritance fashion. Breast cancer 1, early onset (*BRCA1*) encoding a nuclear phosphoprotein that plays a role in maintaining genomic stability is one of the heritable cancer risk factors hitherto identified. Women bearing a mutated *BRCA1* allele are at high risk for both breast and ovarian cancers through their lifespan. According to the recent estimations, average cumulative risks in *BRCA1*-mutation carriers by age 70 years are 65% (95% confidence interval 44%–78%) for breast cancer and 39% (18%–54%) for ovarian cancer [13]. Thus, disease penetrance is incomplete, albeit rather high, in the mutated-*BRCA1* carriers. The *BRCA1* gene maps to human chromosome 17q21, where frequent loss of heterozygosity (LOH) is observed in both familial and sporadic breast cancers. Although tumors developed in the *BRCA1*-mutation carriers are homozygous for the defective *BRCA1* allele via LOH mechanisms, sporadic cases rarely show mutation in the *BRCA1* gene [14]. The *BRCA1* gene may rather undergo inactivation via epigenetic mechanisms such as DNA methylation in sporadic tumors.

Unlike the *BRCA1* case, tumor susceptibility is expressed in a non-Mendelian inheritance manner, because multiple genes with incomplete penetrance participate in the phenotype. Moreover, tumor susceptibility alleles may occasionally express genetic interaction, i. e., epistasis that hides or enhances the effect of some alleles at some susceptibility loci with the effect of other alleles at other susceptibility loci [15]. Despite growing number of association studies localizing tumor susceptibility loci exploiting SNPs in humans [16, 17], validation of these loci in human population with miscellaneous variations in the genetic background might be an intractable task.

Several strains of mice with different susceptibility to lymphomagenesis so far reported might be useful in the study of tumor susceptibility. Using genetic crosses between BALB/cHeA (refer to as BALB/c, hereafter) and STS/A (refer to as STS) mice with different tumor susceptibility, and between the BALB/c and recombinant congenic CcS/Dem strains of mice with 12.5% STS and 87.5% BALB/c allele in the genome, we mapped three loci controlling susceptibility to radiation-induced apoptosis of thymocytes to chromosomes 16, 9 and 3 [18, 19], and two loci for susceptibility to lymphomagenesis to chromosome 4 [20]. We identified the protein kinase, DNA activated, catalytic polypeptide (*Prkdc*) as a candidate for the apoptosis susceptibility gene mapped to chromosome 16, which was also associated with susceptibility to radiation lymphomagenesis [21]. As indicated by our studies, susceptibility to apoptosis as well as lymphomagenesis is controlled by multiple genes. To analyze the effect of one gene involved in such multigenic traits, congenic animals are ordinarily used. We are currently analyzing the genes controlling susceptibility to lymphomagenesis on chromosome 4 by the use of congenic animals.

In this chapter, we initially review recent advances in the research of tumor susceptibility, in particular, susceptibility to radiation lymphomagenesis in mice, and show that two loci controlling radiation lymphomagenesis map to chromosome 4. Then, we show that two types of allele loss, i. e., loss common to lymphoma and parental strain-specific loss, occur in radiation-induced lymphomas from various F_1 hybrids between strains with different lymphoma susceptibility. We show that LOH on chromosome 4 in F_1 hybrids between BALB/c and STS occurs in a strain-specific manner and exhibits a bias towards the STS allele loss. At the close, by exploiting congenic strains of mice containing different segments of chromosome 4 from the donor strain STS on the BALB/c background, we present a concordance between the allele loss region and a lymphoma susceptibility locus area on chromosome 4, where the BALB/c mouse harbors a hypomorphic allele of *Cdkn2a*. Significance of the strain-specific allele loss in probing tumor susceptibility loci will be discussed.

2. Mouse strain difference in susceptibility to radiation-induced lymphomagenesis

In laboratory strains of mice irradiated by ionizing radiation according to a well-established protocol, development of lymphomas starts around three months after the exposure to radiation and is terminated around ten months. Radiation-induced lymphomas are mostly of thymic origin. Several laboratory strains of mice such as BALB/c and C57BL reside in *Mus musculus musculus*, and are known to be highly susceptible to radiation-induced lymphomagenesis, while other strains STS and MSM/Ms (refer to as MSM) are not [22, 23]. The BALB/cHeA and STS/A strains of mice are originally provided by Dr. J. Hilgers at the Netherlands

Cancer Institute [22], and maintained more than twenty generations at the animal facility of Osaka Prefecture University. The BALB/cHeOpu mouse is the direct descendant of the BALB/cHeA mouse [20]. The MSM/Ms strain of mice belongs to a subspecies *Mus musculus molossinus*. Its progenitor was trapped in Mishima-city, Shizuoka, Japan and established as an inbred strain at the National Institute of Genetics (Mishima, Japan). Mice were exposed to 4 x 1.7 Gy of X rays using a well-established protocol for radiation-induced lymphomagenesis. Tumor development was observed during one year. The results were summarized in Table 1. BALB/cHeA mice developed lymphomas at high frequency (33/43, 77%), while STS/A mice develop tumors at less than 10% of frequencies [22]. The onset of tumor development was around three months in both strains. On the other hand, one of 30 MSM/Ms mice developed lymphoma with more than ten months of latency. Lymphomas occurred at high frequency (30/35, 86%) in BALB/cHeOpu mice subjected to X-ray irradiation using the same protocol. Thus, the pattern of tumor development in BALB/cHeOpu mice showed good concordance with that in their progenitor BALB/cHeA mice [20].

Strain of mice	Number of irradiated[a]	Number of affected (%)[b]	Reference
BALB/cHeA	43	33 (77%)	[22]
BALB/cHeOpu	35	30 (86%)	[20]
STS/A	60	5 (8%)	[22]
MSM/Ms	30	1 (3%)	[23]

[a]Only females.

[b]Cumulative incidence of lymphomas within one year in BALB/cHeA, STS/A and MSM/Ms, and within ten months in BALB/cHeOpu.

Table 1. Strain difference in susceptibility to radiation lymphomagenesis

3. Current status of the studies on tumor susceptibility in mice

Numerous tumor susceptibility loci have been mapped by analyzing genetic crosses between strains of mice exhibiting different tumor susceptibility [17]. Several genes responsible for tumor susceptibility have been identified, some of which are validated by supporting evidences: *Pla2g2a* encoding phospholipase A2, group IIA (platelets, synovial fluid), for the modifier of *Min1* (*APC*^{Min})-induced intestinal tumors (*Mom1*) identified in the distal portion of chromosome 4 [24–26]; cyclin-dependent kinase inhibitor 2A (*Cdkn2a*) encoding a tumor suppressor p16, for pristen-induced plsmacytoma resistance1 (*Pctr1*) mapped in the middle of chromosome 4 [27, 28]; protein tyrosine phosphatase, receptor type, J, (*Ptprj*), for susceptibility to colon cancer 1 (*Scc1*) on chromosome 2 [29, 30]. LOH occurs at *PTPRJ*, the human homolog of mouse *Ptprj* (*Scc1*), in the early stage of human colorectal cancer [31]. Hence, *PTPRJ* may play a role in tumor suppression in humans. The biological function of *Pla2g2a* (*Mom1*) differs from other tumor susceptibility genes so far identified. *Pla2g2a* plays a role in physiological processes such as anti-bacterial defense, inflammation and eicosanoid generation, which are preferable targets of medical controls for cancer prevention.

Despite the availability of strains of mice with obvious difference in susceptibility to radiation lymphomagenesis, it is much difficult to analyze such traits as to be expressed in a binominal fashion (tumor-free survivals of animals after exposure to radiation). However, there is one successful case: a suggestive linkage near D4Mit12 at 57.8 centimorgan (cM) position on chromosome 4 with susceptibility to radiation lymphomagenesis, which was detected in the genetic cross between BALB/c and MSM, is confirmed by exploiting congenic mice with the MSM allele at D4Mit12 on the BALB/c background [32, 33]. Because BALB/c mice had a hypomorphic allele at the *Mtf1* locus, they reported the metal-responsive transcription factor-1 (*Mtf1)* gene as the candidate gene for the susceptibility locus near D4Mit12 [32, 34]. *Mtf1* activates expression of metallothionein I and II genes as well as gamma-glutamylcysteine synthetase, a key enzyme for glutathione biosynthesis, and metallothionein and glutathione are involved in detoxification processes, such as scavenging reactive oxygen intermediates generated by ionizing radiation. Reduced reactivity of Mtf-1 retains an increased level of ROS in the BALB/c thymus [35].

4. Mapping of lymphoma susceptibility loci on mouse chromosome 4 using genetic crosses between BALB/c and STS strains of mice

We so far showed that the protein kinase, DNA activated, catalytic polypeptide (*Prkdc*) gene was a candidate for the apoptosis susceptibility gene on chromosome 16, and also responsible for susceptibility to radiation lymphomagenesis [21]. DNA-PK is a key enzyme for DNA double-stranded-break repair as well as V(D)J recombination of T- and B-cell receptors. Because BALB/c mice carry a *Prkdc* variant allele that causes lower DNA-PK activity, resultant hypersensitivity to radiation may raise frequency of cell death in the thymus and promote lymphomagenesis possibly via illegitimate recombination mechanisms. However, strain difference between BALB/c and STS in susceptibility to radiation lymphomagenesis has not been fully explained by the variations in *Prkdc*. According to M. Okumoto *et al.* [22, 23], cumulative incidence of lymphomas in (BALB/c x STS)F_1 exposed to 4 x 1.7 Gy of X-ray irradiation was in between those in parental BALB/c and STS, while (BALB/c x MSM)F_1 developed lymphomas at high frequency similar to BALB/c. The data suggest that strain difference in tumor susceptibility is controlled by multiple genes that influence onset, latency and frequency of tumorigenesis.

Previously, M. Okumoto *et al.* reported a suggestive linkage of susceptibility to radiation-induced lymphomagenesis, named lymphoma resistance *(Lyr)* (Mouse Genome Informatics, MGI: 96893) in the middle area of chromosome 4 using a series of recombinant inbred (RI) CXS strains of mice whose genome was constituted of 50% STS and 50% BALB/c genes [36]. It is worthwhile to test whether the *Lyr* locus is segregated in a genetic cross using BALB/c and STS. We performed genome-wide screen for microsatellite markers linked to lymphoma susceptibility using siblings from (BALB/c x STS)F_1 backcrossed to BALB/c or STS. We detected significant linkage disequilibrium in the middle area of chromosome 4 by the use of 219 siblings from (BALB/c x STS)F_1 backcrossed to BLB/c [20]. No significant linkage was detected by using another backcross. The primary locus with a conspicuous effect existed in an approximately 10 cM segment spanning D4Mit302 (37.6 cM) and D4Mit255 (48.5 cM) in the middle range of chromosome 4 (χ^2=19.3, genome-widely corrected p=0.0075). This locus was likely identical to the *Lyr* locus localized between tyrosinase-related protein 1 (*Tyrp1*) (38 cM) and interferon alpha (*Ifna*) (42.6 cM) [36]. The secondary locus with a weaker effect was detected near D4Mit17 (χ^2=16.0, genome-widely corrected p=0.034), a marker approximately 10 cM proximal to D4Mit302. The STS allele at these loci was associated with resistance to lymphomagenesis. *Mtf1*, a candidate susceptibility gene for radiation lymphomagenesis so far identified by other investigators, is located near D4Mit12 (57.8 cM), more than 10 cM distal to the critical regions containing these loci [32, 33]. Effect of *Prkdc,* which we identified as a lymphoma susceptibility gene by exploiting congenic mice [21], was not detected in tumor-free survivals in these crosses.

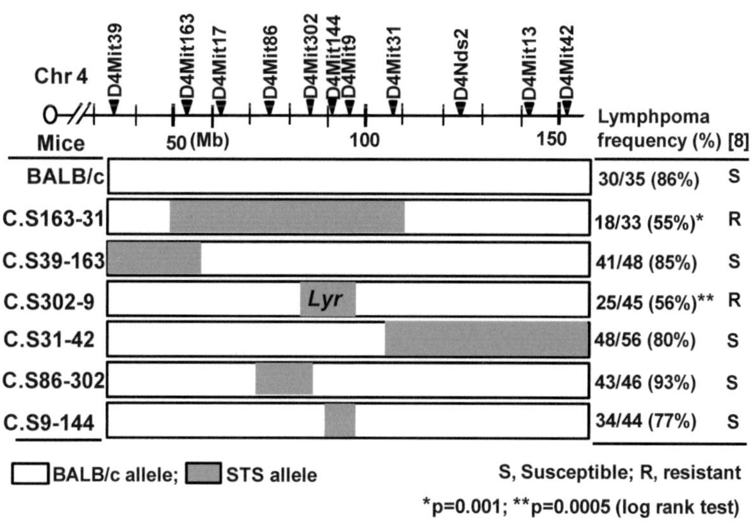

Figure 1. The *Lyr* locus exists between D4Mit302 and D4Mit144 on chromosome 4.

To narrow down the tumor susceptibility gene regions, we generated congenic strains of mice with different portions of STS-derived chromosome 4 on the BALB/c background by back-crossing (BALB/c x STS)F_1 mice to the BALB/c. Establishment of the congenic lines was facilitated by positive and negative selections with typing of microsatellite markers on chromosome 4 and markers distributed in the whole genome [20]. Because the *Lyr* locus was so vicinal (10 cM distance, approximately) to the secondary locus, we selected several strains with or without the STS allele at the critical markers D4Mit17, D4Mit302 and other markers near these markers, and compared their tumor-free survivals with that of BALB/cHeOpu exposed to X-ray irradiation (data shown in Table 1). A part of the results in [20] is represented in Figure 1. In this figure, the strain names of the C.S congenic mice are abbreviated by hyphened two Arabic numbers that represent STS allele-bearing microsatellite (Mit) markers at the proximal and distal end of the chromosomal segment. For instance, C.S163–31 represents a congenic line with the STS allele in the segment spanning D4Mit163 and D4Mit31. The order and megabase (Mb) positions of the markers are indicated by arrowheads on chromosome 4, which is represented by a line at the top of the figure. The primary lymphoma susceptibility locus *Lyr* exists between D4Mit302 (85.2 Mb) and D4Mit9 (94.7 Mb). Although the secondary locus was not detectable by a simple comparison of the tumor-free survival of congenic lines with that of BALB/c, linkage was reconfirmed by crossing congenic lines (data not shown here).

5. Loss of heterozygosity (LOH) in radiation-induced lymphomas from various F_1 hybrids: common loss and cross-dependant loss

Tumor suppressors frequently undergo loss of heterozygosity (LOH) in a variety of tumors in humans and mice. We previously reported that frequent LOH (more than 20%) occurred on chromosomes 4, 12 and 19 in radiation-induced lymphomas from (BALB/c x STS) F_1 mice, with incidences 27% (20 of 74 lymphomas), 57% (42 of 74 lymphomas) and 50% (37 of 74 lymphomas) on chromosomes 4 (at D4Mit31), 12 (at D12Mit17) and 19 (at D19Mit11), respectively [37] (Table 2). Importantly, STS allele-specific loss occurred on chromosome 4. The bias was confirmed using reciprocal F_1 hybrids between BALB/c and STS [37].

Mice[a]	Number of tumors	Chr	Marker (Mb)[b]	LOH (%)	Reference
(CXS)F_1	74	4	D4Mit31 (106.8)	20 (27%)	[37]
		12	D12Mit17[c]	42 (57%)	
		19	D19Mit11 (42.5)	37 (50%)	
(SXM)F_1	20	4	D4Mit54 (137.4)	5 (25%)	[39]
		12	D12Mit233 (109.5)	12 (60%)	
(CXM)F_1 [d]	81	12	D12Mit181 (110.0)	53 (65%)	[40]
		16	D16Mit122 (74.5)	38 (45%)	

[a]Abbreviations used are BALB/c, C; MSM, M; STS, S.

[b]Megabase (Mb) positions of markers are according to Mouse Genome Informatics (MGI) 5.10.03. (http://www.informatics.jax.org/).

[c]Physical position not assigned.

[d]F_1 hybrids between BALB/c and MSM hemizygous for *Trp53*.

Table 2. LOH in radiation-induced lymphomas from various F_1 hybrids.

In these crosses, allele loss involved almost entire chromosomes 4 and 19, without showing any peaks in LOH frequencies. Cytogenetic analysis showed that allele loss in such large areas was not caused by chromosomal deletion, but ascribable to mitotic recombination [38]. In lymphomas from (STS x MSM)F_1 mice, LOH occurred on chromosomes 4 and 12 with incidences 25% (5 of 20 lymphomas) and 60% (12 of 20 lymphomas) on chromosome 4 (at D4Mit54) and chromosome 12 (at D12Mit233), respectively [39]. In these lymphomas, LOH on chromosome 19 was infrequent (1/20, 5% at D19Mit63). In radiation-induced lymphomas from (BALB/c x MSM)F_1 mice, allele loss frequently occurred on chromosomes 12 (53/81, 65% at D12Mit181) and 16 (38/81, 45% at D16Mit122) [40].

Interestingly, LOH on chromosome 12 commonly occurred in radiation-induced lymphomas from these three F_1 hybrids, while LOH frequencies on chromosomes 4 and 19 markedly varied. Frequent LOH was detected on chromosome 4 in lymphomas from (STS x MSM)F_1 mice, but not (0/20 at D4Mit13) in lymphomas from (BALB/c x MSM)F_1 mice [40]. LOH on chromosome 19 was infrequent (0/20 and 1/20, at D19Mit63 and D19Mit123) in lymphomas from (STS x MSM)F_1 mice. In the context of LOH on chromosome 19, results were similar in lymphomas from the (BALB/c x MSM)F_1 hybrid. Thus, LOH on chromosomes 4 and 19 occurred in a cross-dependent manner. This suggests that LOH frequencies on these chromosomes are controlled by genetic interaction, possibly between putative tumor suppressors, the locations of which are indicated by LOH, and by genetic variations in the background. Moreover, the situation of LOH on chromosome 4 is somewhat different from that on chromosome 19. We present allele loss frequencies at several markers on chromosome 4 in these lymphomas in Table 3.

Mice[a]	Number of tumors	Marker (Mb)[b]	LOH (%)	References
(CXS)F_1	47	D4Mit17 (63.0)	14 (30%)	[37]
		D4Mit9 (94.7)	14 (30%)	
		D4Mit13 (142.0)	14 (30%)	
(SXM)F_1	20	D4Mit9 (94.7)	1 (5%)	[39]
		D4Mit54 (137.4)	5 (25%)	
(CXM)F_1 [c]	20	D4Nds2 (124.4)	0 (0%)	[40]
		D4Mit13 (142.0)	0 (0%)	
(CXM)F_1	43	D4Mit9 (94.7)	3 (7%)	Unpublished data
	51	D4Mit13 (142.0)	4 (8%)	[41]

[a]Abbreviations used are BALB/c, C; MSM, M; STS, S.

[b]Megabase (Mb) positions of markers are according to Mouse Genome Informatics (MGI) 5.10.03. (http://www.informatics.jax.org/).

[c]F_1 hybrids between BALB/c and MSM hemizygous for *Trp53*.

Table 3. Variation of LOH frequencies at microsatellite markers on chromosome 4 in radiation-induced lymphomas from various F_1 hybrids.

Notably, LOH frequency at D4Mit9 was reduced compared to that at D4Mit54, a marker in the proximity of D4Mit13 and approximately 43 Mb distal to D4Mit9, in lymphomas from (STS x MSM)F_1 hybrid mice. Using lymphomas from (BALB/c x MSM)F_1 mice, we reconfirmed that allele loss at markers D4Mit9 and D4Mit13 on chromosome 4 was very infrequent (3/4 and 4/51 [41], respectively). Because D4Mit9 is located very close to cyclin-dependent kinase inhibitor 2A (*Cdkn2a*) encoding tumor suppressors p16 and p19Arf, *Cdkn2a* is excluded as the putative tumor suppressor for lymphomagenesis in the (STS x MSM)F_1 and (BALB/c x MSM)F_1 backgrounds. Frequent LOH on chromosome 4 and 19 were also reported by other investigators in lymphomas from (C57BL/6JxBALB/cJ)F_1 and (C57BL/6J x RF/J) F_1 hybrid mice [42, 43]. According to the data in [43], strain-specific allele elimination is not found in the LOH on chromosome 4 from (C57BL/6JxBALB/cJ)F_1 mice.

The LOH frequencies at markers on chromosome 12 formed a sharp peak near telomere [41], and a putative tumor suppressor B cell leukemia/lymphoma 11B (*Bcl11b*) was later cloned from the peak [44]. The *BCL11B* tumor suppressor is also involved in human T cell acute lymphoblastic lymphomas [45]. Some of the lymphomas used for the genome-wide screen of LOH were generated in *Trp53* hemizygous (BALB/c x MSM)F_1 mice [40]. Because LOH frequencies at markers on chromosome 16 were markedly varied depending on the status of *Trp53* in (STS x MSM)F_1 mice [39], the high frequency of the LOH on chromosome 16 observed in lymphomas from (BALB/c x MSM) F_1 mice may likewise be explained.

6. The STS allele-specific loss occurred in the Lyr region on chromosome 4

Allele loss on chromosome 4 was significantly biased towards loss of the STS allele in lymphomas from (BALB/c x STS)F_1 mice [37]. It is of interest to examine whether putative tumor susceptibility genes on chromosome 4, which we identified in different regions of chromosome 4, are associated with the strain-specific allele loss on chromosome 4 by using congenic strains of mice with various regions of chromosome 4 from the donor strain STS on the background strain BALB/c, namely the C.S congenic series. LOH was studied in lymphomas generated in (BALB/c x C.S163–31)F_1 and (BALB/c x C.S302–9)F_1 mice. Both C.S163–31 and C.S302–9 strains of mice showed resistance to lymphomagenesis as shown in Figure 1. The C.S163–31 strain harbors the STS allele at two tumor susceptibility loci, one locus near D4Mit17 and the other, *Lyr* in the D4Mit302–D4Mit9 segment. The results are shown in Table 4.

Frequent allele loss at markers in the chromosome 4 segments was detected in lymphomas from (BALB/c x C.S163–31)F_1 (cross A) and (BALB/c x C.S302–9)F_1 (cross B) with incidences 11/34 (32%) and 10/34 (29%), respectively. The LOH frequencies in these F_1 hybrids were concordant with the original data in (BALB/c x STS)F_1 ([37] in Table 3). The STS-allele loss ratios were 9/11 (D4Mit302) and 10/11 (D4Mit9) in the cross A; 8/10 (D4Mit302) and 9/10 (D4Mit9) in the cross B. Because the STS-allele loss occurred with similar ratio in both crosses, we combined the data from crosses A and B (presented as A + B in Table 4). Analysis of the combined ratios 17/21 (D4Mit302) and 19/21 (D4Mit9) indicate that the distortions are significant at both markers D4Mit304 and D4Mit9 ($\chi 2$ values were 8.0 and 13.7, $p<0.005$, degree of freedom = 1, respectively). The data indicating the STS-allele specific loss (D4Mit31) in lymphomas from reciprocal (BALB/c x STS)F_1 and (STS x BALB/c)F_1 hybrids are also presented ([37] in Table 4). Thus, the skewed allele loss that was originally observed in a wide area of chromosome 4 in (BALB/c x STS)F_1 and (STS x BALB/c)F_1 hybrids is reproducible in the limited segments of the STS-derived chromosome 4. Our results suggest that tumor suppressor(s) associated with susceptibility to lymphomagenesis exist in the limited areas of chromosome 4. Since C.S39–86 mice carry the STS allele in the vicinity of D4Mit17, i. e., the secondary locus controlling susceptibility to lymphomagenesis, we further examined allele loss at markers D4Mit7 (67.7 Mb), a marker in the vicinity of D4Mit17, and D4Mit86 using 25 lymphomas from (BALB/c x C.S39–86)F_1 x BALB/c mice [20]. Allele loss at these markers was detected in only one of 25 tumors (less than 5%). In this case the BALB/c allele was lost. Hence, approximately 40 Mb of the D4Mit39–86 segment, to which the secondary locus for tumor susceptibility was localized, was excluded from the skewed loss region. Analysis on congenic strains strongly suggest that the STS-strain specific loss is ascribable to the D4Mit302–D4Mit9 segment of chromosome 4, which harbors a putative tumor susceptibility gene *Lyr*.

Mice[a]	Number of tumors	Marker (Mb)[b]	LOH (%)	S loss	C loss
A. (C x C.S163–31)F_1	34	D4Mit17 (63.0)	11 (32%)	9	2[c]
		D4Mit302 (85.2)	11 (32%)	9	2[c]
		D4Mit9 (94.7)	11 (32%)	10	1[c]
		D4Mit31 (106.8)	11 (32%)	10	1[c]
B. (C x C.S302–9)F_1	34	D4Mit302 (85.2)	10 (29%)	8	2[d]
		D4Mit9 (94.7)	10 (29%)	9	1[d]
A + B	68	D4Mit302 (85.2)	21 (31%)	17	4
		D4Mit9 (94.7)	21 (31%)	19	2
(C x S)F_1	39	D4Mit31 (106.8)	11 (28%)	10	1[e]
(S x C)F_1	35	D4Mit31 (106.8)	9 (26%)	7	2[e]
(C x S)F_1 + (S x C)F_1	74	D4Mit31 (106.8)	20 (27%)	17	3[e]

[a]BALB/c and STS mice are abbreviated as C and S.

[b]Megabase (Mb) positions of markers are according to Mouse Genome Informatics (MGI) 5.10.03. (http://www.informatics.jax.org/)

[c]Unpublished data.

[d] Data in [20].

[e]Data in [37].

Table 4. LOH at markers on chromosome 4 in lymphomas induced by radiation in F_1 hybrids between BALB/c and STS or C.S congenic lines.

In the *Lyr* region between D4Mit302 and D4Mit144, three known tumor suppressors *Cdkn2a*, cyclin-dependent kinase inhibitor 2B (*Cdkn2b*) encoding p15INK4B and methylthioadenosine phosphorylase (*Mtap*) exist (Figure 2). Involvement of *Cdkn2a* and *Cdkn2b*, specifically in acute lymphoblastic lymphomas (ALL) in humans and thymic lymphomas in mice has been reported [46–49]. Mtap is a key enzyme in purine and polyamine metabolism and regulation of transmethylation reactions and frequently inactivated in human tumors such as lymphomas by large homozygous deletion of the 9p21 region [50]. Since these deletions inactivate *CDKN2A/ARF* and *CDKN2B* as well as *MTAP* [51], it has been hypothesized that loss of *MTAP* in tumors is a result of co-deletion. However, a recent study showed that mice heterozygous for the targeted *Mtap* gene were affected with T-lymphocyte hyperproliferation followed by T-cell lymphomas late in their lives [52]. In these lymphomas, as shown by the study, expression of *Mtap* was markedly reduced, while expression of *Cdkn2a* was not. The results suffice the criteria for tumor suppressors, indicating that *Mtap* is a candidate tumor suppressor distinct from *Cdkn2a*. It has also been reported that the *Cdkn2b* gene is particularly inactivated by allele loss and hypermethylation of the remainder allele in radiation-induced lymphomas in mice [53]. BALB/c mice carry a hypomorphic variant allele at *Cdkn2a*, which is shown to be causative in the sensitivity to plasmacytomagenesis [28]. STS mice and most of strains other than BALB/c have the wild-type allele at the *Cdkn2a* locus ([28], DNA sequences we confirmed). Although *Cdkn2a* is a potential candidate for tumor susceptibility gene *Lyr*, *Cdkn2b* and *Mtap* are at present not ruled out as candidates for the tumor susceptibility gene. Analysis for allele loss, sequences and expression levels of these tumor susceptibility genes in BALB/c and C.S302–9 congenic mice is currently underway.

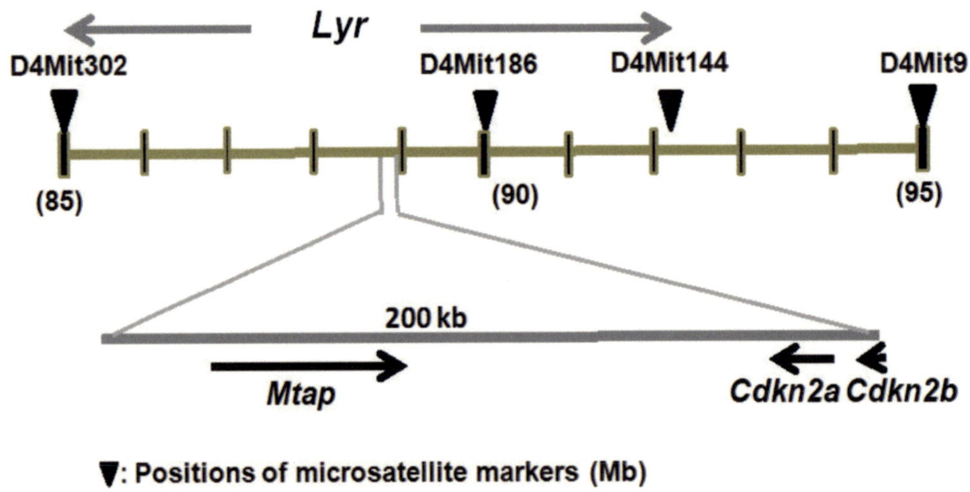

Figure 2. Locations of tumor suppressors in the lymphoma susceptibility *Lyr* gene region.

7. Conclusion

Frequent LOH occurs on chromosomes 4, 12 and 19 in radiation-induced lymphomas from various F_1 hybrid mice. These allele losses are classified into two groups: common loss and cross-dependent loss. The putative tumor suppressor harbored in common loss on chromosome 12 might be a key player in radiation-induced lymphomgenesis. Cross-dependent allele loss such as those on chromosomes 4 and 19 reflects genetic interaction between tumor suppressors harbored in the LOH region and the genetic background. BALB/c and STS strains of mice are susceptible and resistant to radiation-induced lymphomagenesis, respectively. Allele loss occurs on chromosome 4 in approximately 30% of lymphomas induced by radiation in (BALB/c x STS)F_1 mice and shows preferential loss of the STS allele. Our analysis of congenic lines with various portions of STS-derived chromosome 4 on the BALB/c background shows a link between the skewed LOH and the tumor susceptibility *Lyr* locus, where tumor suppressors *p16Ink4a/Arf*, *p15Ink4b* and *Mtap* genes are localized. Although the *Lyr* gene is as yet unidentified, *p16Ink4a/Arf* may be one of the potential candidates. Studying cross-specific LOH and distorted allele loss may lead to better understanding of variable pathways of radiation lymphomagenesis.

http://dx.doi.org/10.5772/55159

Acknowledgements

We thank Ms. Yuko Mitaki and Ms. Ikuko Kinoshita for their contribution to tumor sampling and genotyping. We thank Emeritus Professor M. Okumoto for his helpful discussion in preparation of this manuscript.

Author details

Nobuko Mori[1*] and Yoshiki Okada[2]

*Address all correspondence to: morin@b.s.osakafu-u.ac.jp

1 Department of Biological Science, Graduate School of Science, Osaka Prefecture University, Sakai-shi, Osaka, Japan

2 Department of Biological Science, Graduate School of Science, Osaka Prefecture University, Sakai-shi, Osaka, Japan

References

[1] Lee AH, Fraser ML, Binns CW. Tea, coffee and prostate cancer. Mol Nutr Food Res. 2009;53(2):256-65.

[2] Lee AH, Fraser ML, Meng X, Binns CW. Protective effects of green tea against prostate cancer. Expert Rev Anticancer Ther. 2006;6(4):507-13.

[3] Dang CV. Links between metabolism and cancer. Genes Dev. 2012;26(9):877-90.

[4] Martinez-Outschoorn UE, Pavlides S, Howell A, Pestell RG, Tanowitz HB, Sotgia F, Lisanti MP. Stromal-epithelial metabolic coupling in cancer: integrating autophagy and metabolism in the tumor microenvironment. Int J Biochem Cell Biol. 2011;43(7):1045-51.

[5] Sotgia F, Martinez-Outschoorn UE, Pavlides S, Howell A, Pestell RG, Lisanti MP. Understanding the Warburg effect and the prognostic value of stromal caveolin-1 as a marker of a lethal tumor microenvironment. Breast Cancer Res. 2011;13(4):213.

[6] Trimmer C, Sotgia F, Whitaker-Menezes D, Balliet RM, Eaton G, Martinez-Outschoorn UE, Pavlides S, Howell A, Iozzo RV, Pestell RG, Scherer PE, Capozza F, Lisanti MP. Caveolin-1 and mitochondrial SOD2 (MnSOD) function as tumor suppressors in the stromal microenvironment: a new genetically tractable model for human cancer associated fibroblasts. Cancer Biol Ther. 2011;11(4):383-94.

[7] Wang D, Dubois RN. The Role of Anti-Inflammatory Drugs in Colorectal Cancer. Annu Rev Med. 2012. [Epub ahead of print]

[8] Wang D, Dubois RN. The role of COX-2 in intestinal inflammation and colorectal cancer. Oncogene. 2010;29(6):781-8.

[9] Liotti F, Visciano C, Melillo RM. Inflammation in thyroid oncogenesis. Am J Cancer Res. 2012;2(3):286-97.

[10] El-Omar EM, Ng MT, Hold GL. Polymorphisms in Toll-like receptor genes and risk of cancer. Oncogene. 2008;27(2):244-52.

[11] Zheng SL, Augustsson-Bälter K, Chang B, Hedelin M, Li L, Adami HO, Bensen J, Li G, Johnasson JE, Turner AR, Adams TS, Meyers DA, Isaacs WB, Xu J, Grönberg H. Sequence variants of toll-like receptor 4 are associated with prostate cancer risk: results from the Cancer Prostate in Sweden Study. Cancer Res. 2004;64(8):2918-22.

[12] Yang Y, Wang F, Shi C, Zou Y, Qin H, Ma Y. Cyclin D1 G870A polymorphism contributes to colorectal cancer susceptibility: evidence from a systematic review of 22 case-control studies. PLoS One. 2012;7(5):e36813.

[13] Antoniou AC, Pharoah PD, Narod S, Risch HA, Eyfjord JE, Hopper JL, Olsson H, Johannsson O, Borg A, Pasini B, Radice P, Manoukian S, Eccles DM, Tang N, Olah E, Anton-Culver H, Warner E, Lubinski J, Gronwald J, Gorski B, Tulinius H, Thorlacius S, Eerola H, Nevanlinna H, Syrjäkoski K, Kallioniemi OP, Thompson D, Evans C, Peto J, Lalloo F, Evans DG, Easton DF. Breast and ovarian cancer risks to carriers of the *BRCA1* 5382insC and 185delAG and *BRCA2* 6174delT mutations: a combined analysis of 22 population based studies. J Med Genet. 2005; 42(7):602-3.

[14] Futreal PA, Liu Q, Shattuck-Eidens D, Cochran C, Harshman K, Tavtigian S, Bennett LM, Haugen-Strano A, Swensen J, Miki Y, Eddington K, McClure M, Frye C, Weaver-Feldhaus J, Ding W, Gholami Z, Soderkvist P, Terry L, Jhanwar S, Berchuck A, Iglehart JD, Marks J, Ballinger DG, Barrett JC, Skolnick MH, Kamb A, Wiseman R. *BRCA1* mutations in primary breast and ovarian carcinomas. Science. 1994;266(5182):120-2.

[15] Demant P. Cancer susceptibility in the mouse: genetics, biology and implications for human cancer. Nat Rev Genet. 2003;4(9):721-34.

[16] Frank SA. Genetic predisposition to cancer - insights from population genetics. Nat Rev Genet. 2004;5(10):764-72.

[17] Bartsch H, Dally H, Popanda O, Risch A, Schmezer P. Genetic risk profiles for cancer susceptibility and therapy response. Recent Results Cancer Res. 2007;174:19-36.

[18] Mori N, Okumoto M, van der Valk MA, Imai S, Haga S, Esaki K, Hart AA, Demant P. Genetic dissection of susceptibility to radiation-induced apoptosis of thymocytes and mapping of Rapop1, a novel susceptibility gene. Genomics. 1995;25(3):609-14.

[19] Mori N, Okumoto M, Hart AA, Demant P. Apoptosis susceptibility genes on mouse chromosome 9 (*Rapop2*) and chromosome 3 (*Rapop3*). Genomics. 1995;30(3):553-7.

[20] Mori N. Two loci controlling susceptibility to radiation-induced lymphomagenesis on mouse chromosome 4: *cdkn2a*, a candidate for one locus, and a novel locus distinct from *cdkn2a*. Radiat Res. 2010;173(2):158-64.

[21] Mori N, Matsumoto Y, Okumoto M, Suzuki N, Yamate J. Variations in *Prkdc* encoding the catalytic subunit of DNA-dependent protein kinase (DNA-PKcs) and susceptibility to radiation-induced apoptosis and lymphomagenesis. Oncogene 2001;20(28):3609-19.

[22] Okumoto M, Nishikawa R, Imai S, Hilgers J. Resistance of STS/A mice to lymphoma induction by X-irradiation. J Radiat Res. 1989;30(1):135-9.

[23] Okumoto M, Mori N, Miyashita N, Moriwaki K, Imai S, Haga S, Hiroishi S, Takamori Y, Esaki K. Radiation-induced lymphomas in MSM, (BALB/cHeA x MSM) F1 and (BALB/cHeA x STS/A) F1 hybrid mice. Exp Anim. 1995;44(1):43-8.

[24] Dietrich WF, Lander ES, Smith JS, Moser AR, Gould KA, Luongo C, Borenstein N, Dove W. Genetic identification of *Mom-1*, a major modifier locus affecting *Min*-induced intestinal neoplasia in the mouse. Cell. 1993;75(4):631-9.

[25] Luongo C, Gould KA, Su LK, Kinzler KW, Vogelstein B, Dietrich W, Lander ES, Moser AR. Mapping of multiple intestinal neoplasia (*Min*) to proximal chromosome 18 of the mouse. Genomics. 1993;15(1):3-8.

[26] MacPhee M, Chepenik KP, Liddell RA, Nelson KK, Siracusa LD, Buchberg AM. The secretory phospholipase A2 gene is a candidate for the *Mom1* locus, a major modifier of ApcMin-induced intestinal neoplasia. Cell.1995;81(6):957-66.

[27] Potter M; Mushinski EB; Wax JS; Hartley J; Mock BA. Identification of two genes on chromosome 4 that determine resistance to plasmacytoma induction in mice. Cancer Res.1994: 54 (4) 969-75

[28] Zhang S, Ramsay ES, Mock BA. *Cdkn2a*, the cyclin-dependent kinase inhibitor encoding p16INK4a and p19ARF, is a candidate for the plasmacytoma susceptibility locus, *Pctr1*. Proc Natl Acad Sci U S A. 1998;95(5):2429-34.

[29] Moen CJ, Groot PC, Hart AA, Snoek M, Demant P. Fine mapping of colon tumor susceptibility (*Scc*) genes in the mouse, different from the genes known to be somatically mutated in colon cancer. Proc Natl Acad Sci U S A. 1996; 93(3):1082-6.

[30] Ruivenkamp CA, van Wezel T, Zanon C, Stassen AP, Vlcek C, Csikós T, Klous AM, Tripodis N, Perrakis A, Boerrigter L, Groot PC, Lindeman J, Mooi WJ, Meijjer GA, Scholten G, Dauwerse H, Paces V, van Zandwijk N, van Ommen GJ, Demant P. *Ptprj* is a candidate for the mouse colon-cancer susceptibility locus *Scc1* and is frequently deleted in human cancers. Nat Genet. 2002;31(3):295-300.

[31] Ruivenkamp C, Hermsen M, Postma C, Klous A, Baak J, Meijer G, Demant P. LOH of *PTPRJ* occurs early in colorectal cancer and is associated with chromosomal loss of 18q12-21. Oncogene. 2003;22(22):3472-4.

[32] Saito Y, Ochiai Y, Kodama Y, Tamura Y, Togashi T, Kosugi-Okano H, Miyazawa T, Wakabayashi Y, Hatakeyama K, Wakana S, Niwa O, Kominami R. Genetic loci

controlling susceptibility to gamma-ray-induced thymic lymphoma. Oncogene. 2001; 20(37):5243-7.

[33] Sato H, Tamura Y, Ochiai Y, Kodama Y, Hatakeyama K, Niwa O, Kominami R. The D4Mit12 locus on mouse chromosome 4 provides susceptibility to both gamma-ray-induced and N-methyl-N-nitrosourea-induced thymic lymphomas. Cancer Sci. 2003; 94(8):668-71.

[34] Tamura Y, Maruyama M, Mishima Y, Fujisawa H, Obata M, Kodama Y, Yoshikai Y, Aoyagi Y, Niwa O, Schaffner W, Kominami R. Predisposition to mouse thymic lymphomas in response to ionizing radiation depends on variant alleles encoding metal-responsive transcription factor-1 (*Mtf-1*). Oncogene. 2005; 24(3):399-406.

[35] Maruyama M, Yamamoto T, Kohara Y, Katsuragi Y, Mishima Y, Aoyagi Y, Kominami R. *Mtf-1* lymphoma-susceptibility locus affects retention of large thymocytes with high ROS levels in mice after gamma-irradiation. Biochem Biophys Res Commun. 2007;354(1):209-15.

[36] Okumoto M, Nishikawa R, Imai S, Hilgers J. Genetic analysis of resistance to radiation lymphomagenesis with recombinant inbred strains of mice. Cancer Res. 1990;50(13): 3848-50.

[37] Okumoto M, Park YG, Song CW, Mori N. Frequent loss of heterozygosity on chromosomes 4, 12 and 19 in radiation-induced lymphomas in mice. Cancer Lett. 1999;135(2): 223-8.

[38] Hong DP, Kubo K, Tsugawa N, Mori N, Umesako S, Song CW, Okumoto M. Generation of large homozygous chromosomal segments by mitotic recombination during lymphomagenesis in F1 hybrid mice. J Radiat Res. 2002;43(2):187-94.

[39] Hong DP, Mori N, Umesako S, Song CW, Park YG, Aizawa S, Okumoto M. Putative tumor-suppressor gene regions responsible for radiation lymphomagenesis in F1 mice with different p53 status. J Radiat Res. 2002; 43(2):175-85.

[40] Matsumoto Y, Kosugi S, Shinbo T, Chou D, Ohashi M, Wakabayashi Y, Sakai K, Okumoto M, Mori N, Aizawa S, Niwa O, Kominami R. Allelic loss analysis of gamma-ray-induced mouse thymic lymphomas: two candidate tumor suppressor gene loci on chromosomes 12 and 16. Oncogene. 1998;16(21):2747-54.

[41] Okumoto M, Song CW, Tabata K, Ishibashi M, Mori N, Park YG, Kominami R, Matsumoto Y, Takamori Y, Esaki K. Putative tumor suppressor gene region within 0.85 cM on chromosome 12 in radiation-induced murine lymphomas. Mol Carcinog. 1998;22(3):175-81.

[42] Santos J, Herranz M, Pérez de Castro I, Pellicer A, Fernández-Piqueras J. A new candidate site for a tumor suppressor gene involved in mouse thymic lymphomagenesis is located on the distal part of chromosome 4. Oncogene. 1998;17(7):925-9.

[43] Santos J, Herranz M, Pérez de Castro I, Pellicer A, Fernández-Piqueras J. A new candidate site for a tumor suppressor gene involved in mouse thymic lymphomagenesis is located on the distal part of chromosome 4. Oncogene. 1998;17(7):925-9.

[44] Wakabayashi Y, Inoue J, Takahashi Y, Matsuki A, Kosugi-Okano H, Shinbo T, Mishima Y, Niwa O, Kominami R. Homozygous deletions and point mutations of the *Rit1/Bcl11b* gene in gamma-ray induced mouse thymic lymphomas. Biochem Biophys Res Commun. 2003;301(2):598-603.

[45] Gutierrez A, Kentsis A, Sanda T, Holmfeldt L, Chen SC, Zhang J, Protopopov A, Chin L, Dahlberg SE, Neuberg DS, Silverman LB, Winter SS, Hunger SP, Sallan SE, Zha S, Alt FW, Downing JR, Mullighan CG, Look AT. The *BCL11B* tumor suppressor is mutated across the major molecular subtypes of T-cell acute lymphoblastic leukemia. Blood. 2011; 118(15):4169-73.

[46] Takeuchi S, Bartram CR, Seriu T, Miller CW, Tobler A, Janssen JW, Reiter A, Ludwig WD, Zimmermann M, Schwaller J, et al. Analysis of a family of cyclin-dependent kinase inhibitors: *p15/MTS2/INK4B*, *p16/MTS1/INK4A*, and *p18* genes in acute lymphoblastic leukemia of childhood. Blood. 1995 Jul 15;86(2):755-60.

[47] Drexler HG. Review of alterations of the cyclin-dependent kinase inhibitor *INK4* family genes *p15*, *p16*, *p18* and *p19* in human leukemia-lymphoma cells. Leukemia 1998;12(6): 845-59.

[48] Boström J, Meyer-Puttlitz B, Wolter M, Blaschke B, Weber RG, Lichter P, Ichimura K, Collins VP, Reifenberger G. Alterations of the tumor suppressor genes *CDKN2A* (p16(INK4a)), p14(ARF), *CDKN2B* (p15(INK4b)), and *CDKN2C* (p18(INK4c)) in atypical and anaplastic meningiomas. Am J Pathol.. 2001;159(2):661-9.

[49] Guo SX, Taki T, Ohnishi H, Piao HY, Tabuchi K, Bessho F, Hanada R, Yanagisawa M, Hayashi Y. Hypermethylation of *p16* and *p15* genes and RB protein expression in acute leukemia. Leuk Res. 2000 ;24(1):39-46.

[50] Nobori T, Takabayashi K, Tran P, Orvis L, Batova A, Yu AL, Carson DA. Genomic cloning of methylthioadenosine phosphorylase: a purine metabolic enzyme deficient in multiple different cancers. Proc Natl Acad Sci U S A. 1996;93(12):6203-8.

[51] Zhang H, Chen ZH, Savarese TM. Codeletion of the genes for *p16INK4*, methylthioadenosine phosphorylase, interferon-alpha1, interferon-beta1, and other 9p21 markers in human malignant cell lines. Cancer Genet Cytogenet. 1996;86(1):22-8.

[52] Kadariya Y, Yin B, Tang B, Shinton SA, Quinlivan EP, Hua X, Klein-Szanto A, Al-Saleem TI, Bassing CH, Hardy RR, Kruger WD. Mice heterozygous for germ-line mutations in methylthioadenosine phosphorylase (*MTAP*) die prematurely of T-cell lympho ma. Cancer Res. 2009;69(14):5961-9.

[53] Cleary HJ, Boulton E, Plumb M. Allelic loss and promoter hypermethylation of the *p15INK4b* gene features in mouse radiation-induced lymphoid - but not myeloid - leukaemias. Leukemia. 1999;13(12):2049-52.

Chapter 2

To Grow, Stop or Die? – Novel Tumor-Suppressive Mechanism Regulated by the Transcription Factor E2F

Eiko Ozono, Shoji Yamaoka and Kiyoshi Ohtani

Additional information is available at the end of the chapter

http://dx.doi.org/10.5772/54510

1. Introduction

Proliferation of mammalian cells is strictly regulated by growth stimulation. Cell proliferation is stimulated not only by normal growth stimulation but also by abnormal growth stimulation originated from oncogenic changes. Such abnormal growth stimulation leads to tumorigenesis, if not properly guarded by appropriate cellular response. Cells are endowed with intrinsic tumor suppressor pathways to protect cells from tumorigenesis upon such oncogenic threat [1]. The tumor suppressor pathways halt cell proliferation either by restraining cell cycle progression or by inducing apoptosis (programmed cell death) in case of being unable to stop aberrant cell cycle progression. Consequently, the cell-fate, whether to grow, stop growing or die, is dependent on the balance between growth-promoting effects originated from oncogenic changes and growth-suppressive effects mediated by the tumor suppressor pathways upon oncogenic changes (Figure 1). When the tumor suppressor pathways are disabled by further oncogenic changes, the balance of cell-fate determination shifts from growth suppression to proliferation, and cells start deregulated proliferation, leading to tumorigenesis. Among the intrinsic tumor suppressor pathways, two major pathways are the RB pathway and the p53 pathway. Both pathways are important for induction of cell cycle arrest or apoptosis [2]. In addition, accumulating evidence indicates that the tumor suppressor TAp73, a member of the p53 family, also plays crucial roles in tumor suppression by inducing apoptosis independent of p53 [3, 4].

The transcription factor E2F, the main target of the RB pathway, plays crucial roles in cell cycle progression by activating growth-promoting genes [5]. In this regard, E2F is thought to mediate growth-promoting effects originating from normal growth stimulation and oncogenic changes. Supporting this notion, E2F could be an oncoprotein [6]. On the other hand, recent studies indicate that E2F also plays crucial roles in activation of the major intrinsic tumor suppressor pathways by sensing oncogenic changes, halting cell proliferation by inducing cell cycle arrest or apoptosis. Supporting this notion, E2F could also be a tumor suppressor [7]. Taken together, these observations indicate that E2F is located at the center of the balance between cell proliferation and cell cycle arrest or apoptosis, determining the cell-fate upon oncogenic changes (Figure 1). E2F can be regarded as a double-edged sword in cell growth control. In this chapter, we will describe the major intrinsic tumor suppressor pathways and activation of the tumor suppressor pathways by E2F upon oncogenic changes. We will focus on how E2F differentially regulates expression of target genes upon normal growth stimulation and oncogenic changes that have completely opposite roles in cell-fate determination, to grow, stop or die.

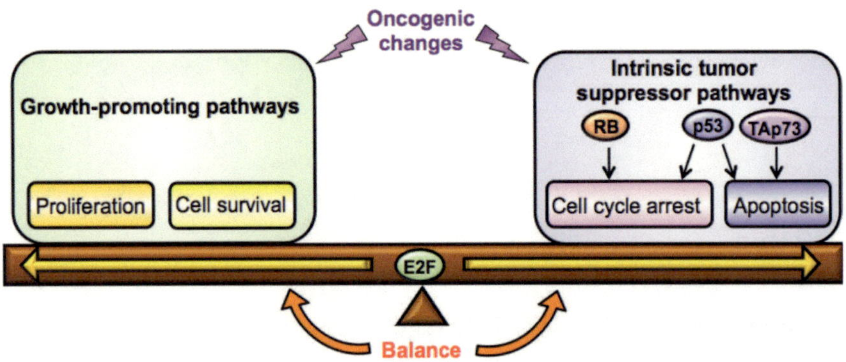

Figure 1. To grow, stop or die? The balance between the activity of growth-promoting pathways and tumor suppressor pathways determines the cell fate upon oncogenic changes. E2F is located at the center of the balance (see also Figure 6).

2. Major tumor-suppressor pathways

2.1. The RB pathway (CDK inhibitor–Cyc/CDK–RB)

The retinoblastoma gene (*RB1*) is the first identified tumor suppressor gene [8]. Individuals with heterozygous deletion or mutation of the *RB1* gene are susceptible to retinoblastoma in early life by additional deletion or mutation of the other allele. The *RB1* gene product pRB, together with its relatives, p107 and p130, comprises a family of pocket proteins [9, 10] (Hereafter we refer all pocket proteins to RB, and the *RB1* gene product to pRB). The main target of RB is the transcription factor E2F, which plays central role in cell proliferation by activating growth-promoting genes including those required for DNA replication and cell cycle progression [9]. In quiescent phase, RB binds to E2F and suppresses transcriptional activity of E2F. Moreover, RB recruits various chromatin-modifying factors, such as histone deacetylases (HDACs) [11] and histone methyl transferase SUV39H1 [12] to actively suppress expression of E2F target genes. Hence the main role of RB in tumor suppression is thought to be suppression of cell proliferation through suppression of E2F target gene expression. Growth stimulation induces expression of cyclins and activates cyclin dependent kinases (CDKs), which are called accelerators and engines in cell cycle progression, respectively. CDKs, in turn, inactivate RB by phosphorylation,

http://dx.doi.org/10.5772/54510

leading to expression of E2F target genes by releasing them from suppression by RB during G1 to S phase cell cycle progression (Figure 2) [13]. In contrast, growth-suppressive signals such as contact inhibition and DNA damage induce expression of CDK inhibitors, which are called brakes in cell cycle progression owing to their ability to inhibit activity of CDKs. Suppression of CDKs keeps RB in hypo-phosphorylated form, which binds to and inhibits E2F. Consequently, the activity of E2F to activate growth-promoting genes is controlled by the activity of RB, which is regulated by CDKs and CDK inhibitors. The pathway converging to RB, including CDKs and CDK inhibitors, is referred to the RB pathway.

Whether a cell progresses one round of the cell cycle or not is determined at the restriction point, which is located at late G1 phase of the cell cycle. Once a cell passed through the restriction point, it is programmed that the cell cycle automatically proceeds to the end of M phase. Thus, whether a cell proliferates or not is determined by whether the cell passes through the restriction point or not. Two major determinants whether a cell passes through the restriction point or not are E2F activity and cyclin dependent kinase activity, which is induced by E2F through activation of the *Cyclin E* (*CycE*) gene [14]. Since E2F plays essential roles in passing through the restriction point, RB can be regarded as a gatekeeper in cell cycle progression by controlling E2F activity. Hence disruption of the RB pathway and consequent activation of E2F is thought to be an essential event for tumorigenesis [2]. Actually, deletion or mutation of the *RB1* gene is observed in about 30% of cancers. Moreover, defects in the RB pathway upstream of RB such as overexpression of CycD and dysfunction of CDK inhibitors such as $p16^{INK4a}$ are frequently observed in other cancers retaining *RB1* [2]. It is predicted that all cancers have at least some defect in the RB pathway.

The RB pathway plays essential roles in tumor suppression by inducing cell cycle arrest or cellular senescence through suppression of the activity of E2F. As described below, disruption of the RB pathway reinforces cell cycle arrest by induction of CDK inhibitor $p21^{Cip1}$ expression through the p53 pathway or $p27^{Kip1}$ expression. When cells failed to induce cell cycle arrest, apoptosis is triggered through the p53 pathway or TAp73, a p53 family member, which activates various pro-apoptotic genes.

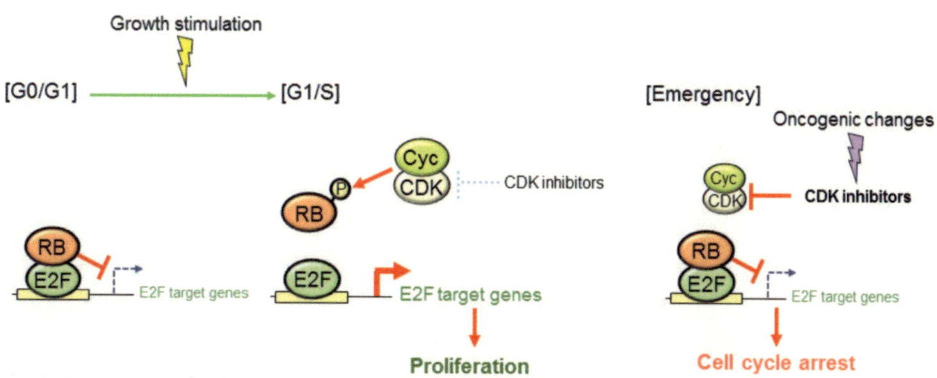

Figure 2. The regulatory mechanism of the transcription factor E2F by the RB pathway.

2.2. The p53 pathway (ARF–p53–cell cycle arrest or apoptosis related effectors)

The tumor suppressor p53 is a transcription factor that is activated by a variety of stress signals, including DNA damage, hypoxia and various oncogenic changes including aberrant activation of E2F [15]. In response to such stress signals, p53 induces either cell cycle arrest or apoptosis. Cell cycle arrest is mainly mediated through activation of the CDK inhibitor *p21*^{Cip1} gene [16], whose product suppresses wide range of CDKs. Apoptosis is mainly mediated through activation of the *Bax* and BH3 only family genes, whose products destabilize mitochondrial membrane to facilitate cytochrome c release, which triggers apoptotic cascades of caspase activation.

Since p53 plays crucial roles in induction of cell cycle arrest or apoptosis, the expression level of p53 is kept low by rapid ubiquitin/proteasome-dependent degradation, mainly caused by Mdm2 (mouse double minute 2, Hdm2 in humans), which is often overexpressed in many cancers [15]. Mdm2 is E3 ubiquitin ligase, which directly binds to p53 and promotes p53 degradation. Mdm2 also inhibits *TP53* mRNA translation [17]. The *Mdm2* gene is a target of p53, forming a negative-feedback loop to control the level of p53 [18]. Another negative regulator of p53 is MdmX, also known as Mdm4. MdmX has recently emerged as a discrete critical negative regulator of p53 [19]. Though MdmX is not a direct target of p53, structure of MdmX is significantly similar to that of Mdm2. MdmX is reported to enhance p53 ubiquitination by altering the substrate preference of the Mdm2, thereby indirectly regulating p53 [20].

Regarding response to oncogenic stresses, a potent activator of p53 is the tumor suppressor ARF (alternative reading frame, known as p14 in humans and as p19 in rodents) [2]. ARF directly binds to and sequesters Mdm2 to nucleoli, stabilizing p53 that leads to the expression of its target genes [21]. The pathway including upstream and downstream of p53 is referred to the p53 pathway. Of note, the *ARF* gene is a direct target of E2F [22] and this E2F-ARF interaction connects the RB pathway and the p53 pathway, enabling efficient tumor-suppressive response. When the RB pathway is disrupted by oncogenic changes, E2F is activated to induce ARF gene expression, leading to activation of the p53 pathway, which inhibits oncogenic cell growth by inducing cell cycle arrest or apoptosis (Figure 3). Therefore, disruption of both of the tumor suppressor pathways powerfully shifts the cell-fate determination balance to proliferation (Figure 1) and is thought to be essential to induce deregulated cell proliferation that leads to tumorigenesis. Indeed, about 50% of cancers carry *TP53* mutations or deletion and most cancers have defects in the p53 pathway including those in upstream and downstream of p53.

2.3. The TAp73 pathway (E2F–TAp73–pro-apoptotic targets)

The tumor suppressor TAp73 is a homologue of p53 and can induce apoptosis independently of p53 [3, 4, 23]. The *TP73* gene encodes two isoforms, TAp73 and DNp73, which are driven by different promoters. DNp73 lacks the transactivation (TA) domain and counteracts TAp73 and p53 [24]. Thus DNp73 is anti-apoptotic. The *TAp73* gene is thought to be a tumor suppressor gene and is known to be a target of E2F1 [3, 4, 25]. Similarly to p53, TAp73 is activated by both oncogenic changes and DNA damage [26, 27] and TAp73 target genes partly overlap with those of p53, such as the *PUMA* and *NOXA* genes [28, 29] that are crucial for induction of apoptosis. Moreover, TAp73 can induce apoptosis in the absence of p53 [3, 4]. Therefore, TAp73 seems to back-up the important tumor suppressive function of p53.

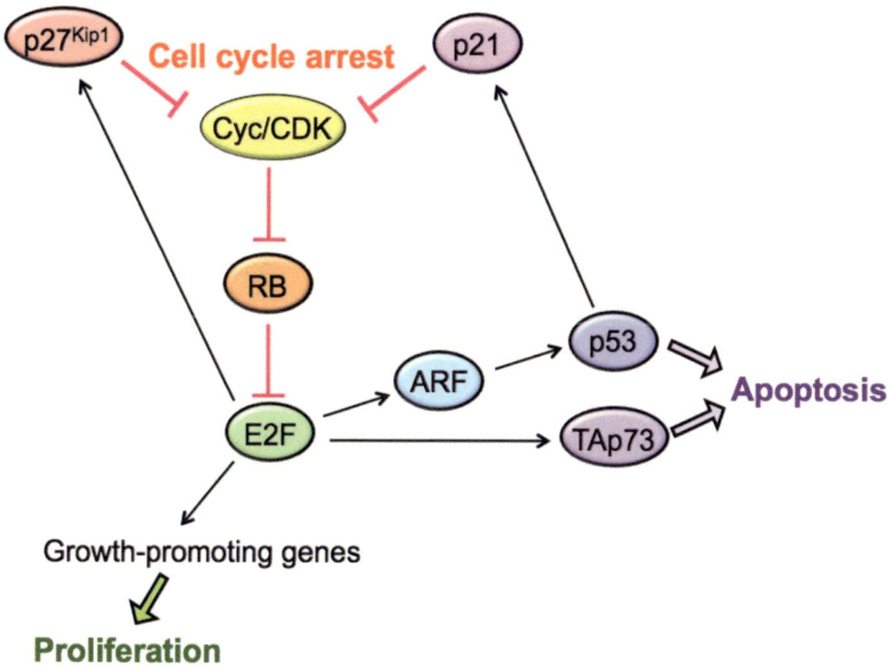

Figure 3. E2F activates major intrinsic tumor suppressor pathways.

3. The transcription factor E2F

The transcription factor E2F was originally identified as a cellular DNA-binding protein, which mediate E1A-dependent activation of the adenovirus E2 promoter [30]. Members of E2F family are downstream targets of the tumor suppressor RB and make repressor complexes with RB that keep cells in quiescent state [9, 10]. Though E2F plays central roles in cell proliferation by activating growth-promoting genes, certain members of E2F can induce apoptosis [15]. In this paragraph, we will describe three points as follows (1) E2F family members, the role of them in cell cycle progression, cell cycle arrest and apoptosis. (2) Regulatory mechanism of E2F. (3) E2F target genes, to better understand the regulatory mechanism of cell-fate determination mediated through E2F.

3.1. E2F family members

E2F consists of eight family members (E2F1-E2F8). E2F1-E2F5 are bound by their repressor RB family proteins through RB binding domain. E2F1-E2F5 activate transcription when free from RB and repress transcription when bound by RB. E2F1-E2F3a are induced at G1/S boundary and activate transcription free from RB. In contrast, E2F3b-E2F5 are expressed all through the cell cycle and play main roles in transcriptional repression in G0/G1 bound by RB. E2F6-E2F8 repress transcription independently of RB. Hence E2F family members are divided into two groups: activator E2Fs (E2F1-E2F3a) and repressor E2Fs (E2F3b-E2F8). Activator E2Fs play major roles in activation of target genes involved in growth promotion, growth suppression and induction of apoptosis. Repressor E2Fs are roughly divided into two groups; E2F3b-E2F5, which make repressor complex together with RB, and recently identified E2F6-E2F8, which function independently of RB.

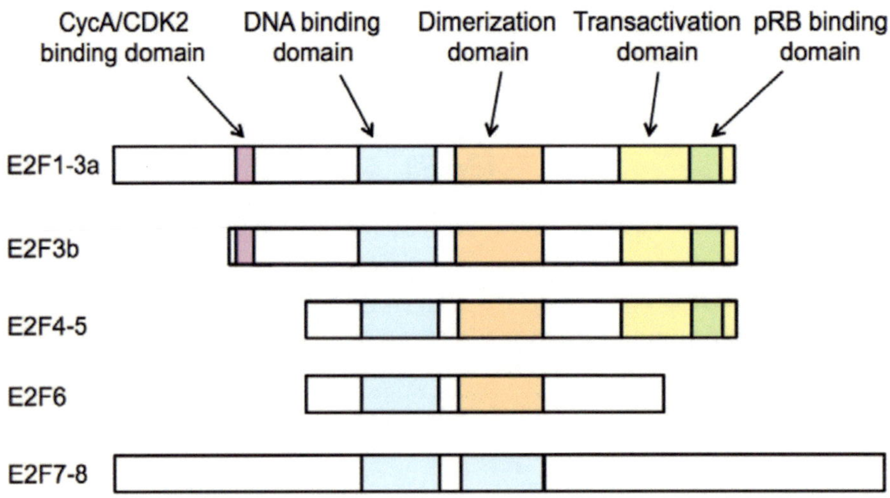

Figure 4. Structure of E2F family members. E2F1-E2F6 bind to their target promoter with binding partner DP proteins through dimerization domain. E2F7 and E2F8 have two DNA binding domains and make homodimers or E2F7/E2F8 heterodimer.

A structural characteristic of activator E2Fs (E2F1-E2F3a) is longer N terminal region, which does not exist in repressor E2Fs (E2F3b-E2F6). Expression of E2F1-E2F3a is induced by E2Fs themselves [31]. When growth stimulation inactivates p130 by phosphorylation through activation of CycD/Cdk4 or CycD/Cdk6, repressor E2Fs (E2F4 and E2F5) are released from p130 and activates the *E2F1-E2F3a* genes. E2F1-E2F3a in turn replace E2F4 and E2F5, and activate growth-promoting genes including E2F1-E2F3a themselves. Each activator E2F has preferential roles in cell cycle progression and induction of apoptosis. p107 preferentially make complexes with E2F4-E2F5 in G1/S to S phases.

E2F1 is generally thought to be the most powerful transcriptional activator of pro-apoptotic genes among activator E2Fs. Overexpression of E2F1 in tissue culture cells alone can induce cell cycle progression in otherwise quiescent fibroblasts [32]. Overexpression of E2F1 can be oncogenic *in vitro* [33, 34] and *in vivo* [6]. In contrast, E2F1 is dispensable for cell cycle progression, since E2F1 knockout mice are viable. On the other hand, over expression of E2F1 in cancer cell lines leads to apoptosis [35]. E2F1 also plays a key role in induction of cellular senescence by activating the ARF-p53 pathway, when overexpressed in human normal fibroblasts [36]. Moreover, E2F1 null mice resulted in tumorigenesis [7]. Taken together, E2F1 seems to play the most important roles in tumor suppression among activator E2Fs by activating genes involved in apoptosis and cell cycle arrest.

E2F2 has 46% overall amino acid sequence similarity to E2F1 [37] and is thought to play roles in both cell proliferation and tumor suppression. E2F2 null mice exhibit increased proliferation of hematopoietic cells and frequently develop autoimmunity and tumors [38, 39]. In tumor suppression, E2F2 plays major roles in suppression of Myc-induced T cell lymphomagenesis. Inactivation of neither E2F1 nor E2F3 had no effect on tumor progression in T cells, and only loss of E2F2 accelerated lymphomagenesis [40].

E2F3 is thought to be the most important activator E2F in cell proliferation. Although E2F1 or E2F2 null mice are viable and tumor-prone, E2F3 null mice are typically embryonic lethal in pure background [41] or show partially penetrant embryonic lethality in mixed background [42]. Although mouse embryonic fibroblasts (MEFs) with combined knockout of E2F1 and E2F2 proliferate, those with combined knockout of E2F1, E2F2 and E2F3 fail to proliferate and can not re-enter the cell cycle [5]. The E2F3 locus encodes two isoforms, E2F3a and E2F3b, a truncated variant of E2F3 in its N-terminus [43, 44]. Although both E2F3s partly overlaps their roles in cell cycle progression [45], E2F3a is expressed in G1/S to S phases and is thought to play crucial roles on cell proliferation, while E2F3b is expressed equivalently in quiescent and proliferating cells, and associates with pRB, representing the predominant E2F-pRB complex in quiescent cells [44, 46]. E2F3b also plays roles in myogenic differentiation by promoting gene expression related to differentiation [47].

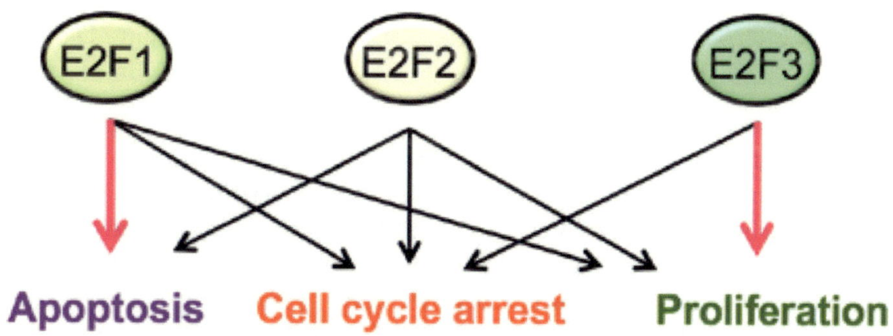

Figure 5. Roles of activator E2Fs in cell-fate determination. E2F1 plays crucial roles in induction of apoptosis. E2F3 is thought be essential for cell proliferation. It is predicted that the character of E2F2 is in the middle of E2F1 and E2F3.

E2F4 and E2F5 were cloned by their association with p107 and p130, and are significantly detected in quiescent cells [48, 49]. Knockout mice of either *E2F4* or E2F5 are viable [50-52]. E2F4 knockout mice are runted and display defects in late stage of maturation. In addition, these mice present reduced thickness of the gut epithelium and developmental craniofacial defects. E2F5 knockout mice develop hydrocephalus after birth apparently due to increased secretion of cerebrospinal fluid by the choroid plexus. E2F4 and E2F5 double knockout mice die before birth because of developmental defects, suggesting that E2F4 and E2F5 have some redundant functions during development [53]. Cells lacking E2F4 and E2F5 are unable to stop cell cycling upon growth-suppressive signals such as TGF-β treatment, suggesting that E2F4 and E2F5 are major repressor E2F in restraining cell cycle progression. These two E2Fs are thought to be important for cell cycle exit and terminal differentiation.

E2F6 does not possess RB binding domain. As predicted from the structure, E2F6 contributes gene silencing independent of RB [54] and is predicted to make repressor complex with polycomb proteins [55]. Consistent with this, *E2f6-/-* animals display overt homeotic transformations of the axial skeleton that are strikingly similar to the skeletal transformations observed in polycomb deficient mice [56]. E2F6 is reported to bind the same E2F-repressor site as E2F4, suggesting E2F6 may partly overlaps its function as transcriptional repressor with E2F4 in S phase [57].

E2F7 and E2F8 are thought to function as transcriptional repressors independently of RB proteins. An important target of E2F7 and E2F8 is the *E2F1* gene. E2F7 and E2F8 knockout mice are embryonic lethal, at least in part, due to apoptosis caused by inability to down regulate expression of E2F1, leading to activation of p53 [58]. Recent report showed that E2F7 and E2F8 promote angiogenesis through transcriptional activation of the VEGFA promoter with hypoxia inducible factor 1 (HIF1) [59].

3.2. Regulatory mechanism of E2F

E2F1 through E2F6 associate with DP family proteins (DP1 or DP2) to form heterodimeric complexes that bind to DNA in a sequence-specific manner (consensus sequence: $TTT^{C}/_{G}{}^{G}/_{C}CGC$). E2F7 and E2F8 have two DNA binding domains and do not require DP proteins for binding to DNA. E2F7 and E2F8 make homodimer or E2F7/E2F8 heterodimer to bind to the target [58].

Transcriptional activity of E2F1 through E2F5 is suppressed by binding of RB family proteins. pRB preferentially binds E2F1 through E2F3, whereas p107 and p130 preferentially bind E2F4 and E2F5. However, pRB can also bind E2F4, depending on the cellular circumstances. E2F6 through E2F8 can repress transcription independently of pRB family proteins [10, 60]. In quiescent phase, E2F3b/pRB, E2F4/p130 and E2F5/p130 repress promoters of E2F target genes. Upon growth stimulation, activated CycD/Cdk4 and CycD/Cdk6 phosphorylate pRB and p130, inhibiting their binding to the E2Fs and allowing accumulation of the free E2Fs [31]. This release from repression conferred by the pRB family proteins is the primary activation step for induction of E2F target genes. In this context, E2F3b, E2F4 and E2F5 mainly act as repressors together with the pRB family proteins during G0/G1 phases. Expression of E2F1, E2F2 and E2F3a is induced at the G1/S boundary by E2F itself, and activate

http://dx.doi.org/10.5772/54510

the growth related genes, including the *CycE* gene. CycE activates Cdk2, whose activity is essential for initiating DNA replication, and drive cells into S phase [31]. In late S phase, CycA/Cdk2 complex represses the transcriptional activity of E2F/DP complex by phosphorylation, which releases the complex from the binding element [61].

3.3. E2F target genes

Classical E2F targets are genes involved in DNA replication and cell cycle progression. In addition to these, recent studies with DNA microarray and chromatin immunoprecipitation (ChIP) identified a variety of E2F targets. These include genes involved in DNA repair, checkpoint, differentiation, development, metabolism, micro RNAs, apoptosis, cell cycle arrest and others.

DNA replication

E2F regulates expression of most of the genes involved in initiation of DNA replication: the *ORC1* (*origin recognition complex1*), *CDC6*, *MCM* (*maintenance of minichromosome*) *2-7*, *ASK* and *CDC45* genes. These factors are assembled into pre-replication complex, which is activated by CycE/Cdk2 to initiate DNA replication [31]. Cdt1 also plays crucial roles in initiation of DNA replication and is negatively regulated by geminin. The *Cdt1* and *geminin* genes are both E2F targets [62]. Genes, which code for machineries responsible for DNA replication, are classical E2F targets. These include the *DHFR* (*dihydrofolate reductase*), *DNA polymerase α*, *thymidine kinase*, *thymidylate synthase* and *PCNA (proliferating cell nuclear antigen)* [53].

DNA repair, checkpoint

E2F target genes related to DNA repair are the *Rad51*, *MSH2* and *MLH1* genes, which are involved in homologous recombination repair and mismatch repair [63]. E2F target genes related to checkpoint are the *ATM*, *Chk1*, *Mad3*, *Bub1*, *Claspin* and *RanBP1* genes [53, 64-67]. It is expected that E2F regulates these DNA repair and checkpoint genes to prepare machineries to quickly respond in case of emergency.

Cell cycle progression

E2F induces expression of genes, which play major roles in induction of S phase, such as the *CycE* and *activator E2F* (*E2F1-E2F3a*) genes themselves [31]. Also, the upstream negative regulators of pRB, the *Emi1* and *Skp2* genes are E2F targets [68, 69]. The genes, which encode CycA, Cdc2 (CDK1), CycB and B-myb that are important for S and G2/M phase progression, are also E2F targets [53]. Repressor E2Fs, *E2F7* and *E2F8* genes are reported to be targets of E2F1 that promote embryonic development by suppressing E2F1-p53 induced apoptosis, forming a negative feedback loop. Although c-Myc is reported as a target of E2F [70], E2F may play a role in suppression of c-Myc expression upon negative growth signals [71].

Development and differentiation

E2F also regulates expression of genes involved in development and differentiation. The *Firizzled homologs1-3*, *Homeobox* and *TGF* genes are shown to be targets of E2F

[53]. It is reported that overexpression of E2F1-3 induced various genes involved in development and differentiation [72]. Interestingly, although E2F7 and E2F8 are generally regarded as transcriptional repressors, a recent study reported the roles of E2F7 and E2F8 in transcriptional activation of genes such as VEGFA as written before [59].

Cellular metabolism

Recent work reported that E2F1 and pRB are required for repression of genes implicated in oxidative metabolism [73]. E2F1 repressed key genes that regulated energy homeostasis and mitochondrial functions in muscle and brown adipose tissue, and E2F1 null mice had a marked oxidative phenotype. Their work suggests a metabolic switch from oxidative to glycolytic metabolism that responds to stressful conditions.

Micro RNAs

Accumulating evidence indicates that expression of E2F is regulated by microRNAs and that E2F also induces expression of microRNAs [74]. One of the major microRNAs regulated by E2F is miR-17~92. miR-17~92 is a negative regulator of E2F1-E2F3 and is also a target of E2F1-E2F3, constructing a fail-safe mechanism to regulate E2F activity. E2F seems to be rigorously controlled by many miRNAs, suggesting that E2F activity must be strictly controlled for appropriate cell cycle progression.

Apoptosis and cell cycle arrest

E2F can activate genes involved in apoptosis and cell cycle arrest, which are inconvenient for cell proliferation. Regulatory mechanism of these growth-suppressive genes by E2F is yet to be elucidated, especially regarding regulation of cell growth versus tumor suppression in response to normal growth stimulation and oncogenic changes. The *caspase3, caspase7* and *Apaf1* genes, which code for apoptotic machineries, are direct targets of E2F [53]. Caspases are expressed as inactive precursors (procaspases), and expression of procaspases and Apaf-1 alone does not necessarily induce apoptosis. These apoptotic machineries require upstream signals to be activated to induce apoptosis. Expression of these two pro-apoptotic E2F targets and all E2F targets mentioned above is induced by growth stimulation. Thus, expression of the pro-apoptotic machinery by growth stimulation is thought to be fail-safe mechanism to induce apoptosis in case of emergency. We refer these E2F target genes, whose expression is induced by growth stimulation, to 'typical E2F targets'. Typical E2F targets are activated by growth stimulation, which physiologically inactivates RB by phosphorylation and activates E2F by release from repression.

In contrast to typical E2F targets, we identified three E2F targets, which are not activated by growth stimulation. These include the tumor suppressor *ARF, TAp73* and CDK inhibitor $p27^{Kip1}$ genes. As described below, these genes are critically important in tumor suppression. We describe the regulatory mechanism of the tumor suppressor *ARF, TAp73* and CDK inhibitor $p27^{Kip1}$ genes by E2F in next paragraph.

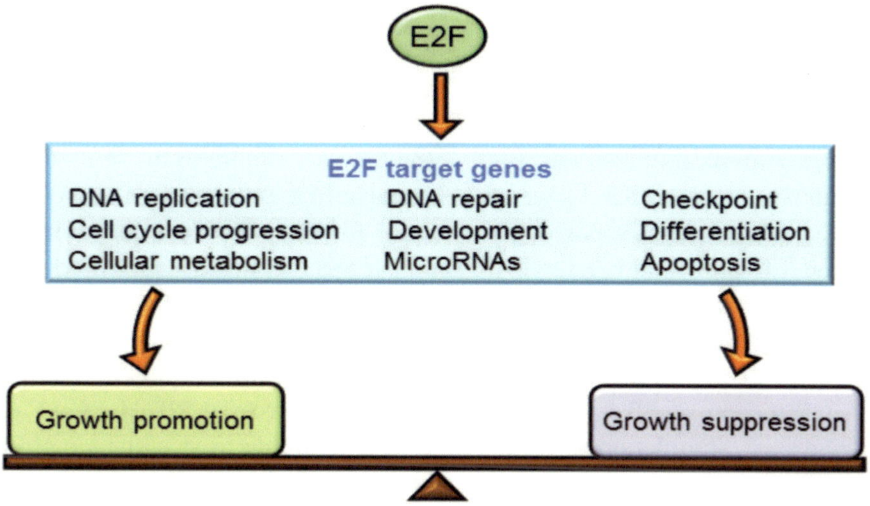

Figure 6. E2F activates both growth-promoting genes and growth- suppressive genes, including E2F itself. It is predicted that the expression levels of E2F targets, related to growth- promotion or growth- suppression, decide the balance of cell-fate determination.

4. Deregulated E2F

Since E2F1-E2F3a are activated by growth stimulation, it is generally thought that their target genes are all activated by growth stimulation. However, E2F1-E2F3a also activate genes involved in cell cycle arrest or apoptotic, which are inconvenient for cell growth. It has yet to be elucidated how E2F regulates genes involved in cell cycle arrest or apoptosis, regarding normal cell growth and tumor suppression.

Since all E2F targets are thought to be activated by growth stimulation, it is surprising that our previous work identified three growth suppressive E2F targets, which were not activated by growth stimulation at all in human normal fibroblasts [75-77]. In contrast, these E2F targets were activated by deregulated E2F activity induced by overexpression of E2F1 or forced inactivation of pRB. Overexpression of E2F1 generates exceeding amount of exogenously introduced E2F1, which becomes out control by RB proteins. Forced inactivation of pRB induces endogenous deregulated E2F activity out control by pRB. We refer these growth suppressive E2F targets, which are not activated by growth stimulation but are activated by deregulated E2F, to 'atypical E2F targets'. These atypical E2F targets include the tumor suppressor *ARF* and *TAp73* genes and the CDK inhibitor $p27^{Kip1}$ gene These three atypical E2F target genes play major roles in tumor suppression. ARF is an upstream activator of the tumor suppressor p53. CDK inhibitor $p27^{Kip1}$ activates the RB pathway by inhibiting CDKs. TAp73 is the tumor suppressor, which can induce apoptosis in dependently of p53. Our observations suggest that E2F activity induced by RB dysfunction, one of major oncogenic changes, has distinct function from that induced by growth stimulation in activating target genes (Figure 7).

E2F activity induced by RB dysfunction activates both typical and atypical E2F targets. In contrast, E2F activity induced by growth stimulation activates only typical E2F targets and interestingly, not atypical E2F targets. Both of the E2Fs are similar in the sense that they are released from repression by RB. However, it is shown that growth stimulation does not totally inactivate pRB and some portion of activator E2Fs is still in complex with pRB [78]. Moreover, the *RB1* gene is an E2F target [79] and expression of pRB is increased in G1/S to S phases [80]. Indeed, the amount of activator E2Fs/pRB complex rather increases in G1/S to S phases, as examined by gel mobility shift assay [78]. This observation indicates that E2F activity induced by growth stimulation is still under control of RB. In contrast, E2F activity induced by dysfunction of RB is thought to be out of control by RB. Here, we refer E2F activated by growth stimulation to 'physiological E2F' and that by dysfunction of RB to 'deregulated E2F'. Our findings indicate that deregulated E2F is functionally different from physiological E2F.

In this paragraph, we describe the regulatory mechanism of atypical E2F targets by deregulated E2F and discuss about the characteristics of deregulated E2F activity regarding its role in tumor suppression.

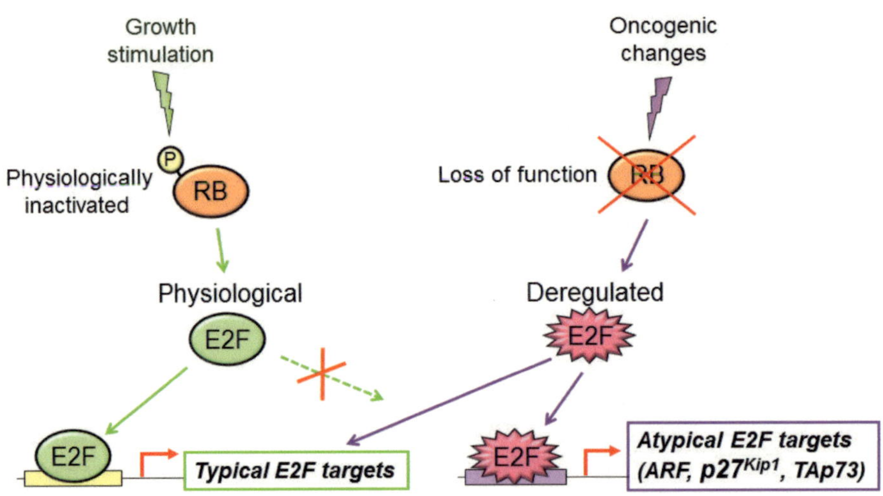

Figure 7. Atypical E2F target genes are specifically activated by deregulated E2F activity through specific E2F responsive elements.

4.1. Atypical E2F targets

The first atypical E2F target identified was the *ARF* gene [75]. ARF is the major activator of p53 pathway and links the RB and p53 tumor suppressor pathways [22], playing crucial roles in tumor suppression. Consistent with this notion, deletion, mutation or silencing of the *ARF* gene is frequently observed in cancers. Moreover, *Arf* null mice are highly prone to tumorigenesis [81]. Various oncogenic signals are able to elicit the activation of the *ARF* gene. Overexpression of adenovirus E1a or E2F1 in primary mouse embryonic fibroblasts (MEFs) rapidly induces ARF gene expression and p53-dependent apoptosis [16]. Myc overexpression and oncogenic mutant Ras are also strong activators of the *ARF* gene and combination of absence of Arf in mice severely impaired the tumor suppressive activity of p53 [82]. Arf promoter seems to monitor these oncogenic signals as shown by ARF promoter-*GFP* transgenic model, in which GFP expression was observed in tumors induced by Myc or Ras but not in normal growing tissues [83]. Other studies elucidated that ARF also restrains cell growth independently of p53, interacting with other factors [84]. Taken together, the *ARF* gene plays crucial roles in tumor suppression through p53-dependent and independent pathways.

The CDK inhibitor $p27^{Kip1}$ is an upstream regulator of the RB pathway and known to contribute to the ability of pRB to induce cell-cycle arrest, differentiation and senescence [85-87]. There is cross regulation between $p27^{Kip1}$ and pRB. $p27^{Kip1}$ enhances pRB growth suppressive function by inhibiting Cyc/CDK, keeping pRB in hypo-phosphorylated form. pRB increases the amount of $p27^{Kip1}$ by sequestrating Skp2, a component of E3 ubiquitin ligase complex, which promotes degradation of $p27^{Kip1}$ [86]. pRB also known to cooperate with APC/C^{cdh1}, another E3 ubiquitin ligase, to induces Skp2 degradation, stabilizing $p27^{Kip1}$ [88]. Taken together, pRB and $p27^{Kip1}$ seem to keep close relationship in the RB pathway to efficiently suppress aberrant cell proliferation. Indeed, our previous study showed that inactivation of RB by adenovirus E1a increased BrdU (bromodeoxyuridine)-positive cells much earlier in $p27^{Kip1}$-/- MEFs than in wild type MEFs [66]. These results support the notion that $p27^{Kip1}$ plays important roles in the RB pathway to suppress cell cycle progression induced by oncogenic changes.

The tumor suppressor *TAp73* gene had been identified as a direct target of E2F using cancer cell lines [4]. We found that the *TAp73* gene was activated by deregulated E2F but not by physiological E2F in human normal fibroblasts [77]. TAp73 is a p53 family member and plays important roles in tumor suppression with its other family member p53 and p63. All of the three genes express differentially spliced isoforms [89, 90]. Two major isoforms are TA isoforms, which retain transactivation (TA) domain, and delta N (DN) isoforms, which lack TA domain. Since these family members activate their targets as tetramers and DN isoforms lack transactivation (TA) domain, DN isoforms have dominant-negative properties [24]. Although p53 is deleted or mutated in half of all human cancers, deletion or mutations of p73 and p63 occur rarely [91, 92]. Rather over expression of DN form of p73 and p63 are commonly observed in many cancers, such as over expression of DNp73 isoform in gliomas and carcinomas of the breast and the colon [93, 94], and that of DNp63 isoform in bladder carcinomas [95]. Each DN isoform can suppress all three types of TA forms [92, 96, 97]. Therefore, it is expected that tumor-suppressive TA isoforms are suppressed by DN isoforms in many cancers. This could explain why deletion or mutations of the *TP73* and *TP63* genes are rare. *TAp73* can induce apoptosis independently of p53. Moreover, TAp73 knockout mice are tumor prone, infertile, sensitive for carcinogen-induced tumorigenesis and defective for maintenance of genomic stability [98]. The *TAp73* gene is activated by various stress signals including deregulated E2F and DNA damage. These observations suggest that TAp73 contributes to tumor suppression in addition to the p53 pathway in response to various oncogenic changes.

Accumulating evidence indicates that the three atypical E2F targets explained above play crucial roles in tumor suppression, by activating the RB pathway, the p53 pathway and the p53 independent pathway. Taken together, deregulated E2F seems to be critically important for activating major intrinsic tumor-suppressor pathways in responding to oncogenic changes to suppress tumorigenesis (Figure 8).

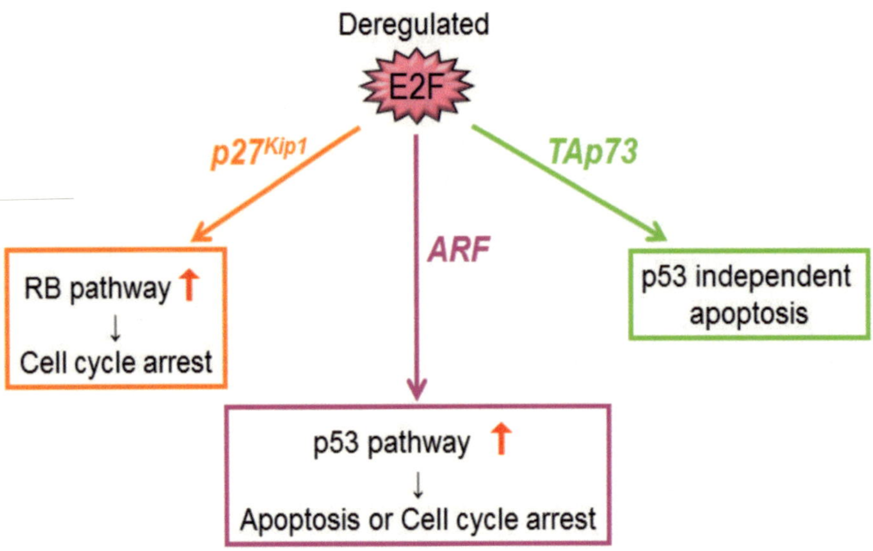

Figure 8. Deregulated E2F plays important roles in activating major tumor suppressor pathways (RB, p53 and TAp73 pathways) by activating the *ARF*, $p27^{Kip1}$ and *TAp73* genes.

4.2. Distinct transcriptional regulatory mechanism mediated by deregulated E2F

ARF, $p27^{Kip1}$ and TAp73 exert its effects when expressed unlike pro-apoptotic targets, which are expressed as inactive precursors such as pro-caspases. Thus the regulation of expression of these genes is critically important for tumor suppression. The finding that these genes are specifically activated by deregulated E2F but not by physiologically activated E2F indicates that there is a mechanism to specifically respond to oncogenic changes to suppress tumorigenesis, while allowing normal cell growth upon normal growth stimulation.

We first identified the tumor suppressor *ARF* gene as an atypical E2F target [75]. In our studies, we used human normal fibroblasts (HFFs or WI-38) to examine the responsiveness of each growth-suppressive E2F target to physiological and deregulated E2F. This is because most of previous studies used cancer cell lines and were unable to examine responsiveness of E2F target genes to normal growth stimulation. To induce physiological E2F, we used serum stimulation, common growth stimulation for fibroblasts. To induce deregulated E2F activity, we used ectopically expressed E2F1 or forced inactivation of RB either by adenovirus E1a, which binds to and inactivates all RB family proteins, or shRNA against *RB1* (shRB), which represses the expression of pRB. The latter (adenovirus E1a and shRB) is expected to induce endogenous deregulated E2F activity.

Ectopically expressed E2F1, adenovirus E1a and shRB induced ARF gene expression in RT-PCR, and activated ARF promoter in reporter assay. However, serum stimulation, which physiologically activates E2F, did not induce ARF gene expression or activate ARF promoter under the condition that CDC6 gene (one of typical E2F targets involved in DNA replication) expression was significantly induced and CDC6 promoter was clearly activated. These results indicate that the *ARF* gene is specifically activated by deregulated E2F but not by physiological E2F. Promoter analyses identified the E2F responsive element of ARF promoter (EREA), which specifically responds to deregulated E2F activity. Interestingly, the sequence of EREA was composed of only GC repeat and lacked T stretch. This is in contrast to that of consensus E2F binding motif, which is composed of T stretch and GC repeat ($TTT^{G}/_{C}{}^{C}/_{G}CGC$) in typical E2F targets. In addition, the location of EREA was far upstream from transcription start site compared to that of typical E2F sites, which is within 100 bp from transcription start site in most cases. Moreover, our gel mobility shift assay and ChIP assay showed that EREA specifically binds ectopically expressed E2F1 but not physiological E2F1 induced by serum stimulation, both *in vitro* and *in vivo*, respectively. Taken together, these observations suggest that the *ARF* gene is specifically activated through EREA by deregulated E2F activity, triggered by ectopically expressed E2F1 or forced inactivation of RB, but not by physiological E2F activity, induced by serum stimulation (Figure 7).

A later study showed that ARF promoter lacking EREA was still activated by overexpression of E2F1 [99]. Our further analyses of ARF promoter identified multiples of EREA-like elements in ARF promoter. It seems that, although EREA is the major E2F responsive element in ARF promoter, it is not the sole responsive element and multiple EREA-like elements co-operate to specifically respond to deregulated E2F activity (manuscript in preparation).

Defects in the RB pathway activate E2F out of control by RB, promoting abnormal cell growth. In response to such oncogenic insults, deregulated E2F activates the *ARF* gene, leading to activation of p53 to protect cell from tumorigenesis. For induction of tumorigenesis, the p53 pathway must be disabled by further oncogenic changes. Indeed, defects in the p53 pathway are observed in almost all cancers. It is expected that the presence of deregulated E2F be tolerated in cancer cells by further inactivation of the p53 pathway. We thus examined the existence of deregulated E2F activity in *RB1* deficient cancer cell lines and in normal growing fibroblasts. When we introduced constitutively active form of pRB into *RB1* deficient cancer cell lines (5637, Saos-2 and C-33 A) and normal growing fibroblasts (WI-38 and HFF), activity of EREA and ARF promoter were decreased in *RB1* deficient cancer cell lines, but not in normal growing fibroblasts. These results showed that deregulated E2F activity specifically exists in *RB1* deficient cancer cell lines but not in normal growing fibroblasts. The presence of deregulated E2F activity may serve as a useful marker to discriminate cancer cells from normal growing cells.

Our search for new E2F targets with subtraction method identified the CDK inhibitor *p27^{Kip1}* gene as an atypical E2F target [66]. p27^{Kip1} plays important roles in cell cycle arrest by inhibiting Cyc/CDKs. Using reporter assay, we showed that EREK (E2F responsive element of p27^{Kip1}) was responsible for specifically sensing deregulated E2F activity in human normal fibroblasts and that EREK was specifically activated in the *RB1* deficient cancer cell lines [76]. Consistent with EREA, the location of EREK was far upstream compared to typical E2F binding sites. Interestingly, the sequence of EREK contained T in addition to GC repeat and is rather similar to that of typical E2F binging site. However, EREK bound deregulated E2F1 but not physiological E2F1 in ChIP assay, showing that its character was similar to EREA. These results suggest that not only the sequence of E2F responsive elements, but also the sequence around the responsive elements may be important for discriminating deregulated E2F activity from physiological E2F activity. There is also a possibility that structure of the whole promoter might also affect the discrimination.

Third atypical E2F target is the tumor suppressor *TAp73* gene [77]. TAp73 promoter specifically responded to deregulated E2F activity through four ERE73s (E2F responsive elements of TAp73), which were specifically activated by deregulated E2F activity. The sequences of ERE73s contained T stretch and were similar to that of typical E2F binding sites. Importantly, our ChIP assay showed that bindings of ectopically expressed 'exogenous' E2F1 and deregulated 'endogenous' E2F1 induced by adenovirus E1a were detected on ERE73s, but not that of physiological E2F1 induced by serum stimulation. Thus, although the sequences of ERE73s were similar to or almost same as that of typical E2F binding sites, the characters of both were completely different. ERE73s were specifically activated by deregulated E2F activity and specifically bound to both 'exogenous' and 'endogenous' deregulated E2F1. These results support the notion that both sequences of the E2F binding site and its flanking region may be important for discriminating deregulated E2F activity from physiological E2F activity. Consistent with EREA and EREK, reporter assay showed that ERE73s were also activated in the *RB1* deficient cancer cell lines and not in normal fibroblasts. Moreover, reintroduction of the constitutive active from of pRB by recombinant adenovirus reduced the expression of the *TAp73* gene in all the cancer cell lines in RT-PCR, indicating that the cancer cell lines harbor deregulated E2F activity that activates the endogenous *TAp73* gene.

Interestingly, our unpublished data suggest that not only the *RB1* deficient cancer cell lines but also cancer cell lines retaining pRB harbor deregulated E2F activity. Activity of ERE73s and expression of the *TAp73* gene were suppressed by introduction of the constitutive active form of pRB in cancer cell lines retaining pRB. These results suggest the possibility that E2F-mediated transcriptional program can sense defects in the RB pathway, not only pRB itself but also upstream regulators of pRB. Taken together, deregulated E2F activity might become a universal means to discriminate cancer cells (may be regardless of the presence of pRB) from normal growing cells.

4.3. Difference between deregulated E2F and physiologically activated E2F

Deregulated E2F and physiologically activated E2F are similar in a sense that both are 'released from RB'. However, there is a functional difference between deregulated E2F and physiologically activated E2F. Deregulated E2F activates both typical and atypical E2F targets. In contrast, physiologically activated E2F activates typical E2F targets but not atypical E2F targets. Why atypical E2F targets are activated by deregulated E2F and not by physiologically activated E2F? What is the difference between deregulated E2F and physiologically activated E2F? Deregulated E2F is totally 'out of control' by RB due to dysfunction of the RB pathway. In contrast, physiologically activated E2F is temporarily released from RB, predicted to be 'under control' of RB. During normal cell growth, activator E2Fs are induced at G1/S boundary of the cell cycle. At the point, it is generally believed that pRB is phosphorylated and inactivated. However, pRB is not totally inactivated at the point. Previous studies indicate that the *RB1* gene is an E2F target [79] and expression of pRB is increased in G1/S to S phases [80]. Moreover, the amount of activator E2Fs/pRB complex rather increases at G1/S boundary as shown by gel mobility shift assay [78], indicating that some portion of pRB is still active and is regulating the activity of activator E2Fs. Thus, physiologically activated E2F is still under control by pRB. It is likely that activity of activator E2Fs is strictly controlled by degree of phosphorylation of pRB dependent on the activity of CDKs, reflecting the strength of growth stimulation. There must be difference between activation of E2F out of control by pRB and activation of E2F under control by pRB. Our studies of regulatory mechanism of atypical target promoters by E2F elucidated the four different points between regulation of typical E2F target promoter and that of atypical E2F target promoter.

1. Although sequences of EREK and ERE73s resemble that of typical E2F binding sites, sequence of EREA is different from that of typical E2F sites. The sequence of consensus E2F binding motif of typical E2F targets is composed of T stretch and GC repeat ($TTT^{G}/_{C}{}^{C}/_{G}CGC$). EREA is composed of only GC repeat and lacks T stretch. Difference in binding sequence suggests the possibility that there may be a difference in factors, which recognize the sequence and bind.

2. In all three cases, location of E2F responsive elements of atypical E2F targets is far from the transcriptional start site compared to that of typical E2F binding sites in typical E2F targets. In the case of typical E2F targets, location of E2F binding sites is within 100 bp from transcriptional start site in most cases. Typical E2F targets are under repression by E2F/RB complex in quiescent phase and are released from repression upon growth stimulation. Close proximity of E2F binding sites to transcriptional start site may be required for this mode of regulation. In contrast, atypical E2F targets are not under repression by E2F/RB complex and literally activated by deregulated E2F. E2F responsive elements of atypical E2F targets behave as enhancer elements, which can function from distance.

3. Deregulated E2F1 bound to EREA, EREK and ERE73s and physiologically activated E2F1 did not bind to these elements as shown by ChIP assay. In the case of EREA, it is also shown that repressor type E2F4 does not bind to EREA. This observation is compatible with the observation that these atypical E2F targets are not under repression by

E2F/RB. The fact that deregulated E2F bind to atypical E2F targets, while physiological E2F does not bind to atypical E2F targets, suggest that there is difference in binding behavior between deregulated E2F and physiologically activated E2F.

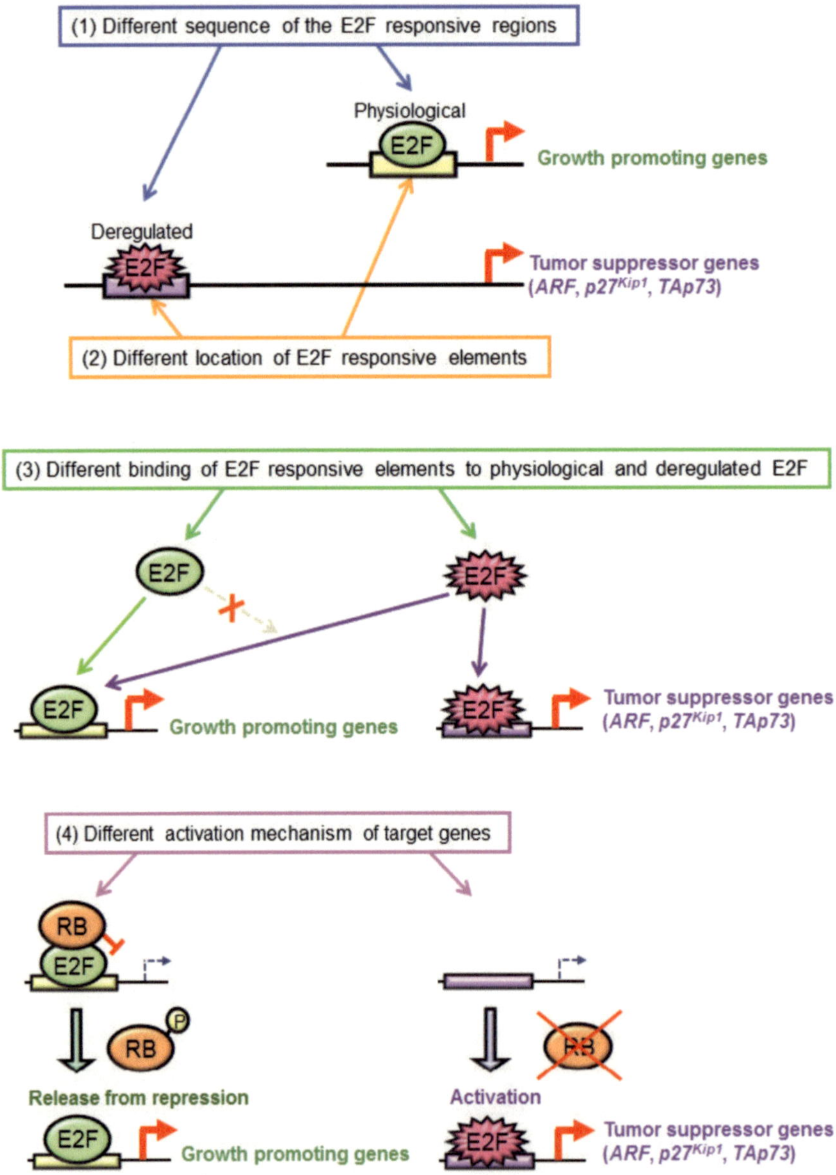

Figure 9. Differences in E2F regulation of target promoters between deregulated E2F and physiological E2F. (1) Sequence of the atypical E2F responsive elements and surrounding regions are different from that of typical E2F targets. (2) E2F responsive elements of the atypical E2F targets locate far upstream from transcriptional start sites compared to that of typical E2F targets. (3) E2F responsive elements of atypical E2F targets specifically bind deregulated E2F and not physiologically activated E2F. (4) Unlike typical E2F targets, promoters of atypical E2F targets are not under repression of RB and are specifically activated by deregulated E2F activity.

http://dx.doi.org/10.5772/54510

4. Regulatory mechanism of promoters is different between typical E2F targets and atypical E2F targets. Promoters of typical E2F targets are repressed by E2F/RB complex. Growth stimulation inactivates RB and releases promoters from the repression by RB. Thus, so-called activation of typical E2F targets by physiological E2F is 'release from repression by RB'. In contrast, activation of atypical E2F targets by deregulated E2F is literally 'activation'. Mutation of EREA, EREK or ERE73s in corresponding full-length promoter did not enhance basal promoter activities, indicating that these three promoters are not under repression through the E2F responsive elements. This is consistent with the observation that binding of physiological E2F to promoters of atypical E2F targets was not observed in ChIP assay, including repressor type E2F4 (Figure 9).

It is generally accepted that the amount of free E2F is important for differential regulation of E2F targets, which have opposite roles in cell-fate determination, as proposed as threshold model [100]. According to this model, when the amount of free E2F (released from repression of RB) is below the threshold, E2F activates only growth-related target genes. When the amount of free E2F exceeds the threshold, E2F activates not only growth-related targets but also pro-apoptotic targets. However, molecular mechanism of how different amount of free E2F differentially regulates target genes is not yet elucidated. The 'quantitative' difference of free E2F seems not sufficient to explain the four differences between deregulated E2F and physiologically activated E2F. The above-mentioned differences strongly suggest the presence of qualitative difference between physiologically activated E2F by 'temporal release' from RB and deregulated E2F induced by 'dysfunction' of the RB pathway [75-77] (Figure 9). The 'qualitative' difference between both of the E2Fs seems to be a useful cue to elucidate how cells discriminate oncogenic growth stimulation from physiological growth stimulation.

5. Conclusion and further research

Accumulating evidence indicates that E2F plays essential roles in cell-fate determination, to grow, stop or die. Together with G1/S gatekeeper RB, E2F governs control of the restriction point, deciding whether to grow or not. Upon normal growth stimulation, physiologically activated E2F facilitates cell proliferation. Upon dysfunction of the RB pathway, deregulated E2F suppresses cell growth by inducing $p27^{Kip1}$ to restrain cells from aberrant cell growth. $p21^{Cip1}$, induced by p53 through activation of the *ARF* gene, may also contribute to suppression of cell growth. When the arrest mechanism failed to stop the aberrant cell cycle progression, E2F induces apoptosis through activation of p53 and TAp73 to protect cells from tumorigenesis. E2F seems to sense and discriminate between normal growth signals and abnormal growth signals originating from various oncogenic changes, asking cells whether to grow, stop or die.

Deregulated E2F activity specifically exists in cancer cell lines but not in normal growing fibroblasts, suggesting that deregulated E2F activity may be a useful means to discriminate abnormally growing cancer cells from physiologically growing normal cells. Since the generation of deregulated E2F activity is expected to be based on the mechanism of oncogenesis, deregulated E2F activity could be a universal marker to discriminate cancer cells from normal growing cells. Analyses of atypical E2F targets suggest that deregulated E2F might be qualitatively different from physiologically activated E2F. One of the most intriguing issues in the future studies would be the molecular nature of deregulated E2F. By elucidating the molecular nature of deregulated E2F, we might be able to specifically approach cancer cells without affecting normal growing cells. For this purpose, qualitative difference between deregulated E2F and physiological E2F is eager to be elucidated.

Acknowledgements

This work is supported by a Grant in-Aid for Science Research from the Ministry of Education, Culture, Sports, Science and Technology of Japan (21570180 to K. Ohtani and 22 2446 to E. Ozono)

Author details

Eiko Ozono[1,2], Shoji Yamaoka[2] and Kiyoshi Ohtani[1]

1 Department of Bioscience, School of Science and Technology, Kwansei Gakuin University, Japan

2 Department of Molecular Virology, Tokyo Medical and Dental University, Japan

References

[1] Lowe SW, Cepero E, Evan G. Intrinsic tumour suppression. Nature. 2004;432(7015) 307-315.

[2] Sherr CJ, McCormick F. The RB and p53 pathways in cancer. Cancer Cell. 2002;2(2) 103-112.

[3] Irwin M, Marin MC, Phillips AC, Seelan RS, Smith DI, Liu W, et al. Role for the p53 homologue p73 in E2F-1-induced apoptosis. Nature. 2000;407(6804) 645-648.

[4] Stiewe T, Pützer BM. Role of the p53-homologue p73 in E2F1-induced apoptosis. Nat Genet. 2000;26(4) 464-469.

[5] Wu L, Timmers C, Maiti B, Saavedra HI, Sang L, Chong GT, et al. The E2F1-3 transcription factors are essential for cellular proliferation. Nature. 2001;414(6862) 457-462.

[6] Pierce AM, Gimenez-Conti IB, Schneider-Broussard R, Martinez LA, Conti CJ, and Johnson DG. Increased E2F1 activity induces skin tumors in mice heterozygous and nullizygous for p53. Proc Natl Acad Sci U S A. 1998;95(15) 8858-8863.

[7] Yamasaki L, Jacks T, Bronson R, Goillot E, Harlow E, Dyson NJ. Tumor induction and tissue atrophy in mice lacking E2F-1. Cell. 1996;85(4) 537-548.

[8] Lee WH, Bookstein R, Hong F, Young LJ, Shew JY, and Lee EY. Human retinoblastoma susceptibility gene: cloning, identification, and sequence. Science. 1987;235(4794) 1394-1399.

[9] Dyson N. The regulation of E2F by pRB-family proteins. Genes Dev. 1998;12(15) 2245-2262.

[10] Cobrinik D. Pocket proteins and cell cycle control. Oncogene. 2005; 24(17) 2796-2809.

[11] Brehm A, Miska EA, McCance DJ, Reid JL, Bannister AJ, Kouzarides T. Retinoblastoma protein recruits histone deacetylase to repress transcription. Nature. 1998;391(6667) 597-601.

[12] Nielsen SJ, Schneider R, Bauer UM, Bannister AJ, Morrison A, O'Carroll D, et al. Rb targets histone H3 methylation and HP1 to promoters. Nature. 2001;412(6846) 561-565.

[13] Cantrup R, Kaushik G, Schuurmans C. Control of Retinal Development by Tumor Suppressor Genes. Tumor Suppressor Genes: In Tech; 2012.

[14] Ohtani K, DeGregori J, Nevins JR. Regulation of the cyclin E gene by transcription factor E2F1. Proc Natl Acad Sci U S A. 1995;92(26) 12146-12150.

[15] Polager S, Ginsberg D. p53 and E2f: partners in life and death. Nat Rev Cancer. 2009;9(10) 738-748.

[16] Sherr CJ. Tumor surveillance via the ARF-p53 pathway. Genes Dev. 1998;12(19) 2984-2991.

[17] Ofir-Rosenfeld Y, Boggs K, Michael D, Kastan MB, Oren M. Mdm2 regulates p53 mRNA translation through inhibitory interactions with ribosomal protein L26. Mol Cell. 2008;32(2) 180-189.

[18] Barak Y, Gottlieb E, Juven-Gershon T, Oren M. Regulation of mdm2 expression by p53: alternative promoters produce transcripts with nonidentical translation potential. Genes Dev. 1994;8(15) 1739-1749.

[19] Marine JC, Jochemsen AG. Mdmx as an essential regulator of p53 activity. Biochem Biophys Res Commun. 2005;331(3) 750-760.

[20] Okamoto K, Taya Y, Nakagama H. Mdmx enhances p53 ubiquitination by altering the substrate preference of the Mdm2 ubiquitin ligase. FEBS Lett. 2009;583(17) 2710-2714.

[21] Weber JD, Taylor LJ, Roussel MF, Sherr CJ, Bar-Sagi D. Nucleolar Arf sequesters Mdm2 and activates p53. Nat Cell Biol. 1999;1(1) 20-26.

[22] Bates S, Phillips AC, Clark PA, Stott F, Peters G, Ludwig RL, et al. $p14^{ARF}$ links the tumour suppressors RB and p53. Nature. 1998; 395(6698) 124-125.

[23] Kaghad M, Bonnet H, Yang A, Creancier L, Biscan JC, Valent A, Minty A, Chalon P, Lelias JM, Dumont X, et al. Monoallelically expressed gene related to p53 at 1p36, a region frequently deleted in neuroblastoma and other human cancers. Cell. 1997;90(4) 809-819.

[24] Stiewe T, Theseling CC, Pützer BM. Transactivation-deficient Delta TA-p73 inhibits p53 by direct competition for DNA binding: implications for tumorigenesis. J Biol Chem. 2002;277(16) 14177-14185.

[25] Lissy NA, Davis PK, Irwin M, Kaelin WG, Dowdy SF. A common E2F-1 and p73 pathway mediates cell death induced by TCR activation. Nature. 2000;407(6804) 642-645.

[26] Zaika A, Irwin M, Sansome C, Moll UM. Oncogenes induce and activate endogenous p73 protein. J Biol Chem. 2001;276(14) 11310-11316.

[27] Flores ER, Tsai KY, Crowley D, Sengupta S, Yang A, McKeon F, et al. p63 and p73 are required for p53-dependent apoptosis in response to DNA damage. Nature. 2002;416(6880) 560-564.

[28] Melino G, Bernassola F, Ranalli M, Yee K, Zong WX, Corazzari M, et al. p73 Induces apoptosis via PUMA transactivation and Bax mitochondrial translocation. J Biol Chem. 2004;279(9) 8076-8083.

[29] Rocco JW, Leong CO, Kuperwasser N, DeYoung MP, Ellisen LW. p63 mediates survival in squamous cell carcinoma by suppression of p73-dependent apoptosis. Cancer Cell. 2006;9(1) 45-56.

[30] Reichel R, Kovesdi I, Nevins JR. Developmental control of a promoter-specific factor that is also regulated by the E1A gene product. Cell. 1987;48(3) 501-506.

[31] Ohtani K, Komori H, Ozono E, Ikeda MA, Iwanaga R. Distinct transcriptional regulation by E2F in cell growth and tumor suppression. Control of Cellular Physiology by E2F Transcription Factors: Research Signpost; 2008.

[32] Johnson DG, Schwarz JK, Cress WD, Nevins JR. Expression of transcription factor E2F1 induces quiescent cells to enter S phase. Nature. 1993;365(6444) 349-352.

[33] Singh P, Wong SH, Hong W. Overexpression of E2F-1 in rat embryo fibroblasts leads to neoplastic transformation. EMBO J. 1994;13(14) 3329-3338.

[34] Johnson DG, Cress WD, Jakoi L, Nevins JR. Oncogenic capacity of the E2F1 gene. Proc Natl Acad Sci U S A. 1994;91(26), 12823-12827.

[35] Shan B, Lee WH. Deregulated expression of E2F-1 induces S-phase entry and leads to apoptosis. Mol Cell Biol. 1994;14(12) 8166-8173.

[36] Dimri GP, Itahana K, Acosta M, Campisi J. Regulation of a senescence checkpoint response by the E2F1 transcription factor and p14^{ARF} tumor suppressor. Mol Cell Biol. 2000;20(1) 273-285.

[37] Ivey-Hoyle M, Conroy R, Huber HE, Goodhart PJ, Oliff A, Heimbrook DC. Cloning and characterization of E2F-2, a novel protein with the biochemical properties of transcription factor E2F. Mol Cell Biol. 1993;13(12) 7802-7812.

[38] Zhu JW, Field SJ, Gore L, Thompson M, Yang H, Fujiwara Y, et al. E2F1 and E2F2 determine thresholds for antigen-induced T-cell proliferation and suppress tumorigenesis. Mol Cell Biol. 2001;21(24) 8547-8564.

[39] Murga M, Fernández-Capetillo O, Field SJ, Moreno B, Borlado LR, Fujiwara Y, et al. Mutation of E2F2 in mice causes enhanced T lymphocyte proliferation, leading to the development of autoimmunity. Immunity. 2001;15(6) 959-970.

[40] Opavsky R, Tsai SY, Guimond M, Arora A, Opavska J, Becknell B, et al. Specific tumor suppressor function for E2F2 in Myc-induced T cell lymphomagenesis. Proc Natl Acad Sci U S A. 2007;104(39) 15400-15405.

[41] Cloud JE, Rogers C, Reza TL, Ziebold U, Stone JR, Picard MH, et al. Mutant mouse models reveal the relative roles of E2F1 and E2F3 in vivo. Mol Cell Biol. 2002;22(8) 2663-2672.

[42] Humbert PO, Verona R, Trimarchi JM, Rogers C, Dandapani S, Lees JA. E2f3 is critical for normal cellular proliferation. Genes Dev. 2000;14(6) 690-703.

[43] Lees JA, Saito M, Vidal M, Valentine M, Look T, Harlow E, et al. The retinoblastoma protein binds to a family of E2F transcription factors. Mol Cell Biol. 1993;13(12) 7813-7825.

[44] Leone G, Nuckolls F, Ishida S, Adams M, Sears R, Jakoi L, et al. Identification of a novel E2F3 product suggests a mechanism for determining specificity of repression by Rb proteins. Mol Cell Biol. 2000;20(10) 3626-3632.

[45] Danielian PS, Friesenhahn LB, Faust AM, West JC, Caron AM, Bronson RT, et al. E2f3a and E2f3b make overlapping but different contributions to total E2f3 activity. Oncogene. 2008;27(51) 6561-6570.

[46] Tsai SY, Opavsky R, Sharma N, Wu L, Naidu S, Nolan E, et al. Mouse development with a single E2F activator. Nature. 2008;454(7208) 1137-1141.

[47] Asp P, Acosta-Alvear D, Tsikitis M, van Oevelen C, Dynlacht BD. E2f3b plays an essential role in myogenic differentiation through isoform-specific gene regulation. Genes Dev. 2009;23(1) 37-53.

[48] Dyson N, Dembski M, Fattaey A, Ngwu C, Ewen M, Helin K. Analysis of p107-associated proteins: p107 associates with a form of E2F that differs from pRB-associated E2F-1. J Virol. 1993;67(12) 7641-7647.

[49] Hijmans EM, Voorhoeve PM, Beijersbergen RL, van 't Veer LJ, Bernards R. E2F-5, a new E2F family member that interacts with p130 in vivo. Mol Cell Biol. 1995;15(6) 3082-3089.

[50] Humbert PO, Rogers C, Ganiatsas S, Landsberg RL, Trimarchi JM, Dandapani S, et al. E2F4 is essential for normal erythrocyte maturation and neonatal viability. Mol Cell. 2000;6(2) 281-291.

[51] Rempel RE, Saenz-Robles MT, Storms R, Morham S, Ishida S, Engel A, et al. Loss of E2F4 activity leads to abnormal development of multiple cellular lineages. Mol Cell. 2000;6(2):293-306.

[52] Lindeman GJ, Dagnino L, Gaubatz S, Xu Y, Bronson RT, Warren HB, et al. A specific, nonproliferative role for E2F-5 in choroid plexus function revealed by gene targeting. Genes Dev. 1998;12(8) 1092-1098.

[53] DeGregori J. The genetics of the E2F family of transcription factors: shared functions and unique roles. Biochim Biophys Acta. 2002;1602(2) 131-150.

[54] Trimarchi JM, Fairchild B, Verona R, Moberg K, Andon N, Lees JA. E2F-6, a member of the E2F family that can behave as a transcriptional repressor. Proc Natl Acad Sci U S A. 1998;95(6) 2850-2855.

[55] Ogawa H, Ishiguro K, Gaubatz S, Livingston DM, Nakatani Y. A complex with chromatin modifiers that occupies E2F- and Myc-responsive genes in G0 cells. Science. 2002;296(5570) 1132-1136.

[56] Storre J, Elsässer HP, Fuchs M, Ullmann D, Livingston DM, Gaubatz S. Homeotic transformations of the axial skeleton that accompany a targeted deletion of E2f6. EMBO Rep. 2002;3(7) 695-700.

[57] Giangrande PH, Zhu W, Schlisio S, Sun X, Mori S, Gaubatz S, et al. A role for E2F6 in distinguishing G1/S- and G2/M-specific transcription. Genes Dev. 2004;18(23) 2941-2951.

[58] Li J, Ran C, Li E, Gordon F, Comstock G, Siddiqui H, et al. Synergistic function of E2F7 and E2F8 is essential for cell survival and embryonic development. Dev Cell. 2008;14(1) 62-75.

[59] Weijts BG, Bakker WJ, Cornelissen PW, Liang KH, Schaftenaar FH, Westendorp B, et al. E2F7 and E2F8 promote angiogenesis through transcriptional activation of VEGFA in cooperation with HIF1. EMBO J. 2012;31(19) 3871-3884.

[60] Attwooll C, Lazzerini Denchi E, Helin K. The E2F family: specific functions and overlapping interests. EMBO J. 2004;23(24) 4709-4716

[61] Krek W, Ewen ME, Shirodkar S, Arany Z, Kaelin WG Jr, Livingston DM. Negative regulation of the growth-promoting transcription factor E2F-1 by a stably bound cyclin A-dependent protein kinase. Cell. 1994;78(1) 161-172.

[62] Yoshida K, Inoue I. Regulation of Geminin and Cdt1 expression by E2F transcription factors. Oncogene. 2004;23(21) 3802-3812.

[63] Iwanaga R, Komori H, Ohtani K. Differential regulation of expression of the mammalian DNA repair genes by growth stimulation. Oncogene. 2004;23(53) 8581-8590.

[64] Berkovich E, Ginsberg D. ATM is a target for positive regulation by E2F-1. Oncogene. 2003;22(2) 161-167.

[65] Carrassa L, Broggini M, Vikhanskaya F, Damia G. Characterization of the 5'flanking region of the human *Chk1* gene: identification of E2F1 functional sites. Cell Cycle. 2003;2(6) 604-609.

[66] Iwanaga R, Komori H, Ishida S, Okamura N, Nakayama K, Nakayama KI, et al. Identification of novel E2F1 target genes regulated in cell cycle-dependent and independent manners. Oncogene. 2006;25(12) 1786-1798.

[67] Fox EJ, Wright SC. The transcriptional repressor gene *Mad3* is a novel target for regulation by E2F1. Biochem J. 2003;370(Pt 1):307-313.

[68] Hsu JY, Reimann JD, Sørensen CS, Lukas J, Jackson PK. E2F-dependent accumulation of hEmi1 regulates S phase entry by inhibiting APC^{Cdh1}. Nat Cell Biol. 2002;4(5) 358-366.

[69] Zhang L, Wang C. F-box protein Skp2: a novel transcriptional target of E2F. Oncogene. 2006;25(18) 2615-2627.

[70] Thalmeier K, Synovzik H, Mertz R, Winnacker EL, Lipp M. Nuclear factor E2F mediates basic transcription and trans-activation by E1a of the human MYC promoter. Genes Dev. 1989;3(4) 527-536.

[71] Chen CR, Kang Y, Siegel PM, Massagué J. E2F4/5 and p107 as Smad cofactors linking the TGFβ receptor to c-myc repression. Cell. 2002;110(1) 19-32.

[72] Müller H., Bracken AP, Vernell R, Moroni MC, Christians F, Grassilli E, et al.E2Fs regulate the expression of genes involved in differentiation, development, proliferation, and apoptosis. Genes Dev. 2001;15(3) 267-285.

[73] Blanchet E, Annicotte JS, Lagarrigue S, Aguilar V, Clapé C, Chavey C, et al. E2F transcription factor-1 regulates oxidative metabolism. Nat Cell Biol. 2011;13(9) 1146-1152.

[74] Emmrich S, Pützer BM. Checks and balances: E2F-microRNA crosstalk in cancer control. Cell Cycle. 2010;9(13) 2555-2567.

[75] Komori H, Enomoto M, Nakamura M, Iwanaga R, Ohtani K. Distinct E2F-mediated transcriptional program regulates $p14^{ARF}$ gene expression. EMBO J. 2005;24(21) 3724-3736.

[76] Ozono E, Komori H, Iwanaga R, Ikeda MA, Iseki S, Ohtani K. E2F-like elements in $p27^{Kip1}$ promoter specifically sense deregulated E2F activity. Genes Cells. 2009;14(1) 89-99.

[77] Ozono E, Komori H, Iwanaga R, Tanaka T, Sakae T, Kitamura H, et al. Tumor suppressor *TAp73* gene specifically responds to deregulated E2F activity in human normal fibroblasts. Genes Cells. 2012;17(8) 660-672.

[78] Ikeda MA, Jakoi L, Nevins JR. A unique role for the Rb protein in controlling E2F accumulation during cell growth and differentiation. Proc Natl Acad Sci U S A. 1996;93(8) 3215-3220.

[79] Gill RM, Hamel PA, Zhe J, Zacksenhaus E, Gallie BL, Phillips RA. Characterization of the human RB1 promoter and of elements involved in transcriptional regulation. Cell Growth Differ. 1994;5(5) 467-474.

[80] Buchkovich K, Duffy LA, Harlow E. The retinoblastoma protein is phosphorylated during specific phases of the cell cycle. Cell. 1989 ;58(6) 1097-1105.

[81] Kamijo T, Zindy F, Roussel MF, Quelle DE, Downing JR, Ashmun RA, et al. Tumor suppression at the mouse *INK4a* locus mediated by the alternative reading frame product p19*ARF*. Cell. 1997;91(5) 649-659.

[82] Junttila MR, Evan GI. p53--a Jack of all trades but master of none. Nat Rev Cancer. 2009;9(11) 821-829.

[83] Zindy F, Williams RT, Baudino TA, Rehg JE, Skapek SX, Cleveland JL, et al. *Arf* tumor suppressor promoter monitors latent oncogenic signals *in vivo*. Proc Natl Acad Sci U S A. 2003;100(26) 15930-15935.

[84] Sherr CJ. Divorcing ARF and p53: an unsettled case. Nat Rev Cancer. 2006;6(9) 663-673.

[85] Alexander K, Hinds PW. Requirement for $p27^{KIP1}$ in retinoblastoma protein-mediated senescence. Mol Cell Biol. 2001;21(11) 3616-3631.

[86] Ji P, Jiang H, Rekhtman K, Bloom J, Ichetovkin M, Pagano M, et al. An Rb-Skp2-p27 pathway mediates acute cell cycle inhibition by Rb and is retained in a partial-penetrance Rb mutant. Mol Cell. 2004;16(1) 47-58.

[87] Thomas DM, Johnson SA, Sims NA, Trivett MK, Slavin JL, Rubin BP, et al. Terminal osteoblast differentiation, mediated by runx2 and $p27^{KIP1}$, is disrupted in osteosarcoma. J Cell Biol. 2004;167(5) 925-934.

[88] Binne UK, Classon MK, Dick FA, Wei W, Rape M, Kaelin WG Jr, et al. Retinoblastoma protein and anaphase-promoting complex physically interact and functionally cooperate during cell-cycle exit. Nat Cell Biol 2007;9(2) 225-232.

[89] Murray-Zmijewski F, Lane DP, Bourdon JC. p53/p63/p73 isoforms: an orchestra of isoforms to harmonise cell differentiation and response to stress. Cell Death Differ. 2006;13(6) 962-972.

[90] Pietsch EC, Sykes SM, McMahon SB, Murphy ME. The p53 family and programmed cell death. Oncogene. 2008;27(50) 6507-6521.

[91] Irwin MS, Kaelin WG. p53 family update: p73 and p63 develop their own identities. Cell Growth Differ. 2001;12(7) 337-349.

[92] Deyoung MP, Ellisen LW. p63 and p73 in human cancer: defining the network. Oncogene. 2007;26(36) 5169-5183.

[93] Domínguez G, García JM, Peña C, Silva J, García V, Martínez L, et al. DeltaTAp73 upregulation correlates with poor prognosis in human tumors: putative in vivo network involving p73 isoforms, p53, and E2F-1. J Clin Oncol. 2006;24(5) 805-815.

[94] Wager M, Guilhot J, Blanc JL, Ferrand S, Milin S, Bataille B, et al. Prognostic value of increase in transcript levels of Tp73 DeltaEx2-3 isoforms in low-grade glioma patients. Br J Cancer 2006;95(8) 1062-1069.

[95] Park BJ, Lee SJ, Kim JI, Lee CH, Chang SG, Park JH, et al. Frequent alteration of p63 expression in human primary bladder carcinomas. Cancer Res 2000;60(13) 3370-3374.

[96] Stiewe T. The p53 family in differentiation and tumorigenesis. Nat Rev Cancer. 2007;7(3) 165-168.

[97] Li Y, Prives C. Are interactions with p63 and p73 involved in mutant p53 gain of oncogenic function? Oncogene. 2007;26(15) 2220-2225.

[98] Tomasini R, Tsuchihara K, Wilhelm M, Fujitani M, Rufini A, Cheung CC, et al. TAp73 knockout shows genomic instability with infertility and tumor suppressor functions. Genes Dev. 2008;22(19) 2677-26791.

[99] del Arroyo AG, El Messaoudi S, Clark PA, James M, Stott F, Bracken A, et al. E2F-dependent induction of $p14^{ARF}$ during cell cycle re-entry in human T cells. Cell Cycle. 2007;6(21) 2697-2705.

[100] Trimarchi JM, Lees JA. Sibling rivalry in the E2F family. Nat Rev Mol Cell Biol. 2002;3(1) 11-20.

Chapter 3

MicroRNAs and lncRNAs as Tumour Suppressors

Emanuela Boštjančič and Damjan Glavač

Additional information is available at the end of the chapter

http://dx.doi.org/10.5772/54701

1. Introduction

Cancer is one of the most serious diseases around the world and it is the third leading cause of death, exceeded only by heart and infectious diseases [1]. There are five major steps for cancer development: initiation, promotion, malignant conversion, progression, and metastasis [2]. Cancer is result of process, where somatic cells mutate and escape the controlled balance of gene expression and cellular networks that maintain cellular homeostasis, which normally prevent unwanted expansion. Perturbations in these pathways results in cellular transformation, where cancer cells differ from their normal counterparts in many characteristics, as is loss of differentiation, increased invasiveness, and decreased drug sensitivity [3-4]. There are six primary hallmarks of cancer: unlimited cell proliferation, autonomous growth without the need of external signals, resistance to growth inhibitory signals, escape from apoptosis the ability to recruit new vasculature and increased tissue invasion and metastasis [5]. The formation of cancer is therefore fundamentally genetic and epigenetic disease requiring accumulation of genomic alterations to inactivate tumour suppressor and activate proto-oncogenes [6]. These results in combined interaction of both tumour suppressors, that are not able to inhibit tumour development and protect cells against mutation that initiate transformation, and cancer inducers, which promotes cancer development as initiators of cellular transformation. When cells exhibit abnormal growth and loss of apoptosis, it usually results in cancer formation [2].

Genetic studies have revealed the mutational and epigenetic alterations of protein-coding genes that control DNA damage response, growth arrest, cell survival and apoptotic pathways [4]. Until recent years ago, the central dogma of molecular biology was that genetic information is stored in protein-coding genes with RNA as an intermediate between DNA sequence and its encoded protein [7]. Recent studies suggest that advanced stages of cancer are possessing more severe molecular perturbations and that this could be due to function of non-coding RNAs (ncRNAs), which were previously known only to have infrastructural functions (as ribosomal RNA, transfer RNA, small nuclear and nucleolar RNA). Eukaryotic genomes are extensively transcribed into thousands of long and short ncRNAs, which are group of endogenous RNAs that also function as regulators of gene expression. They are involved in developmental, physiological as well as pathological processes [7,8].

However, in this review, the following characteristics of ncRNA in human cancers will be summarized: (i) the current understanding of the critical role that lncRNAs and miRNAs may play in cancer as tumour suppressors; (ii) outline current knowledge about some specific lncRNA and miRNAs and their target genes in cancer; (iii) highlight their potential as biomarkers for patho-histological subtype classification; and (iv) highlight their potential as biomarkers and as circulating biomarkers and therapeutic targets in cancer.

Since the majority of research regarding regulatory ncRNAs as tumour suppressor was performed on miRNAs and in lesser extend on lncRNAs/lincRNAs, will this review further focused on these two groups of ncRNAs.

2. Brief overview of non-coding RNAs (ncRNAs)

Classification

Of transcribed eukaryotic genomes, only 1-2 % encode for proteins, whereas the vast majority are ncRNAs that are in more or less functional transcripts. The regulatory ncRNAs are important regulators of gene expression in many eukaryotes and are involved in a wide range of functions in eukaryotic biology [8,9].

Based on their function, ncRNAs can be divided into two groups. First is infrastructural group, with ribosomal RNAs (rRNAs), transfer RNAs (tRNAs), small nuclear RNAs (snRNAs) and small nucleolar RNAs (snoRNAs). Second is regulatory group, with microRNAs (miRNAs), piwi-interacting RNAs (piRNAs), small-interfering RNAs (siRNAs), long non-coding RNAs (lncRNAs), large intergenic non-coding RNAs (lincRNAs), promoter-associated small RNAs (PARs), repeat-associated short interfering RNAs (rasiRNAs) and enhancer RNAs (eRNAs) [8,9]. Recent findings suggest that some structural ncRNAs (e.g. snoRNAs) not only have infrastructural function but have regulatory as well [8].

Based on length, the regulatory ncRNAs can be divided in two groups: larger than 200 nucleotides (nt) are lncRNA, lincRNA, eRNA, whereas the others are smaller than 200 nt, with exception of PARs that are 16-30 nt long or up to 200 nt. Distinct classes of small RNAs are distinguished by their origins, and these are: snRNAs, snoRNAs, miRNAs, piRNAs, siRNAs, and rasiRNAs [8,9].

miRNAs and snoRNAs share similarities in processing pathways and protein interaction partners, genomic organization and location, as well as levels of conservation. However, similarities in sub-cellular localization have been also observed, since large proportion of human mature miRNAs have been detected in the nucleus as well as a subset of small RNAs derived from snoRNAs have been detected in the cytoplasm [10].

Functional role

The most widely studied and characterized of all the regulatory ncRNAs are miRNAs. The roles of regulatory ncRNAs, other than miRNAs, in the mediating transcriptional regulation, chromatin remodelling, post-transcriptional regulation, and other processes are less well understood. The contexts of gene regulation by ncRNAs in non-human systems provided insights into how these processes could function in human cells. Regulatory ncRNAs are involved in diverse cellular pathways, such as development and stem cell maintenance, response to stress and environmental stimuli, regulating chromatin structure and remodelling, chromosome architecture and genome integrity, transcription (positive or negative impact), and post-transcription processing (splicing, transport) and most commonly mRNA stability (translation, degradation) [8,9]. Some ncRNAs trigger different types of gene silencing that are collectively referred to as RNA silencing or RNA interference [11].

RNA interference (RNAi)

RNAi is RNA-guided regulation of gene expression, historically known by other names, including post-transcriptional gene silencing. It is believed to be an evolutionary conserved mechanism in response to presence of foreign dsRNA in the cell. A key step in this silencing pathway is the processing of dsRNAs into short RNA duplexes of characteristic size and structure. The enzyme Dicer, which initiates the RNAi pathway, cleaves dsRNA to short double-stranded fragments of 20–25 base pairs (bp), named siRNAs. siRNAs usually possess perfect complementarity to the mRNA of target gene, thus causing its degradation. When the dsRNA is exogenous, coming from infection by a virus with RNA genome or laboratory manipulations, the RNA is imported directly into the cytoplasm where it is cleave by the Dicer. On other hand, the initiating dsRNA could be result of endogenously expressed RNA-coding genes from the genome. Some of small regulatory RNAs are processed in a similar way or with components of RNAi pathway [11].

2.1. Brief introduction to miRNAs

Genomic organization

miRNAs are endogenously expressed small (~22 nt), single-stranded ncRNAs. It is predicted that they constitute ~1-5 % of human genes [1,12] and in an update from August 2012, miRBase v19 was released with a list of 2019 unique mature human miRNAs. miRNAs are encoded as a single gene or gene clusters, with some of miRNA clusters being co-regulated and co-transcribed. Intergenic miRNAs are transcribed as an independent transcription unit, as a monocistronic, bicistronic or polycistronic primary transcripts [13]. Up to 60 % of currently known miRNAs are proposed to be from intronic sequences of either protein coding or non-coding transcription units and suggestion has been made that some miRNAs are also encoded in antisense DNA, which is not transcribed to the mRNA. Intronic miRNA are preferentially transcribed in the same orientation as the host gene and are together with their host transcripts co-regulated and co-transcribed from the same promoter. They are processed from introns, as are many snoRNA. Within the genome, there might be more than one copy of particular miRNA [13,14].

Biogenesis

miRNAs expression is determined by intrinsic cellular factors and diverse environmental variables [1]. As for protein-coding genes it is known, that regulation of miRNA transcription and expression depends on transcription factors and epigenetic mechanisms (e.g. p53, Myc, and myogenin). In general, from genes encoding miRNAs is transcription guided by RNA-polymerase II (Pol II). Resulting primary transcript (several hundred bases to several kilobases), named *pri-miRNA*, forms distinctive hairpin-shaped stem-loop secondary structure and contains poly-A tail and a cap, similarly to protein-coding mRNA. *pri-miRNA* is processed in the nucleus by Drosha, an RNase III enzyme. The resulting 70-nt stem-loop structure called *pre-miRNA* with a 5' phosphate and 3' 2-nt overhang is imported into the cytoplasm by a transporter protein, Exportin 5. The double-stranded RNA portion of *pre-miRNA* is bound and cleaved by Dicer, another RNase III enzyme, which produces a miRNA:miRNA* duplex (a transient intermediate in miRNA biogenesis, 20–25 nt). One of the two strands of each fragment is together with proteins argonaute (Ago), incorporated into a complex called the miRNA-containing ribonucleoprotein complex (miRNP). It is believed that the *guide strand* is determined on the basis of the less energetically stable 5' end. The resulting complex base-pair with complementary 3'-UTR mRNA sequences. The other strand, miRNA* is presumably degraded, although there are increasing evidence that either or both strands may be functional [2,9]. The schematic overview of canonical miRNA biosynthesis pathway has been represented elsewhere [15].

Numerous alternative pathways differing from canonical miRNA biogenesis pathway have been described recently and subset of several diverse longer non-coding RNAs can serve as precursors for miRNAs [10,16]. As an example, intronic miRNAs presumably bypass Drosha cleavage, since through *pre-mRNA* splicing/debranching machinery is produced an approx. 60-nt hairpin precursor miRNA (*pre-miRNA*) that enter biogenesis pathway at the step of Exportin 5 [14]. However, some of the post-transcriptional mechanism include miRNA editing, which is mechanism mediated by adenine deaminase of alteration of adenines to inosines, and not yet thoroughly studied regulations of miRNA, such as export step from nucleus or miRNAs turnover rate [17].

miRNAs mechanism

The functional role of miRNA varies, but the primary mechanism of miRNA action in mammals is believed to be base-pairing to 3'-UTR of target mRNA followed by inhibition of mRNA translation (when base pairing between these two molecules is incomplete) or deadenylation and degradation (perfect complementarity of miRNA:mRNA binding) [9]. Especially in animals, the primary mechanism of miRNA action is reducing mRNA translation and each miRNA can inhibit the translation of as many as 200 target genes. In addition, mRNA can be regulated by more than one miRNA. The cooperative action of multiple identical (multiplicity) or different miRNPs (cooperativity) appears to provide the most efficient translational inhibition. Additional mechanism to increase the specificity of miRNAs is combinatorial control of gene expression, which may be also provided by a set of co-ordinately expressed miRNAs. Proteins or mRNA secondary structures could restrict miRNP accessibility to the UTRs, or may facilitate recognition of the authentic mRNA targets [12,13,18,19].

There is the prospect that some miRNA might specify more than just post-transcriptional repression [9,13]. miRNAs may also target promoter to regulate transcription through epigenetic mechanism. miRNAs have been paradoxically also shown to up-regulate gene expression by enhancing translation under specific conditions [9].

Biological function

Translational repression, as major mechanism of miRNAs, may in normal cell conditions occur in different ways: as switch off the targets, that is for mRNAs that should not be expressed in a particular cell type, the protein production is reduced to inconsequential levels; as fine-tuners of target expression, that is when miRNAs can adjust protein output for customized expression in different cell types; as neutralizers of target expression, that is when miRNAs act as bystanders, where down-regulation by miRNAs is tolerated or reversed by feedback processes [13]. Role of miRNA can be further divided in three paradigms: combinatorial control (defined as cooperativity), cell-to-cell variation, specific (tissue-specific and/or cell-type specific) and housekeeping functions [20].

Despite the large number of identified miRNAs, the scope of their roles in regulating cellular gene expression is not fully understood [11]. It is believed that miRNAs through negative gene regulation influence at least 50 % of genes within the human genome [9]. Expression profiling of many miRNAs in various normal and diseased tissues have demonstrated unique spatial and temporal expression patterns. Many miRNAs are important at distinct stages of development and have been found to regulate a variety of physiological and pathological processes [11]. miRNAs are involved in a numerous biological processes, such as stem cell division and developmental timing, proper organ formation, embryonic pattering and body growth, proliferation and differentiation, apoptosis, epithelial-mesenchymal-transition (EMT), cholesterol metabolism and regulation of insulin secretion, resistance to viral infection and oxidative stress, immune response etc. [2,11]. All these effects may occur by regulating or being regulated by the expression of signalling molecules, such as cytokines, growth factors, transcription factors, pro-apoptotic and anti-apoptotic genes [21]. With all different genes and expression patterns, it is reasonable to propose that every cell type at each developmental stage might have a distinct miRNA expression profile.

Defining miRNA targets and databases

Up to date, over 2000 human miRNAs have been identified and this number is still growing. All annotated miRNAs are collected in miRBase [22]. The first step in miRNA target identification is usually defining reciprocally regulated miRNA-mRNA or miRNA-protein. Since miRNAs target mRNA mainly by incomplete base-pairing, many computational methods have been recently developed for further identifying potential miRNA targets [23]. Most of these methods search for three criteria in predicting miRNA target genes: first, multiple conserved regions of miRNA complementarities within 3'-UTR of target mRNA (evolutionary conservation); second, interaction between seven consecutive nucleotides in the target mRNAs 3'-UTR and the 1-8 nt ("seed sequence") at the 5' miR-

NA end; third, stability of base pairing and predicted binding energy. Further complicating target site prediction in mammals is the fact that not all 3'-UTR sites with perfect complementarities to the miRNA seed nucleotides are functional. Moreover, mRNAs sites with imperfect seed complementarities can themselves be very good miRNA targets [24,25]. Bioinformatics is therefore much noisier and more prone to false positive and false negative predictions. Among many available programs for predicting mRNA targets for specific miRNA, none of these programs can be used as an independently approach for validating the targets, and all predicted targets must be validated *in vitro* and/or *in vivo.* Thus the gold standard for miRNA target identification is the experimental demonstration that a luciferase reporter fused to the 3'-UTR of the predicted target is repressed by over-expression of the miRNA and that this repression is abrogated by point mutation in the target sequences in 3'-UTR [26,27]. Finaly, expression profiling in human disease gives the starting point for target verification/validation and association to disease prognosis and pathogenesis. All identified disease related miRNAs are listed in The human microRNA disease database (HMDD), where you can search for specific miRNA, for tissue expression of annotated miRNAs, and for disease related miRNAs [28].

2.2. Brief introduction to lncRNA

Genomic organization

LncRNAs are those longer than 200 nt, and many of them can also act as primary transcripts for the production of short RNAs [9]. It is estimated that total number of lncRNA transcripts, including new unexplored, is approx. 15000. Thousands of protein-coding genes in humans harbour natural antisense transcripts (approx. 61 % of transcribed regions show antisense transcription) belonging to the lncRNA, and majority of known lncRNAs in some way overlap protein-coding loci. All these data are giving the importance to lncRNA annotation [29].

Classification

LncRNAs can be classified according to their proximity to protein coding genes. There are five categories of lncRNAs: sense, antisense, bidirectional, intronic, intergenic. Just to mention a few of them, lincRNAs, a class of ncRNAs, exhibit a high conservation between different species; they both up- and down- regulate hundreds of gene expression and participate in the establishment of cell type-specific epigenetic states [9]. Further, ncRNAs were found expressed at enhancer regions, suggesting that some enhancer RNA is also transcribed with an average size of 800 nt; these transcripts are termed eRNAs. Studies propose a possible role for eRNAs as transcriptional activators, however, question remains whether such eRNAs are in fact a subset of the activating lncRNAs. Similar to eRNA, a novel diverse class of ncRNAs has been linked to the promoters, called PARs, ranging from 16-36 nt to 200 nt. It is suggested that they participate in the transcriptional regulation [9]. Most lncRNAs are characterized by low expression levels, low level of sequence conservation, by composition of poly-A tail and without poly-A tail as well as by spliced and un-spliced forms. They are believed to have nuclear localization, but can also accumulate in cytoplasm of cells [3].

Function

lncRNAs may act through diverse molecular mechanisms, and play regulatory as well as structural roles in different biological processes [3]. Many of the identified lncRNAs show spatial- and temporal-specific patterns of expression. Almost every step in the life cycle of genes – transcription, mRNAs splicing, RNA decay, and translation – can be influenced by lncRNAs. Generally lncRNAs have been implicated in gene-regulatory roles, such as chromatin dosage-compensation, imprinting, epigenetic regulation, cell cycle control, nuclear and cytoplasmic trafficking, cell differentiation etc. [7]. A number of studies suggest that lncRNAs are key components of the epigenetic regulatory network [4]. Two general modes of lncRNAs regulation seem to be important: interaction with chromatin remodelling complexes that promote silencing of specific genes; and modulation of splicing factors. Chromatin remodelling guided by ncRNAs contributes to the establishment of chromatin structure and to the maintenance of epigenetic memory. Various ncRNAs have been identified as regulators of chromatin structure and gene expression [30]. Additional mechanisms of action are yet to be revealed [3].

Database

The lncRNA database provides sequence, structural, and conservation evidence for multi-species lncRNAs, together with a list of lncRNAs that are experimentally known to interact with coding mRNAs, harbouring other short ncRNAs and other characteristics of specific lncRNA [29].

3. Involvement of ncRNA in cancer

Three major mechanisms are known to give rise to deregulated ncRNAs function, genetic alterations, epigenetic alterations, and in case of miRNAs, an aberrant miRNA biogenesis machinery. Since brief overview of first two mechanisms is described below, will be here mentioned only aberrant machinery of miRNA processing. Proteins involved in miRNA biogenesis (Drosha, Dicer, Ago) are deregulated in several cancers. Co-factors involved in miRNA biogenesis can be mutated causing consequently deregulation of Dicer; Exportin 5, mediating pre-miRNA nuclear export, is often mutated and truncated, leaving pre-miRNAs within nucleus [31,32].

3.1. Mutations, SNPs and epigenetics of ncRNAs

Cancer cells have different genetic and epigenetic changes from their normal counterparts and the role of ncRNAs in mediating these differences is beginning to emerge. Specific genetic polymorphisms are associated with the risk of developing several types of cancer [7-9]. Multiple studies have identified small-scale and large-scale mutations and genomic alterations affecting also noncoding regions of the genome. Some of these mutations are structural alterations, rearrangements and chromosomal translocation, amplification, loss of heterozigocity and copy-number variation, nucleotide expansion, and single-nucleotide polymorphisms (SNPs), and they are linking distinct types of mutations in ncRNA genes with diverse diseases [7]. First, lncRNA have already been implicated in human diseases such as cancer and neurodegeneration [33]. Second, approx. half of miRNA genes are encoded in genomic region prone to cancer-associated rearrangements or in fragile chromosomal sites (amplified, deleted or rearranged) that are often associated with cancer, such as ovarian and breast carcinomas, and melanomas [8,11]. Third, presence of SNPs in miRNAs, where disruption of miRNA target interaction either in the miRNA gene or its target site (3'-UTR mRNA) can lead to complete gain or loss of the miRNA function or target gene thus causing disease [34,35]. In contrast to the miRNA target sites in mRNA transcripts, where the potential of variation is huge, variants identified in miRNA precursor sequences tend to be rarer [36]. The presence of SNPs in *pri-miRNA* or *pre-miRNA* can in addition affect the processing of miRNAs, their expression and/or binding to target mRNA [27,37]. Forth, recent advances in miRNA research have provided evidence of a miRNA association with epigenetic mechanisms activated in diseased human tissues [38]. Heritable changes in gene expression that do not involve coding sequence modification are referred as epigenetics. Gene regulation by ncRNAs was considered as an epigenetic mechanism, but ncRNAs can be regulated by the same mechanism in which they participate [39]. DNA methylation, one of the two major epigenetic mechanisms, leads to gene silencing, and serves as an alternative mechanism of gene inactivation. The aberrant DNA methylation of gene promoters has been shown to result in the inactivation of tumour suppressor genes [40]. For an example, *miR-34* family is a family of tumour suppressors' genes, with *miR-34a* being deregulated by DNA methylation in both epithelial and haematological cancers. *miR-34* is an important component of the p53 tumour suppressor network, and p53 is a predicted target for members of the *miR-34* family. *miR-34a* reinforces the tumour suppressor function of p53, transactivation of *miR-34a* by p53 was also shown to promote apoptosis [11,41-44]. Another example is that *miR-29s* could target two enzymes of methylation process, DNMT3A and DNMT3B [39]. Antisense ncRNAs have been recently showed to be implicated in the silencing of tumour suppressor genes through epigenetic remodelling events [30]. All these miRNAs abnormalities suggest that they play a broad role in cancer pathogenesis.

3.2. Promising role of ncRNAs in cancer: As cancer-subtype classifiers and detection in body fluids

ncRNAs have been recognized as gene-specific regulators. They are similar in activity to a large number of protein transcription factors that are known to be critical in the transformation of cells to a malignant state. Majority of research has been involved in defining the role of miRNAs in cancer; however, lincRNAs have been shown to play role in tumour development by promoting the expression of genes involved in metastasis and angiogenesis [9]. Genome-wide analyses have shown that ncRNAs have distinct signatures specific for a certain cancer type. Importance of combining ncRNAs with other biomarkers for cancer detection and prognosis would improve cancer risk assessment, detection, and prognosis. Thus, there is a need to combine genomic mutations with ncRNA markers to develop marker panels for more accurate risk assessment and early diagnosis [7-9].

Most cancers are diagnosed in advance stages, leading to poor outcome. Intense investigation is going on seeking specific molecular changes that are able to identify patients with early cancer or precursor lesions [1]. Genome-wide expression profiling has examined miR-

NAs in preneoplasia or their usefulness to predict progression from preneoplasia to cancer. Several lines of evidence suggest the potential usefulness of ncRNAs, particularly miRNAs: first, as signature of early events in carcinogenesis and as biomarkers in early cancer detection, second, as differential indicators of benign tumours, preneoplasia, and neoplasia, and third, that miRNAs and perhaps other ncRNAs might be useful in determining which preneoplastic lesions are likely to progress to cancer. Distinguishing benign diseases and certain non-precancerous lesions from precancerous lesions and metastatic tumours would improve patient outcomes, survival, and reduce patient discomfort [45]. Characterization of ncRNAs involved in the development or maintenance of oncogenic states may therefore define ncRNAs as early biomarkers for the emergence of cancer, and could have have an impact on the development of tools for disease diagnosis and treatment [30].

miRNAs are believed to be promising potential biomarkers for cancer diagnosis, prognosis and targets for therapy. As potential markers for diagnosis are better classification factors than mRNAs. miRNAs seems to be evolutionarily selected gene regulatory molecules, their expression profiles might therefore be rich in gene regulatory information. Only small percentage of the 16000 genes on the mRNA-expression arrays are regulatory molecules. This difference may be responsible for more efficient microRNA expression arrays in classifying cancer than mRNA-expression arrays [21,45,46]. Some of the key features of miRNAs that make them useful as potential biomarkers can be briefly summarized. First, expression patterns of miRNAs in human cancers appear to be tissue specific. Second, miRNA profiles appear to reflect developmental lineage and differentiation state of the tumours. Third, miRNAs can successfully classify poorly differentiated tumours with high accuracy (~70 %). In contrast, mRNA profiles in the same set of specimens had an accuracy of only 6 %. Therefore, a combination of both miRNA and mRNA profiling data has the potential of enhancing accuracy of tumour classification. Forth, miRNAs can also be profiled and quantitatively measured in formalin-fixed paraffin-embedded tissues. And last, miRNAs are stable in human body fluids of plasma and serum and can be quantitatively measured in microliter quantities of human sera or plasma using qPCR. [45-47].

Highly stable cell-free circulating nucleic acid (cfCNA), both RNA and DNA, has been discovered in the blood, plasma, and urine in humans. Since there is good correlation between tumours and genetic, epigenetic and/or transcriptomic changes and alterations in cfCNA levels, it gives a usefulness of cfCNA as biomarkers for clinical applications. Release of cfCNA in body fluids is probably related to apoptosis and necrosis. Circulating RNAs are stable in serum and plasma in spite of high amounts of RNAase in blood of cancer patients [48]. They are packed in microparticles, of which the most analyzed are in recent years exosomes [3,49,50]. Tumour derived exosomes are small membrane vesicles of endocytic origin released by the tumour and found in peripheral circulation. Several recent reports showed that exosomes could be an important resource of cf-lncRNA/cf-miRNA in serum or plasma [51]. Small size, relative stability and resistance to RNAase degradation make the miRNAs more superior molecular markers than mRNAs [1]. Using non-invasive diagnostic procedures, the extraction and reliable determination of cf-miRNAs, circulating in body fluids like plasma, serum, and others, could serve as circulating tumour biomarkers [52,53].

LncRNAs show greater tissue specificity compared to protein-coding mRNAs, making them attractive in the search of novel diagnostics and/or prognostics cancer biomarkers in body fluid samples. For an example, lncRNA PCA3 was initially identified as over-expressed in prostate tumours relative to benign prostate hyperplasia and normal epithelium. It was latter showed that is very specific prostate cancer gene, whose mechanism is not yet identified, but it can be detected in urine samples and has been shown to improve diagnosis of prostate cancer [3].

3.3. ncRNAs can act as both tumour suppressor genes and oncogenes

There are different ways in which miRNAs appear to be involved in cancer: as tumour suppressors, as oncogenes, or as agents involved in affecting genome stability. Below is discussed role of miRNAs acting both, as tumours suppressor and as oncogenes, since are much more investigated in this field than are lncRNAs. Care must be taken in assigning oncogenic or tumour suppressor activity to a miRNA, since miRNA expression patterns are highly specific for cell-type and cellular differentiation status. The same miRNA can function as tumour suppressor in one cell type and as potential oncogene in other cell type. Some of the aberrant miRNA expression observed in tumours may also be a secondary consequence of the loss of normal cellular function that accompanies malignant transformation. Up- or down-regulation of a miRNA in a given tumour type is not obvious a causative role in tumorigenesis [6].

The increased expression of oncogenic miRNAs appears to act in a manner analogous to an oncogene. Over-expression of oncogenic miRNAs are presumed to function by down-regulating the levels of protein product of target tumour suppressor gene or by reduction of tumour suppressor processes, such as apoptosis [2,6,20]. A loss of expression of tumour suppressor miRNA may lead to elevated levels of the protein products of target oncogenes [6], activation of an oncogenic processes, such as proliferation [2,20]. MicroRNAs with anti-proliferative and pro-apoptotic activity are likely to function as tumour suppressors and thus may be under-expressed in cancer cells. Figure 1 represents schematic overview of miRNAs acting as tumour suppressors or oncogenes in comparison to non-cancerous cells.

There should be at least four type of evidence before assigning tumour suppressor function to ncRNAs: (i) data about widespread deregulation in diverse cancer, (ii) gain or loss of function in tumours owing to deletion, amplification or mutation, (iii) direct documentation of tumour suppressing activity using cell line or animal models, (iv) the identification and verification of cancer relevant targets that define mechanisms through which miRNAs participate in oncogenesis [6].

4. ncRNAs as potential therapeutic targets in cancer

4.1. RNAi in therapeutic applications

Using RNAi approaches, ncRNAs may in future serve as therapeutic targets. For ncRNA that is under-expressed and possess tumour suppressor function, re-introduction of the mature

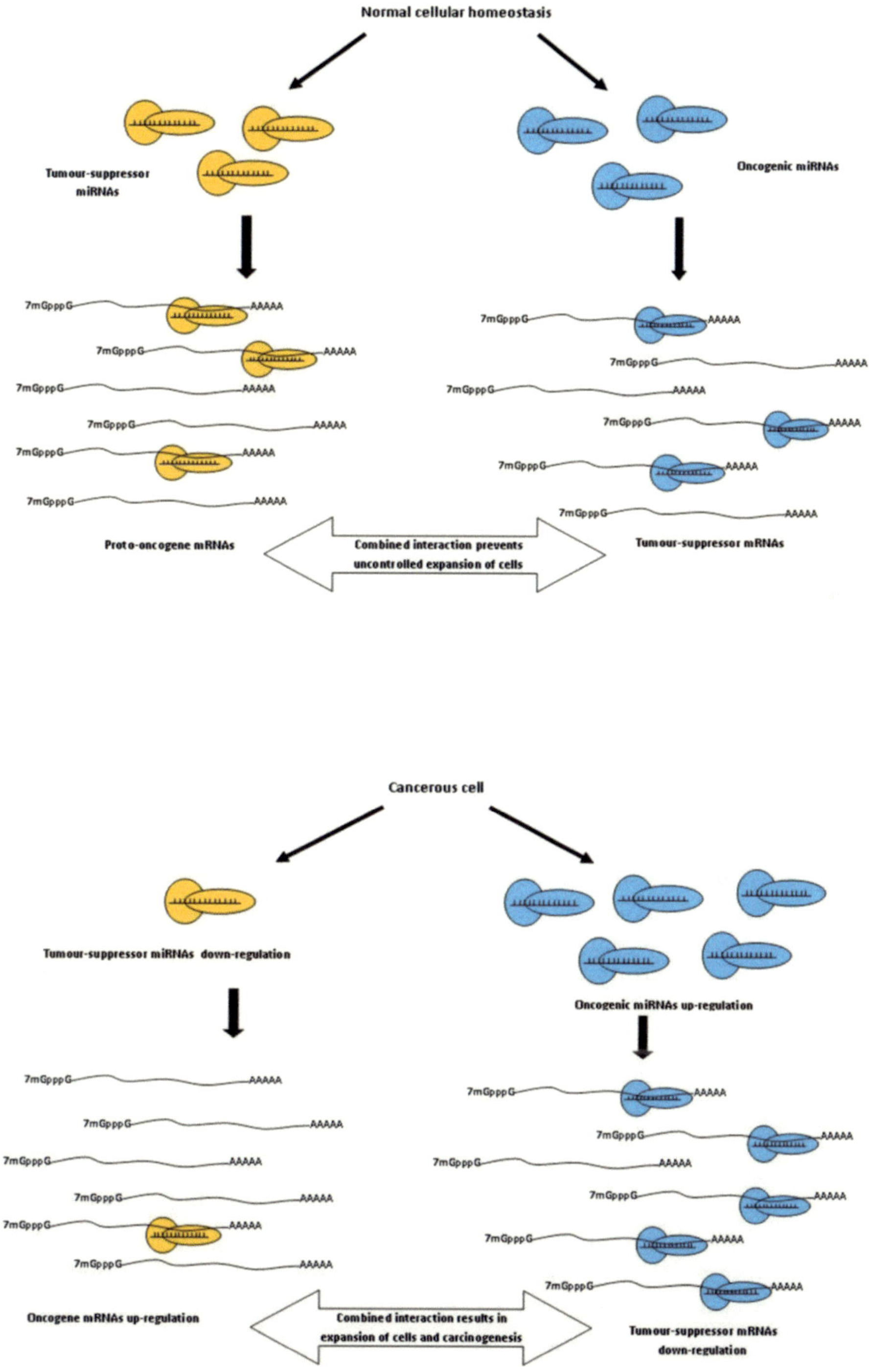

Figure 1. Schematic overview of miRNAs acting as tumour suppressors or oncogenes.

ncRNA into the affected tissue would restore the regulation of the target gene. By contrast, over-expressed ncRNA with oncogenic function could be down-regulated by reducing mature ncRNA level by its direct targeting [54].

Due to the interferon response it is difficult to introduce long dsRNAs into mammalian cells, however, the use of RNAi as a therapeutic approach has been successfully used. Among the first applications to reach clinical trials were in the treatment of macular degeneration and respiratory syncytical virus infection, reversal of induced liver failure in mouse models, antiviral therapies, neurodegenerative diseases, and cancer. Cancer was treated by silencing up-regulated genes in tumour cells or genes involved in cell division. A key area of research in the use of RNAi for clinical applications is the development of a safe delivery method, which to date has involved mainly viral (lentivirus, adenovirus, adeno-associated virus) and non-viral (nanoparticles, aptamers, stable nucleic-acid-lipid particle, e.g.) vector systems similar to those suggested for gene therapy [55].

4.2. ncRNAs with tumour suppressor function as therapeutic targets

Replenishing small RNAs/miRNAs

Pharmacological manipulation of miRNAs is still in its infancy; however, the correlation between the expression of miRNAs and their effects on target oncogenes, on tumorigenesis, and on the proliferation of cancer cells has gained experimental support. miRNAs are small molecules, making their *in vivo* delivery feasible. It has been shown that miRNAs can be delivered systematically, and can reduce invasion, proliferation and growth as well as induce radio-sensitivity and resistance. miRNAs may therefore serve as therapeutic targets in the future.

For miRNA that is under-expressed, re-introduction of the mature miRNA into the affected tissue would restore regulation of the target gene. For this purpose, artificial miRNA (miRNA-mimic) have been developed to enhance the expression of beneficial miRNAs or the introduction of short hairpin duplex, similar to *pre-miRNA,* into the cell. This suggests that individual miRNAs are potential therapeutic agents, provided that their expression or delivery can be targeted to appropriate tissue. Most of the developed protocols have used local administration in easily accessible tissue; systemic delivery has also give some promising results; the major challenge remains tissue and cell-type specific targeting [56].

miRNA mimic can only last a couple of days and the long term biological effects were not observed very effectively. To overcome this, the cells were infected with a lentivirus that expressed mature miRNAs. This generated stable cell expressing miRNAs. miRNA mimics and lentiviral miRNAs showed great potential in restoring tumour suppressor miRNAs. However, viral and non-viral delivery systems have been developed. Viral vector-directed methods show high gene transfer efficiency, but have some limitations. However, non-viral gene transfer vectors have been also developed: cationic liposome mediated gene transfer system, lipoplexes, neutral lipid emulsion, etc. [57].

Expression of miRNA-mimic would simultaneously suppress many gene targets. miRNAs-mimic would be useful in conjunction with standard chemotherapy or radiotherapy, by influencing drug resistance or enhancing responsiveness to therapy. Current limitation is need for improvement of efficiency of delivery to target tissue, for systemic drug administration, potential inhibition of non-target genes ("off-target effect"), redundancy among miRNAs efficacy, potential toxicity and immunogeneic responses. However, studies introducing miRNAs strategies to inhibit cancer propagation in animal models are showing promising results [32].

Examples for miRNA

Therapeutic delivery to animal models was demonstrated using miRNA-mimics of the tumour suppressor miRNAs, *miR-34a* and *let-7a*, both of which are often down-regulated or lost in lung cancer. It has been shown that re-introduction of *let-7* directly represses cancer growth in the lung [58] and that development of chemically synthesized therapeutic *miR-34a* and lipid-based delivery vehicle block tumour growth in mouse models of non-small-cell lung cancer (NSCLC) [59]. Systemic treatment of these mice led to significant decrease in tumour burden. Mice treated with *miR-34a* displayed a 60 % reduction in tumour area compared to mice treated with a miRNA control. Similar results were obtained with the *let-7* mimic [60].

Targeting lncRNAs

Successful inhibition of lncRNAs seems to be more difficult than inhibition of miRNAs. Our growing knowledge of other ncRNAs might exploit in future to develop new therapeutic strategies not only against cancer, but also for other diseased states. The findings regarding lncRNA and Alzheimer disease are attracting the attention of pharmaceutical and biotechnology industries [8]. Therapy using small RNAs that targets ncRNA transcripts, such as eRNAs or PARs, may represent a new way to treat disease conditions caused by epigenetic changes [9].

Targeting both, lncRNAs and miRNAs

Another possible approach for manipulation of ncRNAs level may also be by altering DNA methylation. As mentioned above, DNA methylation is a crucial mechanism associated with epigenetic regulation. It has been shown that in cancer cells treated with DNA demethylating agent reactivation of certain miRNAs occurs [40]. ncRNAs mediated therapy may also be useful in combination with DNA methyltransferase inhibitors that are other way toxic [39].

5. ncRNAs as tumour suppressor in different types of cancers

5.1. miRNAs as tumour suppressors

In the following section, down-regulated miRNAs will be describe and miRNAs with suggested tumours suppressive roles in different types of cancer. However, down-regulation does not ncessary mean that miRNA is tumours uppressor.

Hematological cancers

Leukaemia. Chronic lymphocytic leukaemia (CLL) is characterized by overexpression of the protein Bcl-2 in B cells and represents the most common human leukaemia. In less than 5 %

of cases, over-expression of Bcl-2 is due to a translocation of the Bcl-2 gene, whereas for the majority of CLL cases no explanation for the deregulation of Bcl-2 has been reported. It has been demonstrated that mutations in genomic regions containing miRNAs were associated with disease progression in a number of CLL patients. In this type of cancer, *miR-15* and *miR-16* expression is often reduced and indeed, one of the first associations between miRNAs and cancer development was observed for *miR-15* and *miR-16* in CLL [46]. Both miRNAs are located in a 30 kb region on chromosome 13 that had been found deleted in more than half of B cell CLL (chromosome 13q14 deletion) [8,46], and *miR-15a* and *miR-16-1* have been shown to be deleted or translocated in approx. 65 % of CLL patients [2,11]. Several papers indicate that miRNA regulates cell growth and apoptosis. Indeed, over-expression of *miR-15* and *miR-16* directly inhibit anti-apoptotic Bcl-2, a key player in many types of human cancers, and thus activate apoptotic processes [2,11]. However, it was further demonstrated that other mutations in miRNA genes are frequent in CLL; many mutations were located in the flanking sequence of *pre-miRNA*, thus cell culture assay indicated that a point mutation of the *miR-16-1* precursor abolishes expression of mature *miR-16* [21]. Few other miRNAs were also recognized as tumour suppressors in CLL. *miR-29a* and *miR-29b* are associated with fragile site FRA7H that is not associated with any known tumour suppressor gene. Over-expression of *miR-29b* may target TCL1 and reduces anti-apoptotic Mcl-1 protein in CLL patients [11]. Another well-known tumour suppressor was analysed. Low expression of *miR-34a* in CLL was found to be associated with p53 inactivation, impaired DNA damage response, apoptosis resistance and chemotherapy-refractory disease irrespective to p53 mutation (cases with CLL with p53 mutation are resistant to chemotherapy). It was latter showed that *miR-34a* is induced by p53. In another type of leukaemia, particularly acute myeloid leukaemia, an inverse correlation between *miR-34b* and CREB expression has been observed. After restoring expression of *miR-34b*, cell cycle abnormalities, reduces growth and altered CREB expression has been observed, suggesting tumour suppressor potential of this miRNA [47].

Lymphoma.miR-142 gene was found at the junction of the t(8;17) translocation, which may contribute to the progression of an indolent lymphoma into aggressive B-cell leukaemia [21].

Breast cancer

Breast cancer is one of the most important cancers in adult females. *miR-125b, miR-145, miR-21* and *miR-155* were significantly reduced in breast cancer tissue and this expression was correlated with specific breast cancer pathologic features, such as tumour stage, proliferation, oestrogen and progesterone receptor expression, and vascular invasion. Some of these miRNAs act as oncogenes (e.g. *miR-21*) in many cancer types, so it is suggested that some miRNAs act as tumour suppressors in one cancer type and as oncogenes in another [2]. *miR-125b-1* is located on a fragile site on chromosome 11q24, which is deleted in a subset of patients with breast cancer [61]. Down-regulation of *mir-221* in breast cancers was detected, whereas germ line mutation in mature *miR-125a* is highly associated with breast cancer tumorigenesis, suggesting its tumour suppressor role. *miR-125a* is also down-regulated in human breast cancer and when over-expressed post-trancriptionally regulates CYP24 resulting in an anti-proliferative effect. Ectopic expression of *miR-30e* suppresses cell growth in breast cancer, probably through targeting Ubc9 [47]. *miR-17-5p* was down-regulated in breast cancer cells, and enhanced expression decreased tumour cell proliferation [11]. Cyclin D1 has been identified as a direct target for *miR-17/20* that functions to suppress proliferation of breast cancer cells. Additional miRNAs have also been shown to be down-regulated and have tumour suppressor function in other types of cancer: *let-7, miR-145, miR-34a, miR-214,* and *miR-205.* MicroRNAs, *miR-31, miR-126, miR-146a/b, miR-206* and *miR-335* have been shown as anti-metastatic miRNAs [62,63].

Colorectal Cancer (CRC)

Colorectal cancer is the third most commonly diagnosed cancer in the world, but it is more common in developed countries. Also in colorectal neoplasia miRNAs expression is associated to the tumour formation. Reduced expression of *miR-143* and *miR-145* have been shown to be a frequent feature of colorectal tumours (adenomatous and cancer stage) when compared to normal mucosa [2,6,64]. A tumour suppressive role of *miR-143* has been elucidated in the epigenetic aberration of CRC with DNMT3A as a target. Restoration of *miR-143* expression in CRC decreases tumour cell growth and down-regulates DNMT3A expression [47]. *let-7* has been implicated in development of colon cancers and progression of colorectal cancers; together with *miR-143* and *miR-18a* was observed to be down-regulated and target KRAS [11, 64]. *miR-145* has been proposed as a tumour suppressor and it has been shown that target IRS-1, and when over-expressed it dramatically inhibits the growth of colon cancer cells. A ubiquitous loss of *miR-126* expression in colon cancer lines was observed and its reconstitution resulted in a significant growth reduction. Also, in a panel of matched normal colon and primary colon tumours, each of the tumours demonstrated *miR-126* down-regulation [64]. Down-regulation of *miR-200* family is a hallmark of EMT as well as up-regulation of ZEB1 transcription factor, and it was shown that in colorectal cells ZEB1 directly suppress transcription of *miR-141* and *miR-200c* [57]. It was found that *miR-192* and *miR-215* was down-regulated in CRC, and their anti-proliferative effect was identified in CRC cell lines. It was further defined that both are regulated by p53 and that their targets are a number of transcripts that regulate cell cycle checkpoints [57]. An inverse correlation between COX-2 and *miR-101* was reported in CRC cell lines, and this was further confirmed in colon cancer tissue and liver metastases derived from CRC patients. *miR-16, miR-125b, miR-31, miR-133b,* and *miR-96* were along with already mentioned miRNAs showed to be down-regulated in colorectal cancer [1].

Aberrant DNA methylation may further induce silencing of specific miRNAs in CRC. While methylation of *miR-129* and *miR-137* CpG islands is frequently observed in CRC, is methylation of *miR-9-1* associated with the presence of lymph node metastasis and expression of *miR-9* in CRC inversely correlated with the methylation of its promoter regions [47]. In human colon cancer cell lines, *miR-34a* was showed to participate in the apoptotic program triggered by p53 activation and loss of *miR-34a* expression occurs frequently in cancer cells. p53 directly binds to the genomic region defined as the *miR-34a* promoter with consequent *miR-34a* targeting genes of cell cycle, DNA repair, mitotic checkpoint, DNA integrity checkpoint, cell proliferation, and angiogenesis. Among the down-regulated targets of *miR-34* family were well-characterized p53 targets, such as CDK4/6, cyclin E2, E2F5, BIRC3 and Bcl-2. These effects were nearly identical irrespective of whether *miR-34a, miR-34b* or *miR-34c* was introduced into cell lines. Another target was identified for *miR-34a,* SIRT1, negative regulator of apoptosis.

miR-34a promoter hyper-methylation was observed in 3 of 23 cases of colon cancer, *miR-34b/c* were found to be epigenetically silenced in 9 of 9 cell lines and in 101 of 111 primary CRC tumours [42,64].

Gastric cancer

Gastric cancer is the fourth most common cancer and the second leading cause of cancer death in the world. It was reported that loss of Ago2, which leads to premature stopping of miRNAs biogenesis and general deregulation of miRNAs expression, was observed in 40 % of human gastric cancer patients with high microsatellite instability. A number of miRNAs were reported to be down-regulated. Among these, *miR-141* was significantly low expressed in 80 % of primary gastric carcinoma compared to non-cancer adjacent tissue. It targets FGFR2 and its down-regulation means proliferative potential and poor differentiation of gastric cancer cells [65]. It appears that down-regulation of *miR-451* is related to the worse prognosis of the gastric cancer patients. Over-expression of *miR-451* in gastric cancer cells regulates the oncogene MIF production, reduces cell proliferation and increases sensitivity to radiotherapy [47]. *miR-101* was down-regulated in gastric cancer cells, its targets are: EZH2, Cox-2, Mcl-1, and Fos [65]. Other potential tumour suppressor miRNAs in gastric cancer are: *miR-181b/c* and *miR-432AS*, and for *miR-181* it was proposed that modulate expression of Bcl-2. Low level or loss of expression in gastric cancer also showed *let-7a, miR-486,* and *miR-449*. However, a proposed role for *miR-107* and *miR-126* is controversial in gastric cancer, either tumour-suppressive or oncogenic [66].

An epigenetic silencing of *miR-512-5p* was observed in gastric cancer cells. As its target is was shown anti-apoptotic protein Mcl-1 and after epigenetic treatment (demethylation), it results in apoptosis of gastric cancer cells [65].

The association between genetic polymorphism of *miR-196a-2* and risk of gastric cancer has been identified; it was found that the variant homozygous genotype of *miR-196a-2* was associated with significantly increased risk of gastric cancer [65].

Pancreatic cancer

Pancreatic cancer is the eighth most common cause of cancer-related deaths worldwide and it has a poor prognosis for all stages. It is usually diagnosed at advent stages, therefore it is an urgent need to find some specific biomarkers and key components of carcinogenesis. Several miRNAs were reported to suppress metastasis. In pancreatic cancer cell lines, *miR-146a* was decreased compared to normal ductal epithelial cell line, and its ectopic expression inhibited invasive capacity of pancreatic cancer cell lines. *miR-96* is believed to be a potential tumour suppressor through targeting KRAS. It is significantly down-regulated in pancreatic cancer, its ectopic expression induces apoptosis, inhibits cell proliferation, migration and invasion. In human clinical samples there is observed inverse correlation between *miR-96* and KRAS. Further, ectopic expression of *miR-520h* has inhibitory effect on pancreatic cancer cell migration and invasion, *miR-20a,* with metastasis-suppressing effect, is reduced in pancreatic cancer and its cell lines. [67]. In pancreatic ductal carcinoma, *miR-345, miR-139,* and *miR-142-p* were the most down-regulated miRNAs in tumour tissue compared to normal tissue [1]. Other potential tumour suppressor miRNAs involved in pancreatic cancer are: *miR-100, miR-181a,* and *miR-15b,* but are as well as *miR-200* family up-regulated [67].

miR-200 family, potential tumour suppressors, is up-regulated in pancreatic cancer. Low expression of *miR-200* family genes and higher expression of their target is common in different cancers. However, most pancreatic cancer cell investigations showed hypo-methylation of *miR-200a/b* and its over-expression, in contrary its targets are hyper-methylated, suggesting that this pathway is not involved in metastases in most pancreatic cancer [67]

Hepatocellular Carcinoma (HCC)

Primary liver cancer mainly refers to HCC, which is one of the most common malignant tumours in liver and accounts for 85-90 % of primary liver cancers. Cyclins D2 and E2 were validated as direct targets for *miR-26a,* which exhibit reduced expression in HCC [68]. In another early study on animal models, down-regulation of number of miRNAs has been detected, including known tumour suppressor miRNAs, such as: *miR-15/16, miR-34a, miR-150* and *miR-195* [69]. *miR-122,* which represents 70 % of all liver miRNAs, was found to be frequently down-regulated in HCCs. Loss of *miR-122* expression in tumour cells segregates with specific gene expression profiles linking to HCC progression. *miR-122* is specifically repressed in a subset of primary HCCs that are characterized by poor prognosis and is therefore suggested as tumour suppressor miRNA [47,68]. As one of *miR-122* targets, cyclin G1 was identified, through its regulation *miR-122* influences p53 protein stability and transcriptional activity. Two other *miR-122* targets, which promote tumorigenesis, are SRF and IGF1R. The cellular mRNAs and protein levels of Bcl-w were also repressed by *miR-122.* Other pro-apoptotic functions were assigned for *let-7* through targeting Bcl-xL, for *miR-101* through targeting Mcl-1, and for *miR-29* through targeting Mcl-1 and Bcl2. *miR-195* was significantly reduced in HCC tissues and cell lines, it suppress tumorigenicity through targeting cyclin D1, CDK6, and E2F3. CDK6 was showed to be also target for *miR-124,* which is silenced through CpG methylation in HCC; *miR-124* in addition targets vimentin, SET, and MYND. *Let-7g* inhibits the proliferation of HCC by down-regulating c-Myc. Methylation of *miR-1* in HCC results in enhanced tumour cell growth, probably through release of its oncogenic targets c-Met, FoxP1, HDAC4. *miR-223* targets Stahmin 1 and is down-regulated in HCC whereas *miR-375* inhibits the proliferation and invasion of HCC cells by targeting Hippo-signalling effectors YAP [68,70]. Research on expression profiling in HCC and adjunct non-tumour tissue defined that *miR-199a**, miR-195, miR-199a, miR-200a,* and *miR-125a* were also under-expressed in HCC tissue [2]. Anti-metastatic functions were showed for *miR-122* by targeting desintegrin and metalloprotease, and *let-7g* by targeting type I collagen A2. c-Met is target for *miR-1, miR-34a, miR-23b* and *miR-199a-3p,* and all of these miRNAs are down-regulated in HCC [68,70].

Lung cancer

Lung cancer is one of the most common cancers of adults and is also leading cause of cancer-related deaths in many economically developed countries [2]. *Let-7* family is a family of miRNAs, whose genes map to different chromosome regions that are frequently deleted in lung cancer [71]. Significantly worse survival was observed among patients with low expression of *let-7a-2* compared to those with opposite expression pattern, independent of disease stage [2]. Over-expression of *let-7* in lung adenocarcinoma cell lines inhibited cancer cell growth and reduced cell cycle progression. These findings reflect that *let-7* mediates tumour suppressive function [2,45]. It has been shown that *let-7* regulates the expression of several oncogenes, RAS, MYC, HMGA2, and cell-cycle progression regulators, CDC25, CDK6, cyclin D2. *let-7* is down regulated in lung tumours and its expression anti-correlated with that of RAS relative to the normal lung tissue [46,71]. In animal models, ectopic *let-7g* expression reduces tumour burden and intranasal administration repress lung adenocarcinoma. A SNP in *let-7* complementary site 6 in 3'-UTR of its target KRAS is significantly associated with increased risk for NSCLC among moderate smokers [71]. The *let-7* family was subsequently found to be deregulated in a large number of tumour types [46].

Mutations in EGFR gene are more frequent in NSCLC patients who never smoked tobacco; a significant down-regulation of *miR-145* has been demonstrated in the cancer tissues of these patients. Restoration of this miRNA can inhibit cancer cell growth in EGFR mutant lung AD [47,71]. *miR-128b* has been also showed as direct regulator of EGFR, whereas *miR-128b* loss-of-heterozigocity is frequently found in NSCLC. *miR-7* is frequently down-regulated in lung cancer, it suppress EGFR and Raf1, it attenuates activation of Akt and ERK, suggesting that is negative regulator of EGFR pathway [71].

It was proposed that *miR-140* regulates PDGF in lung cancer development, but this has not yet been thoroughly investigated. *miR-29* is down-regulated in lung cancers, its targets are Mcl-1, DNMT3A and DNMT3B, suggesting that it is pro-apoptotic and that it has role in regulating epigenetic DNA methylation. Further, in lung cancer cells induction of *miR-34* results in apoptosis; miRNA profiling revealed that *miR-34a/b/c* are directly correlated with expression of p53. Decreased expression of *miR-126* and increased expression of VEGFA was found in various lung cancer cell lines. Introduction of *miR-126* down-regulates VEGFA, inhibits growth, and reduces average tumour weight. Mouse model of lung adenocarcinoma showed that *miR-200* family possess anti-metastatic abilities [5]. *miR-125b-1* is located on a fragile site on chromosome 11q24 which is deleted in a subset of patients with lung cancer [61].

Human brain cancer

The phrase "brain tumours" describes an inhomogeneous collection of various tumours of the brain, which represents primary tumours of nervous central system or metastases. Glioblastomas (belongs to family of gliomas) are the most frequent occurrence and malignant form of primary brain tumors in contrary to medulloblastomas, which have a better prognosis [2,11]. Several articles have described the effects of ectopic miRNA modulation on medulloblastoma cell proliferation and growth.

Rescued expression of *miR-9* and *miR-125a* were shown to promote medulloblastoma cell growth arrest and apoptosis by targeting TrkC, whereas *miR-29* has been shown to be down-regulated in neuroblastoma and brain tumour [47,72]. Further, *miR-34a* induces apoptosis in neuroblastoma cells, possibly by targeting the transcription factor E2F3 [11]. Transient transfection in medullablastoma cells with *miR-34a* strongly inhibited cell proliferation, cell cycle progression, cell survival and cell invasion. *miR-34a* was shown to inhibit c-Met, Notch-1,

Notch-2 and CDK6. Ectopic up-regulation of *miR-124* was shown to inhibit cell proliferation, and it was demonstrated that *miR-124* target and regulates CDK6. *miR-125b, miR-324-5p,* and *miR-326* over-expression inhibit medulloblastoma cell growth by targeting Hedgehog signalling pathway. *miR-128* also inhibits growth by targeting Bmi-1 oncogene. Another miRNA, *miR-199b-5p* negatively regulates proliferation and cell growth [72].

In contrary to medulloblastomas, are gliomas the most common and deadly primary human brain tumours, and its subtype glioblastomas are highly invasive, very aggressive, and one of the most incurable [2]. *miR-181a, miR-181b* and *miR-181c* were originally identified as down-regulated in glioblastoma cells and tumours when compared to normal brain controls. *miR-181a* and to a greater extent *miR-181b* were subsequently described as tumour suppressors that inhibit growth and induce apoptosis of glioma cells. *miR-181a* over-expression down-regulates Bcl-2. Several other miRNAs have been implicated in glioma malignancy as tumour suppressors. *miR-15b* was suggested to target CCNE1, the gene encoding cyclin E1, however, a direct link between CCNE1 down-regulation by *miR-15* and cell cycle regulation was not demonstrated. *miR-146b* was shown to inhibit glioma cell migration and invasion, and was identified as one of miRNAs that is significantly deregulated in human glioblastoma tissue. *miR-146b* over-expression or knock-down did not affect the growth of human glioblastoma cell line, while it significantly reduced the migration and invasion of one glioblastoma cell line. MMP16 was identified as one of the downstream targets of *miR-146b*. *miR-125b* was shown to induce cell cycle arrest and inhibits CDK6 and CDC25A expression in glioma cell lines. However, another study suggested oncogene function for *miR-125b*. *miR-153* decrease cell proliferation and increased apoptosis (pro-apoptotic miRNA), it inhibited Bcl-2 and Mcl-1. *miR-17* and *miR-184* were identified as two miRNAs with reduced expression in higher grades of glioblastomas. Their over-expression inhibited viability, proliferation, invasion and decreased expression of AKT 2 and several other genes. *let-7* over-expression effect was investigated in glioma cells, its transfection reduced expression of RAS oncogenes, proliferation in vitro, migration of the cells, and reduced the size of tumours generated [73].

Head and Neck Squamous Cell Carcinoma (HNSCC)

Head and neck tumours are a heterogenous group with different behaviour at the various sites arising from anatomical factors, cell-type variation, and differences in exposure to risk factors including tobacco, alchocol, and viruses [74]. Head and neck squamous cell carcinoma is represented by epithelial cancers of the oral cavity, pharynx, nasal cavity, paranasal sinuses, salivary glands and larynx [75]. Studies were made in expression profiling regarding different sites of head and neck tumours, tongue, tonsil, larynx, hypopharynx, nasopharynx, saliva, oral cavity, salivary gland and animal models. Several miRNAs were identified as down-regulated, and for some of their target genes were validated [74].

Low expression of *miR-205* is significantly associated with local-regional recurrence independent of disease severity at diagnosis and treatment. Combined low expression of *let-7d* and *miR-205* is significantly associated with poor survival. In nasopharengyal carcinoma down-regulation of *miR-34* family, *miR-145* and *miR-143* was also observed [47]. Tumour suppressive role in HNSCC has been suggested for *let-7* family, *miR-125a/b, miR-200, miR-133a/b,* and *miR-100* [75]. General down-regulation of *miR-1, miR-133a* and *let-7b* was also observed [74].

Reduced expression for majority of members of the *let-7* family (except *let-7i*) was observed in HNSCC. KRAS and HMGA2 have been characterized as targets for *let-7* [75]. Notable among down-regulated was also *miR-98*. However, another group identified *miR-98* as another regulator of an oncogene HMGA2 [74]. A possible molecular mechanism of *miR-125a/b* down-regulation might be through targeting ERBB2, since its higher level of expression was observed in oral SCC. Other target were also suggested for *miR-125*, namely KLF13, CXCL11 and FOXA1 [74,75]. Down-regulation of *miR-133a/b* in primary HNSCC may further contribute to increased cell proliferation and decreased apoptosis. PKM2 has been validated as cellular target for both, *miR-133a* and *miR-133b*, and increased expression of PKM2 has been associated with cancer progression. Finally, *miR-100* has been observed at suppressed levels in primary HNSCC and derived cell lines. A few of its targets are known, namely FGFR1, MMP13, ID1, FGFR3, EGR2. The exact role has to be investigated yet, but suggestion has been made that down-regulation of *miR-100* means higher rate of cell proliferations.

Deregulation of miRNAs in cancer can occur through epigenetic changes (promoter CpG island hyper-methylation in the case of *miR-200* family) [8]. Suppressed *miR-200a* was detected in primary oral SCC. Members of *miR-200* family inhibit EMT by directly targeting ZEB1/ZEB2, suppressed levels of *miR-200a* may promote EMT. At last, it was observed that *miR-137* and *miR-193a* could be also silenced via hyper-methylation, CDK6 and E2F6 has been suggested as their major targets. It was also shown that *miR-137* hyper-methylation is associated with poorer average survival [74,75].

Urological tumours

Renal cell carcinoma (RCC). The VHL tumour suppressor signalling pathway is the most important deregulated pathway in clear cell RCC, the dominant subtype of kidney cancers. The VHL gene can be spontaneously deleted or hyper-methylated. The regulation of the VHL pathway by miRNAs has not been well studied in RCC. The interactions have been proposed but a direct relation between miRNAs and VHL or HIF1A were not proven. A subset of miRNAs has been identified as regulated by VHL pathway, 3 miRNAs were up-regulated and 6 were down-regulated. The second commonly deregulated pathway is VEGF signalling pathway, which is transcriptionally regulated by HIF (after hypoxia) or due to loss of VHL. Interaction between miRNAs and VEGF has not been well studied in RC, and only *miR-29b* has been shown to indirectly regulate VEGF. However, strong inverse correlation between *miR-200* family and VEGFA has been observed with suggestion that VEGFA is direct target of these miRNAs [76]. Down-regulation of *miR-141* was found in malignant compared to matched non-malignant tissue samples [47].

Bladder cancer. Mutation or over-expression of the FGFR3 gene occurs in approx. 80 % of all patients with low-grade non-invasive urothelial carcinomas. Some well-established deregulated miRNAs in bladder cancer are predicted to target FGFR3, and regulation of FGFR3 by *miR-99* and *miR-100* of four predicted miRNAs has been experimentally validated. Family *miR-200* is associated with an epithelial phenotype; its ectopic expression in bladder cancer cell lines induces up-regulation of epithelial and down-regulation of mesenchymal markers. Up-regulation of *miR-143* is accompanied by down-regulation of RAS [76]. Transfection of bladder cancer cell lines with *pre-miR-129* exerts significant growth inhibition and induces cell death. *mir-129* is shown to target GALNT1 and SOX4. Transfection of *miR-30-3p, miR-133a* and *miR-199a** results in decrease of tumour cell growth in bladder cancer. *miR-101* inhibits cell proliferation in bladder transitional cell carcinoma by targeting EZH2 and altering global chromatin structure. Down-regulation of *miR-145* in bladder cancer has been also observed [47].

Prostate cancer. A major signal transduction pathway in prostate cancer is PI3K/Akt signalling pathway that is hyper-activated in approx. 30-50 % of prostate cancer. Many of the predicted miRNAs that are predicted to target proteins of this pathway are differentially regulated in prostate cancer. Second pathway is androgen receptor (AR) pathway. *miR-125b* is an androgen-sensitive miRNA, which has been shown to regulate apoptosis through inhibition of BAK1. *miR-101* has been shown to be up-regulated in human prostate cancer and it seems that through inhibition of EZH2 reduce invasion and induce morphological changes in prostate cancer cells [76]. *miR-221* is found to be progressively down-regulated in aggressive forms of prostate cancer. Down-regulation of *miR-221* is linked to cancer progression and recurrence in a high risk prostate cohort. Progressive miR-221 down-regulation also hallmarks metastasis [47,76]. *miR-499a* has been shown to be down-regulated in prostate cancer tissue. Its introduction into prostate cancer cells results in cell cycle arrest and apoptosis, where it regulates cell growth and viability in part by repressing the expression of HDAC1 [47]. *miR-15a-miR-16-1* cluster, located at chromosome 13q14, is deleted in most cases of prostate cancer [21].

Melanomas

Melanoma is the most aggressive type of skin cancer, and it is resistant to therapy in its advanced stages [77]. Abnormalities in several signal transduction pathways, which are important for normal melanocyte development, only partly explain molecular mechanism directly linking UV radiation to the development of melanoma. miRNAs are emerging as important causal factors to melanoma initiation and progression [78]. In 45 primary cultured melanoma cell lines, there was observed that many genomic loci containing miRNAs are frequently affected (85.9 %) by copy number abnormalities. For an example, copy number losses of the region containing *miR-218-1* and SLIT2 were shown in 33 % of all investigated melanoma lines [79]. Proteins involved in miRNA biogenesis, Drosha, Dicer and Ago, are over-expressed in melanoma [32].

However, deregulation of miRNAs expression is not always explained. *let-7* plays a role in melanoma development and progression, its targets are many cancer-promoting molecules, such as NRAS, Raf, c-myc, cyclin D1/D3, and CDK4 [77]. Analyzing 10 melanocytic nevi and 10 primary melanomas, it was revealed that five members of the *let-7* family were significantly down-regulated in melanoma [80]. Over-expression of *let-7b* leads to inhibition of cell cycle progression and inhibition of its targets (cyclin D1/D3/A and CDK4) [77]. A direct interaction of *let-7b* with the cyclin D1 3'-UTR was showed. *Let-7a* was thus demonstrated to regulate expression of integrin beta3 and RAS oncogene, which is highly related to melanoma progression.

Another family, *miR-34*, has tumour suppressive role. Expression of *miR-34a* is silenced due to an aberrant CpG methylation in 43.2 % of melanoma cell lines and in 62.5 % of primary melanoma samples. The tumour suppressive function of *miR-34a* has not yet been investigated, however, there was shown a reduced expression of *miR-34b/c* and *miR-199a** and it was proposed that their target is MET oncogene [80]. Another miRNA with potentially tumour suppressor role is embedded in CpG island and epigenetically regulated in melanoma, this is *miR-370* [79].

miR-203 has also an important tumours suppressor role in a lot of cancers, and it is often lost due to deletion or due to promoter CpG hyper-methylation. *miR-203* target p63, and their relationship might be relevant also in melanoma, since it is known that *miR-203* functions as switch between epidermal proliferation and differentiation [81].

Gynaecological tumours

miR-125b-1 is located on a fragile site on chromosome 11q24 which is deleted in a subset of patients with ovarian and cervical cancer [61].

Cervical cancer. Cervical cancer aetiology is strongly linked to HPV infection, and involvement of virus protein E6 and E7 in pathogenesis is well established. The exact pathway from infection to tumorigenesis has not been elucidated yet. However, down-regulation was observed for: *let-7b/c, miR-23b, miR-196b, miR-143,* and *miR-145. miR-143* and *miR-145* were equally down-regulated in all cell lines of cervical cancer (HPV infected and HPV not infected), whereas miR-*218* was the unique miRNA down regulated only in HPV-16 and HPV-18 positive cell lines. Down-regulation of *miR-214* is related to the ability of this miRNA to inhibit HeLa cells proliferation through targeting MEK3 and JNK1 transcripts. HPV protein E6 induces destabilization of p53, down-regulation of *miR-34a* and increased proliferation of pre-malignant HPV infected cervical cancer cell lines [82].

Endometrial cancer. Reciprocal association between down-regulation of *miR-192-2* and SOX4 expression was determined; it was further established that restoration of *miR-192-2* induced a decrease in SOX4 expression and this resulted in diminished cell proliferation. Decreased expression for *miR-152* and *miR-101* was found to consist of an independent risk factor for disease free survival. Restoration of those miRNAs by transfection in cell lines lead to diminished cell proliferation. Down-regulation of *miR-101* was correlated with strong positive immunoreactivity of COX2, which was previously shown to be associated with worse prognosis. To date, no data are available for relationship between miRNAs and oestrogen response in endometrial cancer [82].

Ovarian cancer. Inconsistencies are observed between results in ovarian cancer studies for well-known tumour suppressors. These could be due to the differences in study populations and methodologies used, due to the choice of control group and type of control. For instance, the number of studies used as control cell lines and another number of studies used whole normal ovaries. The existence of significant discrepancies in expression profiles of certain miRNAs indicate the need of further and more in-depth research that would establish those results [82].

5.2. lncRNAs as tumour suppressors

lncRNAs are known to mediate epigenetic modifications of DNA by recruiting chromatin complexes to specific loci [8]. Only a handful of lncRNAs have been characterized, and their involvement in control of gene expression [3]. We therefore presented four lncRNAs with proposed tumour suppressor function in cancer.

MEG3

MEG3, located in chromosome 14q32, is maternally expressed imprinted gene, which represents lncRNA, but also hosted miRNAs and snoRNAs. It plays role in cell proliferation, and its expression is under epigenetic control. *MEG3* and its hosted miRNAs and snoRNA could represent a tumour suppressor gene, since aberrant CpG methylation (promoter hypermethylation, and hyper-methylation of the intergenic region) has been observed in several types of cancer, as well as their gene copy number loss [9].

MEG3 ncRNA might modulate binding of p53 on the promoter of its target genes [9]. It was later verified that *MEG3* was associated with p53 and that this association was required for p53 activation, further suggesting tumour suppressor role for *MEG3*. It was demonstrated that *MEG3* expression is markedly decreased in glioma tissues compared to adjunct normal tissues. Ectopic expression of *MEG3* inhibited cell proliferation and promoted cell apoptosis in glioma cell lines [83]. Growth inhibition is partially due to apoptosis induced by *MEG3*, which induces accumulation of p53, stimulates transcription from p53-dependent promoter and regulates p53 target gene expression. Loss or significantly reduction of *MEG3* expression has been further found in other cancer cell lines examined, bladder, bone marrow, breast, cervix, colon, liver, lung, meninges, and prostate, as well as in other primary tumours, neuroblastoma, hepatocellular carcinoma, and meningioma. It has been suggested that DNA methylation plays a major role in silencing the *MEG3* gene in tumours [84].

GAS5

LncRNA *GAS5* is highly expressed in cells that have arrested growth and can sensitize a cell to apoptosis by regulating activity of glucocorticoids in response to nutrient starvation. It has been linked with breast cancer. *GAS5* transcript levels are significantly reduced compared to un-affected normal breast epithelia, suggesting that could act as tumour suppressor. *GAS5* maintain sufficient caspase activity to activate appropriate apoptotic response in diseased cells. Chromosomal translocation affecting 1q25 locus that contains the *GAS5* gene has been detected in melanoma, B-cell lymphoma, prostate and breast cancer [7,85]. *GAS5* regulates expression of a critical subset of genes with tumour suppressive consequences [4].

LincRNA-21

LincRNA-p21 is required for the global repression of genes that interfere with p53 function regulating cellular apoptosis; it physically interacts with a protein hnRNP-K, allows it localization to promoters of genes that need to be repressed in a p53-dependent manner [4]. In response to DNA damage, lncRNAs are induced by the p53 tumour suppressor pathway. *lincRNA-p21* plays an important role in cellular response to apoptotic signal, it is induced by p53 and act as an inhibitor of the p53-dependent transcriptional response by repressing the transcription of genes that interfere with apoptosis (guidance of hnRNP-K to the promoters of genes repressed by p53). *LincRNA-p21* has not been directly associated with disease yet, but loss of function of *lincRNA-p21* might be involved in cancer initiation since functions to trigger cell death through the induction of apoptosis program [7,9,85].

CCND1

It is involved in the regulation of Cyclin D1 gene expression. Cyclin D1 is a cell cycle regulator often mutated, amplified and over-expressed in various types of cancer. After binding of this lncRNA on RNA-binding protein, consequently inhibition of enzymatic activities of the histone acetyltransferases occurs, leading to silencing of cyclin D1 gene. These studies suggest that this lncRNA is a tumour suppressor RNA, which can be rapidly induced by cellular stress to regulate it sense gene expression [85].

6. Conclusion

The rest of ncRNAs, other than miRNAs, in regulation biological functions are more or less unexplored, and this should be further investigated in future research. Regarding therapeutic approaches, we still need more knowledge concerning which miRNAs to target, how to produce and stabilize them, how to direct them to the target tissue. The specificity of drug-like oligonucleotides is important, because of the off-target effect. The off-target effect is also a significant challenge, especially considering that miRNA-mediated repression often requires a homology of only six to seven nucleotides in the seed region of the miRNA and mRNA target site. Toxicity due to chemical modifications, which is used to facilitate cellular uptake and prevent degradation, should be take into account. However, only recently was described the possibility of using exososmes and exosomal tumour-suppressive miRNAs as novel cancer therapy [86].

Author details

Emanuela Boštjančič and Damjan Glavač

Department of Molecular Genetics, Institute of Pathology, Faculty of Medicine Ljubljana, University of Ljubljana, Slovenia

References

[1] Paranjape T, Slack FJ, Weidhaas JB. MicroRNAs: tools for cancer diagnostics. Gut 2009;58(11): 1546-54.

http://dx.doi.org/10.5772/54701

[2] Zhang B, Pan X, Cobb GP, Anderson TA. microRNAs as oncogenes and tumor suppressors. Dev Biol 2007;302: 1-12.

[3] Reis EM, Verjovski-Almeida S. Perspectives of Long Non-Coding RNAs in Cancer Diagnostics. Front Genet 2012;3: 32.

[4] Huarte M, Rinn JL Large non-coding RNAs: missing links in cancer? Hum Mol Genet 2010;19(R2): R152-61.

[5] Du L, Pertsemlidis A.microRNAs and lung cancer: tumors and 22-mers. Cancer Metastasis Rev 2010;29(1): 109-22.

[6] Kent OA, Mendell JT. A small piece in the cancer puzzle: microRNAs as tumor suppressors and oncogenes. Oncogene 2006;25(46): 6188-96.

[7] Wapinski O, Chang HY. Long noncoding RNAs and human disease. Trends Cell Biol 2011; 21: 354-61.

[8] Esteller M. Non-coding RNAs in human disease. Nat Rev Genet 2011;12: 861-74.

[9] Kaikkonen MU, Lam MT, Glass CK. Non-coding RNAs as regulators of gene expression and epigenetics. Cardiovasc Res 2011;90: 430-40.

[10] Scott MS, Ono M. From snoRNA to miRNA: Dual function regulatory non-coding RNAs. Biochimie 2011;93: 1987-92.

[11] Williams AE. Functional aspects of animal microRNAs. Cell Mol Life Sci 2008;65: 545-62.

[12] Pillai RS. MicroRNA function: multiple mechanisms for a tiny RNA? RNA 2005;11: 1753-61.

[13] Bartel DP. MicroRNAs: genomics, biogenesis, mechanism, and function. Cell 2004;116: 281-97.

[14] Kim YK, Kim VN. Processing of intronic microRNAs. EMBO J 2007;26, 775-83.

[15] Boštjančič E, Glavač D. MicroRNAs as possible molecular pacemakers. Rijeka: InTech, 2011.

[16] Miyoshi K, Miyoshi T, Siomi H. Many ways to generate microRNA-like small RNAs: non-canonical pathways for microRNA production. Mol Genet Genomics 2010;284: 95-103.

[17] Blow MJ, Grocock RJ, van Dongen S, Enright AJ, Dicks E, Futreal PA, Wooster R, Stratton MR. RNA editing of human microRNAs. Genome Biol 2006;7(4): R27.

[18] Pillai RS, Bhattacharyya SN, Filipowicz W. Repression of protein synthesis by miRNAs: how many mechanisms? Trends Cell Biol, 2007;17: 118-26.

[19] Ying SY, Chang DC, Lin SL. The microRNA (miRNA): overview of the RNA genes that modulate gene function. Mol Biotechnol, 2008;38: 257-68.

[20] Perera RJ, Ray A. MicroRNAs in the search for understanding human diseases. BioDrugs 2007;21: 97-104.

[21] Chen CZ. MicroRNAs as oncogenes and tumor suppressors. N Engl J Med 2005;353(17): 1768-71.

[22] Griffiths-Jones S, Saini HK, van Dongen S, Enright AJ. miRBase: tools for microRNA genomics. Nucleic Acids Res 2006;36: D154-8.

[23] Ioshikhes I, Roy S, Sen CK. Algorithms for mapping of mRNA targets for microRNA. DNA Cell Biol 2007;26: 265-72.

[24] Min H, Yoon S. Got target? Computational methods for microRNA target prediction and their extension. Exp Mol Med 2010;42(4): 233-44.

[25] Xia W, Cao G, Shao N. Progress in miRNA target prediction and identification. Sci China C Life Sci 2009;52(12): 1123-30.

[26] Kuhn DE, Martin MM, Feldman DS, Terry AV Jr, Nuovo GJ, Elton TS. Experimental validation of miRNA targets. Methods 2008;44: 47-54.

[27] Barnes MR, Deharo S, Grocock RJ, Brown JR, Sanseau P. The micro RNA target paradigm: a fundamental and polymorphic control layer of cellular expression. Expert Opin Biol Ther 2007;7: 1387-99.

[28] Lu M, Zhang Q, Deng M, Miao J, Guo Y, Gao W, Cui Q. An analysis of human micro-RNA and disease associations. PLoS One 2008;3: e3420.

[29] Amaral PP, Clark MB, Gascoigne DK, Dinger ME, Mattick JS. lncRNAdb: a reference database for long noncoding RNAs. Nucleic Acids Res 2011;39(Database issue): D146-51.

[30] Malecová B, Morris KV. Transcriptional gene silencing through epigenetic changes mediated by non-coding RNAs. Curr Opin Mol Ther 2010;12(2): 214-22.

[31] Wilmott JS, Zhang XD, Hersey P, Scolyer RA. The emerging important role of micro-RNAs in the pathogenesis, diagnosis and treatment of human cancers. Pathology 2011;43(6): 657-71.

[32] Bell RE, Levy C. The three M's: melanoma, microphthalmia-associated transcription factor and microRNA. Pigment Cell Melanoma Res 2011;24(6): 1088-106.

[33] Derrien T, Guigó R, Johnson R. The Long Non-Coding RNAs: A New (P)layer in the "Dark Matter". Front Genet, 2011;2: 107

[34] Sethupathy P, Collins FS. MicroRNA target site polymorphisms and human disease. Trends Genet 2008;24(10): 489-97.

[35] Chen K, Song F, Calin GA, Wei Q, Hao X, Zhang W. Polymorphisms in microRNA targets: a gold mine for molecular epidemiology. Carcinogenesis 2008;29(7): 1306-11.

[36] Saunders MA, Liang H, Li WH. Human polymorphism at microRNAs and microRNA target sites. Proc Natl Acad Sci U S A 2007;104(9): 3300-5.

[37] Borel C, Antonarakis SE. Functional genetic variation of human miRNAs and phenotypic consequences. Mamm Genome 2008;19(7-8): 503-9.

[38] Chuang JC, Jones PA. Epigenetics and microRNAs. Pediatr Res 2007;61(5 Pt 2): 24R-29R.

[39] Wang Y, Liang Y, Lu Q. MicroRNA epigenetic alterations: predicting biomarkers and therapeutic targets in human diseases. Clin Genet 2008;74(4): 307-15.

[40] Lujambio A, Esteller M. CpG island hypermethylation of tumor suppressor microRNAs in human cancer. Cell Cycle 2007;6(12): 1455-9.

[41] He X, He L, Hannon GJ. The guardian's little helper: microRNAs in the p53 tumor suppressor network. Cancer Res 2007;67(23): 11099-101.

[42] Chang TC, Wentzel EA, Kent OA, Ramachandran K, Mullendore M, Lee KH, Feldmann G, Yamakuchi M, Ferlito M, Lowenstein CJ, Arking DE, Beer MA, Maitra A, Mendell JT. Transactivation of miR-34a by p53 broadly influences gene expression and promotes apoptosis. Mol Cell 2007;26(5): 745-52.

[43] Beckman M. MicroRNAs found cavorting with p53. J Natl Cancer Inst 2007;99(22): 1664-5.

[44] Hermeking H. The miR-34 family in cancer and apoptosis. Cell Death Differ 2010;17(2): 193-9.

[45] Jay C, Nemunaitis J, Chen P, Fulgham P, Tong AW. miRNA profiling for diagnosis and prognosis of human cancer. DNA Cell Biol 2007;26: 293-300.

[46] Blenkiron C, Miska EA. miRNAs in cancer: approaches, aetiology, diagnostics and therapy. Hum Mol Genet 2007;16: R106-13.

[47] Cho WC. MicroRNAs: potential biomarkers for cancer diagnosis, prognosis and targets for therapy. Int J Biochem Cell Biol 2010;42: 1273-81.

[48] Mitchell PS, Parkin RK, Kroh EM, Fritz BR, Wyman SK, Pogosova-Agadjanyan EL, Peterson A, Noteboom J, O'Briant KC, Allen A, Lin DW, Urban N, Drescher CW, Knudsen BS, Stirewalt DL, Gentleman R, Vessella RL, Nelson PS, Martin DB, Tewari M. Circulating microRNAs as stable blood-based markers for cancer detection. Proc Natl Acad Sci U S A 2008;105(30):10513-8.

[49] Zomer A, Vendrig T, Hopmans ES, van Eijndhoven M, Middeldorp JM, Pegtel DM. Exosomes: Fit to deliver small RNA. Commun Integr Biol 2010;3(5): 447-50.

[50] Iguchi H, Kosaka N, Ochiya T. Secretory microRNAs as a versatile communication tool. Commun Integr Biol 2010;3(5): 478-81.

[51] Kosaka N, Ochiya T. Unraveling the Mystery of Cancer by Secretory microRNA: Horizontal microRNA Transfer between Living Cells. Front Genet 2011;2: 97.

[52] Kosaka N, Iguchi H, Ochiya T. Circulating microRNA in body fluid: a new potential biomarker for cancer diagnosis and prognosis. Cancer Sci 2010;101: 2087-92.

[53] Chen X, Ba Y, Ma L, Cai X, Yin Y, Wang K, Guo J, Zhang Y, Chen J, Guo X, Li Q, Li X, Wang W, Zhang Y, Wang J, Jiang X, Xiang Y, Xu C, Zheng P, Zhang J, Li R, Zhang H, Shang X, Gong T, Ning G, Wang J, Zen K, Zhang J, Zhang CY. Characterization of microRNAs in serum: a novel class of biomarkers for diagnosis of cancer and other diseases. Cell Res 2008;18(10): 997-1006.

[54] Soifer HS, Rossi JJ, Saetrom P. MicroRNA in disease and potential therapeutic applications. Mol Ther 2007;15: 2070-9.

[55] Kim DH, Rossi JJ. Strategies for silencing human disease using RNA interference. Nat Rev Gen 2007;8: 173-84.

[56] van Rooij, E. The art of microRNA research. Circ Res 2011;108: 219-34.

[57] Zarate R, Boni V, Bandres E, Garcia-Foncillas J. MiRNAs and LincRNAs: Could They Be Considered as Biomarkers in Colorectal Cancer? Int J Mol Sci 2012;13(1): 840-65.

[58] Kumar MS, Erkeland SJ, Pester RE, Chen CY, Ebert MS, Sharp PA, Jacks T. Suppression of non-small cell lung tumor development by the let-7 microRNA family. Proc Natl Acad Sci U S A 2008;105: 3903-8.

[59] Wiggins JF, Ruffino L, Kelnar K, Omotola M, Patrawala L, Brown D, Bader AG. Development of a lung cancer therapeutic based on the tumor suppressor microRNA-34. Cancer Res 2010;70: 5923-30.

[60] Trang P, Wiggins JF, Daige CL, Cho C, Omotola M, Brown D, Weidhaas JB, Bader AG, Slack FJ. Systemic delivery of tumor suppressor microRNA mimics using a neutral lipid emulsion inhibits lung tumors in mice. Mol Ther 2011; 19: 1116-22.

[61] Esquela-Kerscher A, Slack FJ. Oncomirs - microRNAs with a role in cancer. Nat Rev Cancer 2006;6(4): 259-69.

[62] Negrini M, Calin GA. Breast cancer metastasis: a microRNA story. Breast Cancer Res 2008;10(2): 203.

[63] Piao HL, Ma L. Non-coding RNAs as regulators of mammary development and breast cancer. J Mammary Gland Biol Neoplasia 2012;17(1): 33-42.

[64] Slaby O, Svoboda M, Michalek J, Vyzula R. MicroRNAs in colorectal cancer: translation of molecular biology into clinical application. Mol Cancer 2009;8: 102.

[65] Wang J, Wang Q, Liu H, Hu B, Zhou W, Cheng Y. MicroRNA expression and its implication for the diagnosis and therapeutic strategies of gastric cancer. Cancer Lett 2010;297(2): 137-43.

[66] Ma YY, Tao HQ. Microribonucleic acids and gastric cancer. Cancer Sci 2012;103(4): 620-5.

[67] Zhang L, Jamaluddin MS, Weakley SM, Yao Q, Chen C.Roles and mechanisms of microRNAs in pancreatic cancer. World J Surg 2011;35(8): 1725-31.

[68] Huang S, He X. The role of microRNAs in liver cancer progression. Br J Cancer 2011;104(2): 235-40.

[69] Kumar A. MicroRNA in HCV infection and liver cancer. Biochim Biophys Acta 2011;1809(11-12): 694-9.

[70] Haybaeck J, Zeller N, Heikenwalder M. The parallel universe: microRNAs and their role in chronic hepatitis, liver tissue damage and hepatocarcinogenesis. Swiss Med Wkly 2011;141: w13287.

[71] Lin PY, Yu SL, Yang PC. MicroRNA in lung cancer. Br J Cancer 2010;103(8): 1144-8.

[72] Hummel R, Maurer J, Haier J. MicroRNAs in brain tumors : a new diagnostic and therapeutic perspective? Mol Neurobiol 2011;44(3): 223-34.

[73] Zhang Y, Dutta A, Abounader R. The role of microRNAs in glioma initiation and progression. Front Biosci 2012;17: 700-12.

[74] Tran N, O'Brien CJ, Clark J, Rose B. Potential role of micro-RNAs in head and neck tumorigenesis. Head Neck 2010;32(8): 1099-111.

[75] Babu JM, Prathibha R, Jijith VS, Hariharan R, Pillai MR. A miR-centric view of head and neck cancers. Biochim Biophys Acta 2011;1816(1): 67-72.

[76] Fendler A, Stephan C, Yousef GM, Jung K. MicroRNAs as regulators of signal transduction in urological tumors. Clin Chem 2011;57(7): 954-68.

[77] Sand M, Gambichler T, Sand D, Skrygan M, Altmeyer P, Bechara FG. MicroRNAs and the skin: tiny players in the body's largest organ. J Dermatol Sci 2009;53(3): 169-75.

[78] Perera RJ, Ray A. Epigenetic regulation of miRNA genes and their role in human melanomas. Epigenomics 2012;4(1): 81-90.

[79] Howell PM Jr, Liu S, Ren S, Behlen C, Fodstad O, Riker AI. Epigenetics in human melanoma. Cancer Control 2009;16(3): 200-18.

[80] Mueller DW, Bosserhoff AK. Role of miRNAs in the progression of malignant melanoma. Br J Cancer 2009;101(4): 551-6.

[81] Aberdam D, Candi E, Knight RA, Melino G. miRNAs, 'stemness' and skin. Trends Biochem Sci 2008;33(12): 583-91.

[82] Torres A, Torres K, Maciejewski R, Harvey WH. MicroRNAs and their role in gynecological tumors. Med Res Rev 2011;31(6): 895-923.

[83] Wang P, Ren Z, Sun P. Overexpression of the long non-coding RNA MEG3 impairs in vitro glioma cell proliferation. J Cell Biochem 2012;113(6): 1868-74.

[84] Zhou Y, Zhang X, Klibanski A. MEG3 noncoding RNA: a tumor suppressor. J Mol Endocrinol 2012;48(3): R45-53.

[85] Qi P, Du X The long non-coding RNAs, a new cancer diagnostic and therapeutic gold mine. Mod Pathol 2012.

[86] Kosaka N, Takeshita F, Yoshioka Y, Hagiwara K, Katsuda T, Ono M, Ochiya T. Exosomal tumor-suppressive microRNAs as novel cancer therapy: "Exocure" is another choice for cancer treatment. Adv Drug Deliv Rev 2012

Chapter 4

Roles of Tumor Suppressor Signaling on Reprogramming and Stemness Transition in Somatic Cells

Arthur Kwok Leung Cheung, Yee Peng Phoon,
Hong Lok Lung, Josephine Mun Yee Ko,
Yue Cheng and Maria Li Lung

Additional information is available at the end of the chapter

http://dx.doi.org/10.5772/55712

1. Introduction

The pioneering landmark, established by Takahashi and Yamanaka (Takahashi et al., 2007; Takahashi and Yamanaka, 2006) in reprogramming somatic cells into induced pluripotent stem (iPS) cells using the four transcriptional factors of Oct4, Sox2, Klf4, and c-Myc, represents one of the most important paradigm shifts in current stem cell biology. This unprecedented discovery could potentially revolutionize regenerative medicine, cell-based therapy and personalized medicine. Despite recent great advancement in cell reprogramming, there are still considerable technical challenges to circumvent restrictions of applications of reprogramming technology (Kawamura et al., 2009; Saha and Jaenisch, 2009). The utilization of over-expressed transcriptional factors, which of many play oncogenic roles, during somatic reprogramming posts the risk of malignant transformation, thus, limiting its clinical applications. Moreover, the reprogramming process using these factors is still inefficient in some of cell types, and is not always successful in other kinds of cells (Kawamura et al., 2009; Marion et al., 2009; Menendez et al., 2012). Therefore, the underlying mechanisms for signaling control of these factors still need to be further explored.

Somatic cell reprogramming is a complicated cellular process that is controlled by many signaling networks. Accumulated evidence indicated that stemness transition can be detected in some tumor cells following the introduction of relevant signal stimulation, and cancer cells or differentiated cells can be changed into stem cell-like cells that go through less-differentiated stages (Chen et al., 2008; Fodde and Brabletz, 2007; Huang et al., 2009; Liu et al., 2009a).

However, stemness transition may not lead to a full reprogramming of treated cells, which is determined by the delicate controls of signaling network activities in living cells. Interestingly, stemness transition may accompany epithelial-mesenchymal transition (EMT) events in cancer cells, and both programs are closely linked to the core stem cell gene network activities. Not surprisingly, multiple signaling pathways have been reported to be involved in EMT events and generation of stem cell-like cells. Wnt/β-catenin and TGF-β signaling are two potent inducers of EMT during embryonic development and cancer progression (Li et al., 2010; Mani et al., 2008; Morel et al., 2008; Scheel et al., 2011). Other involved pathways in these cellular activities may include BMP/Activin/Nodal, Notch, Hedgehog, Fibroblast growth factor signaling, and others (Chen et al., 2008; Huang et al., 2009; Kang and Massague, 2004; Natalwala et al., 2008; Thiery, 2002, 2003; Wu and Zhou, 2008).

The Wnt/β-catenin signaling pathway, highly conserved among various species and composed of a large family of proteins that control many biological properties (Fodde and Brabletz, 2007; Kikuchi et al., 2009; ten Berge et al., 2008b), may play a central role in the control of reprogramming and stemness process. This pathway includes more than two hundred genes and plays a critical role in modulating the delicate balance among stemness, proliferation, and differentiation in certain stem cell niches and tumor cells (Gu et al., 2010; Katoh, 2007; Lowry et al., 2005; Reya and Clevers, 2005). The established evidence reveals that various levels of Wnt/β-catenin signaling are likely to contribute to distinct cellular activities such as stemness transition, differentiation, carcinogenesis, and the EMT program. Therefore, the cellular activities and fate decisions are determined by this signaling activity in both dosage-dependent and tissue-dependent fashions (Anton et al., 2007; Kikuchi et al., 2009; Lluis et al., 2008; Reya and Clevers, 2005; Slack et al., 1995; Tapia and Scholer, 2010a; ten Berge et al., 2008a; Vermeulen et al., 2010). However, whether and how this signaling pathway has its direct influence on pluripotency gene networks and EMT events is largely unexplored.

As mentioned previously, cell fate decisions are controlled by both positive and negative forces in human cells. It has been well-established that tumor suppressor genes (TSGs) are important regulators to control cell proliferation, differentiation and cell death. Not surprisingly, these genes also play important roles in programming, reprogramming, and stemness transition in human cells. The well-studied TSGs, such as *p53*, *p16*, and *RB1*, serve as key regulators for the cell programming (Bonizzi et al., 2012; Hong et al., 2009; Liu et al., 2009b; Marion et al., 2009; Molchadsky et al., 2010; Wenzel et al., 2007). There are a number of reports on *p53* / p21 pathway that are involved in the reprogramming process and stemness transition in somatic cells. It should be noted that Wnt signaling was linked to the *p53* pathway a long time ago, suggesting that both signaling pathways may play interactive and critical roles in cell fate determination (Damalas et al., 1999; Kinzler and Vogelstein, 1996; Lee et al., 2010). Recent findings demonstrated that several mechanisms play a limiting role in somatic reprogramming and cell stemness transition (Figure 1) (Kawamura et al., 2009; Menendez et al., 2012; Menendez et al., 2010; Takahashi, 2010; Tapia and Scholer, 2010b). In most situations, these genes serve as active players or barriers for cell reprogramming. However, many essential questions on the roles of TSGs in cell fate decision remain unclear. For example, whether *p53*-induced inhibition in reprogramming is transient or just in the early stage is still in question (Cox and Rizzino,

http://dx.doi.org/10.5772/55712

2010; Krizhanovsky and Lowe, 2009; Wahl, 2011). Also, it was reported that the loss of *RB1* is critical for the expansion of the stem cell populations (Liu et al., 2009a; Wenzel et al., 2007). Undoubtedly, there is an urgent need to further elucidate the molecular mechanism and signaling pathways in regulating and controlling the process of somatic reprogramming and stemness transition.

Epigenetic regulation is one of the important mechanisms in the regulation of TSG activities. Recently, epigenetic modification has been shown to influence the reprogramming process, suggesting that many known TSGs may be involved in these cellular activities. Some reports illustrated that a dedifferentiation process of somatic cells to iPS cells involves dynamic epigenetic remodeling. In addition, there seem to be interactions between reprogramming transcription factors and epigenetic modifiers during these cellular activities (Takahashi, 2010).

In this chapter, the role of TSGs in cell reprogramming and stemness process, and regulation of these genes during stem cell renewal will be discussed, as described in Figure 1. We will review the role of TSG-mediated pathways and epigenetics as a barrier in cell fate determinations.

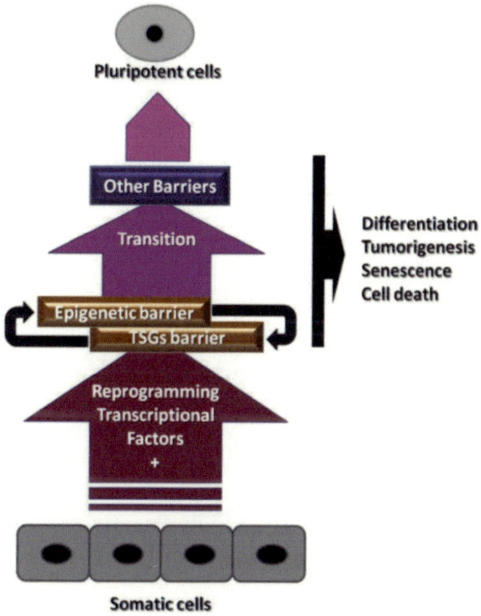

Figure 1. Schematic representative of somatic reprogramming. The reprogramming efficiency is markedly influence by TSG-mediate pathways and epigenetic modifications.

1.1. *CDKN2A* ($p16^{INK4A}$ *and* $p14^{ARF}$) gene

The *CDKN2A* (*INK4/ARF*) locus encodes two important TSGs, the $p16^{INK4A}$ (or *p16*) and $p14^{ARF}$. They are important regulators for two other critical tumor suppressive signaling pathways for controlling cell proliferation, namely *RB1* and *p53*. Utikal et al. reported that secondary murine embryonic fibroblasts (MEFs) were capable of generating iPS cells at early passage, but the efficiency decreased after serial cell culture passaging and the concomitant onset of cellular senescence (Utikal et al., 2009). This phenomenon was mainly correlated with accumulation of molecular changes in the late passage senescent MEFs (Utikal et al., 2009). Indeed, up-regulation of *p16*^{INK4A} (*INK4A*), *p14*^{ARF} (*ARF*), and *p21*^{CIP} was concurrently observed in the late passage of the MEFs (Utikal et al., 2009). Deficiency and knockdown of *INK*, *ARF*, and *p53* expression resulted in higher efficiency of iPS cell formation. Interestingly, when MEFs were cultured in low oxygen condition (4%), both the expression of *INK4A* and *p53* were reduced. Most importantly, the efficiency of the iPS reprogramming was increased in the low oxygen condition. This further supports the role of *CDKN2A* and *p53* in inhibiting the reprogramming process (Utikal et al., 2009).

Concurrently, Li et al. also worked on the role of *INK4/ARF* locus which encodes three TSGs, *p16*^{IN4A}, *p14*^{ARF}, and *p15*^{INK4B} on the reprogramming of differentiated cell into iPS cells. They showed that the locus is completely silenced in iPS and embryonic stem cells. The three transcription factors, Oct4, Klf4, and Sox2 repressed the gene expression of *p16*^{INK4A}, *p14*^{ARF}, and *p15*^{INK4B} with concomitant appearance of iPS cells. In addition, genetic knockdown of the *INK4/ARF* locus improved the efficiency of iPS cell generation. In mouse cells, *ARF* played more significant role as compared to *INK4A*. In contrast, the *INK4A* function was more prominent than the *ARF* in human cells (Li et al., 2009). Interestingly, ageing up-regulated the gene expression of the three genes at the *INK4/ARF* locus and, in turn, led to less efficient reprogramming in cells from old organisms; this defect can be rescued by genetically inhibiting the *INK4/ARF* locus. Taken together, these findings provide strong evidence that supports the role of *CDKN2A* in regulating cell reprogramming in iPS cells.

The epidermis is a tissue that undergoes continual and rapid self-renewal, and which is dependent on the presence of stem cells and transient amplifying keratinocytes. In primary human keratinocytes, *INK4A* also plays an important role in regulation of their stemness properties (Maurelli et al., 2006). The *INK4A* inactivation enabled the primary human keratinocytes to escape replicative senescence and blocked clonal evolution and maintained keratinocytes having the stemness phenotypes. A persistent *INK4a* inactivation is necessary for maintenance of immortalization of the keratinocytes, which was accompanied by reactivation of B cell-specific Moloney murine leukemia virus site 1 *(Bmi-1)* expression and telomerase activity. *Bmi-1* expression is necessary to maintain the immortalization induced by *INK4a* inactivation. In contrast, the *INK4a* inactivation in the transient amplifying keratinocytes did not undergo immortalization but senescence. Thus, *INK4a* inactivation appears to selectively inhibit clonal conversion in highly proliferative somatic cells. Interestingly, inactivation of *INK4a* up-regulated the *ARF/p53/p21*^{Waf1} pathway but this up-regulation of the *p53* pathway was unlikely to suppress the cell proliferation. The *p53* pathway was necessarily inactivated during immortalization of human keratinocytes. This study clearly indicates the regulation of keratinocyte clonal evolution by *INK4a* regulation and its inactivation in epidermal stem cells is necessary for maintaining the stemness phenotypes (Maurelli et al., 2006).

1.2. *RB1* gene

RB1 (*pRB1* family members: *RB1, RBL1,* and *RBL2*) was identified as a TSG in patients with inherited retinoblastoma. It is one of the well-studied TSGs. It involves in cell cycle G_1/S transition regulation and binds to an important transcription factor family, E2F. Based on the Knudson two-hit hypothesis, loss of single copy of *pRB1* gene is not sufficient to induce tumor formation, loss of another copy is necessary for inducing tumor formation (Knudson, 1971). Mouse *pRB1*was found to be crucial during embryonic development; loss of two copies of *RB1* gene in mouse embryo is lethal (Clarke et al., 1992; Jacks et al., 1992; Lee et al., 1992; Wu et al., 2003). Trophoblasts are cells forming the outer layer of a blastocyst, which provide nutrients to the embryo and develop into a large part of the placenta. Specific loss of mouse *pRB1* gene in trophoblast stem cells resulted in an overexpansion of trophoblasts, profound placental abnormalities, and eventually fetal death (Wenzel et al., 2007). Loss of *pRB1* resulted in an increase of *E2F3* expression and the combined depletion of *pRB1* and *E2F3* in trophoblast stem cells rescued the *pRB1* mutant phenotypes by restoration of placental development and by extending the lifespan of embryos. As can be seen, the *pRB1* pathway plays a critical role in the maintenance of a mammalian stem cell population for proper development of both extra-embryonic and fetal tissues.

Humans and other mammalians are unable to regenerate large portions of lost limbs or other internal organs after traumatic injury or surgical excisions. In contrast, lower vertebrates are able to regenerate entire limbs, the lens of the eye, and portions of the heart (Brockes and Kumar, 2008; Poss et al., 2002; Tanaka and Weidinger, 2008). The difference can be explained in part by the observation that inactivation of *pRB1* alone in lower vertebrates was sufficient to induce skeletal muscle regeneration by reversing differentiation and post-mitotic arrest in the muscle cells (Tanaka et al., 1997). In mammalian muscle cells, suppression of *pRB1* alone was not sufficient to reverse the post-mitotic arrest and terminal differentiation (Camarda et al., 2004; Huh et al., 2004; Pajcini et al., 2010). The tumor suppressor *ARF* which is present in mammals, but absent in regenerative vertebrates, is a regeneration suppressor in addition to *pRB1* (Pajcini et al., 2010). Concurrent inactivation of both *ARF* and *pRB1* resulted in mammalian muscle cell cycle re-entry cell proliferation and dedifferentiation (Pajcini et al., 2010). These results indicate that suppression of both *pRB1* and *ARF* will result in the ability of skeletal muscle cells to lose their differentiated characters, and the skeletal muscle cells will then proliferate and dedifferentiate in a manner that mimics the regenerative lower vertebrate cells. Furthermore, *pRB1* is not only restricted to serve as a cell cycle regulator, but also to impact differentiation and tissue-specific gene expression directly by binding histone deacetylase 1 (HDAC1) and promoting activation of muscle genes such as the myogenic activator MyoD (Puri et al., 2001).

The *pRB1* gene family plays an important regulatory role in neuronal differentiation (Slack et al., 1995). When treated with retinoic acid, the embryonal carcinoma p19 cells were induced to differentiate into cultures primarily consisting of neurons and astrocytes. During this neuroectodermal differentiation, a dramatically increase of pRB1 protein levels was observed. When the pRB1 family proteins in the p19 cells were inactivated by the E1A mutant, the differentiating p19 cells underwent apoptosis. The dying cells were those committed to the neural lineages because neurons and astrocytes were lost from the differentiating cell culture. The results suggest that the pRB1 family proteins are essential for the neural lineage development and the absence of functional pRB1 activities will trigger cell death of the differentiating neuroectodermal cells.

The *pRB1* pathway is also critical for inducing the cellcycle arrest that mediates cell-cell contact inhibition in fibroblasts; when all three *pRB1* family members, *RB1, RBL1,* and *RBL2,* were inactivated by triple knockouts (TKOs), the fibroblasts escaped from contact inhibition and grew into 3D colonies or stacks in cell culture (Dannenberg et al., 2000; Sage et al., 2000). The outgrowth of TKO MEFs into spheres triggered reprogramming to produce cells with cancer stem cell properties. Whereas the fibroblasts with a single *pRB1* mutation retained contact inhibition, when this inhibition was bypassed by forcing the cells to form outgrowth spheres, the fibroblasts were reprogrammed to generate cells with a cancer stem cell phenotype (Liu et al., 2009a). These findings suggest a potential mechanism for generation of cancer stem cells from differentiated somatic cells as a result of tumor outgrowth.

1.3. *p53* gene

p53, as the "guardian of the genome" (Lane, 1992), plays a pivotal role in regulating the delicate balance of cell proliferation and cell death (Molchadsky et al., 2010). Since its discovery more than three decades ago, the role of *p53* in suppressing tumor initiation and progression is well established. It is, therefore, not surprising that *p53* is lost, inactivated, or mutated in the majority of cancers. In respond to external stress stimuli, *p53* prevents cancer development by inducing cellcycle arrest, DNA repair, senescence, and apoptosis.

Researchers have newly identified roles played by *p53* including regulation of pluripotency and dedifferentiation, as a potent barrier in reprogramming. (Hong et al., 2009). Undoubtedly, the function of *p53* is now far more complex than just simply playing the role as the classical tumor suppressor (Bonizzi et al., 2012; Kawamura et al., 2009; Marion et al., 2009; Menendez et al., 2010; Molchadsky et al., 2010; Tapia and Scholer, 2010a; Wahl, 2011; Zhao and Xu, 2010). This provides us with a new insight on the complexity of *p53* signaling in controlling cell fate decisions. Despite accumulating effort in deciphering the diversified roles played by *p53,* the cellular and molecular mechanism underlying the acquisition of "stemness" involved in the *p53* signaling is still largely unexplored.

During somatic cell reprogramming, the *p53* pathway is activated, thus disrupting iPS reprogramming (Kawamura et al., 2009). *p53* may act as a limiting factor in the iPS reprogramming efficiency. Inhibition of the *p53* pathway either by mutating, deleting or knocking down *p53* or its target gene , *p21,* further enhances the reprogramming efficiency (Kawamura et al., 2009; Liu et al., 2009b; Marion et al., 2009; Tapia and Scholer, 2010b).

The *p53/p21* pathway was reported to suppress the iPS cell generation. Suppression of *p53* increased the efficiency of the generation of iPS cells (Hong et al., 2009). A dominant negative *p53* mutant, P275S, was used to study the effect of *p53* on regulating the iPS cell generation. Results suggested that inhibition of *p53* function by introducing the dominant negative *p53* mutant into the MEFs increased GFP-positive colonies in the *p53*-heterozygous MEFs (Hong et al., 2009). Similar experiments were also performed in terminally differentiated somatic cells (T-lymphocytes from Nanog-GFP reporter transgenic mice with either *p53* wild-type or null genotype). In this study, the four important stem cell reprogramming factors, *Oct4, Sox2, Klf4,* and *c-Myc* were introduced into the T-lymphocytes. No GFP-positive colony can be observed in the *p53* wild-type T lymphocytes (Hong et al., 2009). On the other hand, GFP-positive colonies can be observed in *p53*-null lymphocytes and the cells were expandable and have a similar morphology with the mouse ES cells (Hong et al., 2009). The increased GFP-positive cells can also be observed in the adult human dermal fibroblasts (HDFs) by introducing the dominant negative *p53* together with the reprogramming factors into the HDFs (Hong et al., 2009), suggesting the importance of *p53* in regulating the iPS cell reprogramming.

The function of *p53* in regulating stem cell multipotency was confirmed in germ-line stem cells (GSCs). Depletion of *p53* function in the GSCs increased the efficiency of reverting GSC multipotency status (Kanatsu-Shinohara et al., 2004). This finding can also be observed in a *p53* knockout mouse study (Lam and Nadeau, 2003). Hanna et al. suggested that depletion of *p53* function in clonal B cells can only enhance the kinetics of reprogramming somatic cells into iPS cells with a higher cell division rate (Hanna et al., 2009). However, it does not regulate the overall efficiency (Hanna et al., 2009). A *p53* mutant, R172H, which induces conformation change of the *p53* protein, was reported to associate with higher reprogramming efficiency than WT *p53* in the MEFs (Lang et al., 2004). Lang et al. showed that reprogramming efficiency in that particular *p53*-mutated MEFs, which was induced by utilizing a two factor system (Oct4 and Sox2), is higher than the *p53* knockout MEFs that was induced by using the three factors system (Oct4, Sox2, and Klf4) (Lang et al., 2004), suggesting the importance of *p53* in regulating the reprogramming process.

Cicalese et al. suggested that the function of *p53* in stem cells is critical to maintain a constant number of stem cells by imposing an asymmetric mode of self-renewing division. In the $p53^{-/-}$ and *ErbB2* tumor mammospheres, up-regulation of Nanog is observed. These studies also revealed the importance of *p53* in regulating the stem cell polarity, and the loss of *p53* induces increased frequency of symmetric division and tumor initiation and growth (Cicalese et al., 2009).

The suppression of the reprogramming efficiency of iPS cells by *p53* can be associated with the maintenance of genomic integrity of iPS cells. Deficient *p53* resulted in shorter telomeres in the reprogramming MEFs (Marion et al., 2009), suggesting the low efficiency of reprogramming in the WT *p53* cells to prevent the spreading of cells upon DNA damage and to ensure iPS cell genomic integrity (Marion et al., 2009).

Another barrier affecting the reprogramming is the *INK4A/ARF* tumor suppressor locus, as described previously. A recent report by Li and colleagues illustrated that the *INK4A/ARF* locus was suppressed during the early stage of reprogramming, leading to inactivation of the *p53* and *pRB1* pathways (Li et al., 2009). Interestingly, cells with $p16^{INK4A}$ knockdown alone are sufficient to enhance the reprogramming efficiency (Li et al., 2009). Together, these observations indicate that both *p53* and *pRB1* may work synergistically as barriers in somatic cell reprogramming (Li et al., 2009; Menendez et al., 2010; Utikal et al., 2009).

In a recent report by Lee K.H. et al., *p53* preferentially targets the Wnt signaling pathway in the murine ESC differentiation program (Lee et al., 2010). Evidently, the crosstalk between *p53* and Wnt signaling pathway plays an integrated role in stemness acquisition. A *p53* downstream phosphatase, *Wip1,* which shows high expression in the intestinal cells, was reported to associate with *p53*-dependent apoptosis of stem cells in the mouse intestine (Demidov et al., 2007). Removal of *Wip1* reduced the polyp formation in the APC^{Min} mice. The $APC^{Min/+}$ mice contain a nonsense mutation in the *APC* gene. Constitutively activated Wnt signaling pathway increased the apoptosis events of intestinal stem cells in the *Wip1*-deficient mice (Demidov et al., 2007). Low level of *Wip1* reduced the threshold of *p53*-dependent apoptosis of stem cells. However, *Wip1* deficiency does not affect the activities of β-catenin in terms of its nuclear localization level. A high level of β-catenin can be observed in the nuclei of polyp cells and this contributes to up-regulation of *c-Myc* and *Cyclin D1* in the *Wip1* null/$Apc^{Min/+}$mice. The β-catenin signaling pathway activation and attenuation of *p53* resulted in increasing efficiency of intestinal stem cell apoptosis (Demidov et al., 2007).

Recently, researchers demonstrated that the p53-miR34-Wnt network is a determinant factor of dichotomy between stem cell properties and tumor progression. miR34, one of the direct downstream targets of *p53*, is found to interact with Wnt and EMT genes, including *β-catenin, AXIN2, LEF1* and *Snail*. With the loss of *p53* due to miR34, the Wnt pathway is activated, which further induced the transformation of EMT (Liu et al., 2011). Therefore, the *p53* gene plays an important role in the controlling EMT.

Chang et al suggested that *p53* induced transcriptional activation of microRNA, miR-200c, through direct binding to its promoter region. The miR-200c was reported to regulate the EMT process through inhibition of transcriptional suppressors of an epithelial marker, E-cadherin (Chang et al., 2011). The miR-200c can target to and suppress ZEB1/2 (Wellner et al., 2009), which is a well-studied E-cadherin transcriptional suppressor and thus, regulates the EMT process. The knockdown of *p53* in MCF12A cells resulted in loss of epithelial phenotype and shows a significant elevation of the $CD24^{-}CD44^{+}$ population. Re-expression of *p53* in TGF-β-treated MCF12A showed inhibition of TGF-β-induced increase of the stem cell population.

p53 is not a sole player in deciding the cell fate determination. In fact, *p53* works as an integrated network, interplaying with other important pathways, depending on the external stimuli and microenviroment. However, there is a great need to further elucidate the roles of the *p53* network in reprogramming, dedifferentiation, self-renewal, and pluripotency.

2. Signaling pathways involved in the reprogramming and stemness transition

2.1. TGF–β signaling pathway

TGF-β signaling pathways play multiple roles in regulating tumorigenesis and other cellular processes, including reprogramming, stemness transition, and EMT events. Many components in this signaling pathway were defined to participate in both oncogenic and tumor suppressive pathways in various tumors. This provides a complicated story for researchers to study the function of TGF-β signaling pathways in stem cells or reprogrammed cells. The ligands of the TGF-β family have multiple functions and can cause opposite effects in different cell types. The TGF-β can regulate cell proliferation, growth arrest, differentiation, survival, cell migration, and also the pluripotency of cells. In cancer, over-expression of TGFβ1 and deregulation of the TGF-β receptor type II (TGFBRII) were reported to associate with skin cancer tumorigenesis and invasiveness (Cui et al., 1996). However, the role of TGF-β signaling in regulating reprogramming is still not well-defined. In a previous report, TGF-β family ligands play an important role in reprogramming of somatic cells into iPS cells, regulating ESCs self-renewal, pluripotency maintenance, and controlling differentiation.

TGF-β signalling may have the ability to induce reprogramming of somatic cells into iPS cells. Treatment of TGF-β/activin inhibitor in partially reprogrammed iPS cells can induce Nanog expression (Ichida et al., 2009; Maherali and Hochedlinger, 2009). Furthermore, the functional role of TGF-β in regulating the reprogramming was defined by utilizing chemical TGF-β inhibitors. Interestingly, inhibition of TGF-β signaling can enhance the mouse fibroblast reprogramming efficiency. A substitute of Sox2 (E-616452) and TGFBR1 kinase (SB-431542) inhibitor, were reported to replace the function of Sox2 in MEFs with Oct4, Klf4, and c-Myc expression (Ichida et al., 2009; Maherali and Hochedlinger, 2009). These results suggest the important roles TGF-β plays in the controlling reprogramming process.

Maintenance of the pluripotencies and self renewal properties are important for both ESCs and iPS cells.The canonical TGF-β signaling pathway may play important regulatory roles in ESCs maintenance and generation of pluripotency. BMP4 together with the LIF protein can induce Oct4 expression (Ying et al., 2003). The BMP activated Smad signaling to support self-renewal properties of stem cells. The inhibition of Smad activities by the Smad6 and Smad7 in the ES cells induced smaller and fewer ES cell colon formation (Ying et al., 2003). Secretion of BMP4 by the feeder cells is necessary for ES cell self-renewal (Ying et al., 2003). Inhibition of the Erk and p38 MAPK pathways can further enhance the BMP4-associated effect on self-renewal of mouse ESCs (Qi et al., 2004). Besides this, bFGF (basic fibroblast growth factor) and activin are also important to maintain the pluripotency in human ESCs (Greber et al., 2010; James et al., 2005). The TGF-β signalling may play multifunctional roles in regulating pluripotency of cells. Smad1 was reported to suppress the expression of Nanog by inhibiting its promoter activities (Jiang and Ng, 2008; Xu et al., 2008). The Smad proteins were reported to bind directly to Nanog promoter (Xu et al., 2008) and this is the major mechanism for Smad proteins to regulate Nanog expression. These results suggest the multiple roles of TGF-β signaling in the regulation of stem cell renewal.

Furthermore, TGF-β also plays a role to control the differentiation of ESCs. One of the TGF-β family members, BMP4, was reported to associate with induction of inhibitor of differentiation (*Id)* gene via interaction with the LIF/Jak-Stat3 and Smad pathways. The *Id* gene is an important factor to block ESC differentiation. The undifferentiated ES cells expressed BMP signaling ligands (Ying et al., 2003) and regulated downstream molecules, the Smads, to control the cell differentiation process (Ying et al., 2003).

Collaborating with Wnt signaling, TGF-β signaling is also involved in the EMT program and both pathways are regarded as the axis of EMT in breast cells (Scheel et al., 2011). The hy-

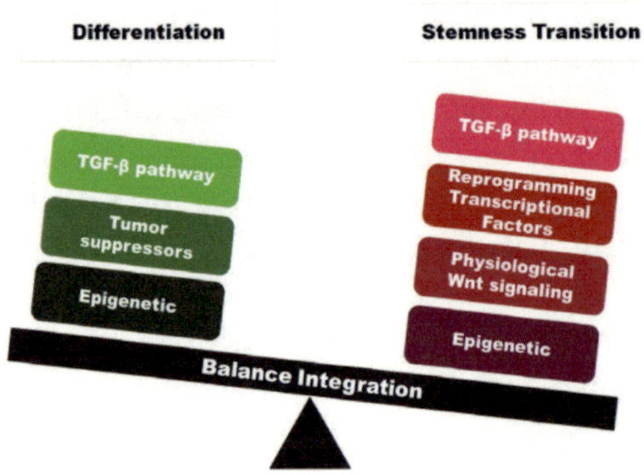

Figure 2. The cell fate determination is delicately controlled by positive and negative forces. Cellular activity balance regulated by both core stem gene-mediated pathways and TSGs is the key determinant in reprogramming process.

pothesis of these two pathways linked to the stem cell networks and TSG pathways is presented in Figure 2.

2.2. Wnt pathway

Cellular reprogramming can be achieved by overexpression of defined transcription factors in somatic cells (Ichida et al., 2009; Takahashi et al., 2007). However, the underlying mechanism of signaling activities that regulated these factors are not fully understood now. Overexpression for certain genes may not be suitable for all pathways, such as *β-catenin*, a mediator of Wnt signaling, because discrete levels of expressed genes are usually needed for maintaining the pluripotent status or direct programming through this pathway (Gu et al., 2010; Lluis et al., 2008; Marson et al., 2008; Merrill, 2008). It still remains unclear the gene-dosage effects of critical factors on somatic cell reprogramming and stem cell renewal. Recent studies revealed that activation of Wnt/β-catenin signaling may directly control reprogramming of fused somatic cells. For example, Wnt stimulators, Wnt3a and BIO, strikingly enhanced reprogramming ability after cell fusion (Lluis et al., 2008; Merrill, 2008). The fusion clones derived from both ESCs and somatic cells had an obvious β-catenin accumulation with increased expression of *Axin2*, a *β-catenin*-dependent gene, suggesting that basic or lower levels of stabilized β-catenin might drive somatic cell reprogramming.

The lower levels of Wnt signaling play a critical role in the control of development of several types of tissues through a dosage-dependent manner, as reported in crypt progenitor cells (Batlle et al., 2002; Korinek et al., 1998), hair follicles (Lowry et al., 2005), and hematopoietic stem cells (Luis et al., 2011). Taken together, observations from both *in vitro* and *in vivo* studies indicated that Wnt/β-catenin signaling was a single dominant force in the control of cell fate determinations in some of tissues, which suggests that basic or physiological levels of Wnt signaling may be required for many cellular activities.

More and more evidence revealed that Wnt signaling plays important roles in maintenance of pluripotency in ESCs and cell self-renewal (Cole et al., 2008; Lluis et al., 2008; Macarthur et al., 2009; Marson et al., 2008; Takao et al., 2007). For example, expression of *β-catenin* was confirmed

http://dx.doi.org/10.5772/55712

to associate with hemtopoietic stem cells and neural stem cell growth (Kalani et al., 2008; Reya et al., 2003). *Wnt3A* activation associated with expression of the stem cell reprogramming markers, Oct4 and Nanog (Ogawa et al., 2007) and maintenance of the pluripotency of mouse ES cells (Hao et al., 2006; Singla et al., 2006). *Wnt3A* induced generation of iPS cells in the absence of *Myc* (Marson et al., 2008). Those cells contained iPS cell properties and were able to form teratomas during subcutaneous injection into SCID mice (Marson et al., 2008). The Wnt signaling pathway is also involved in regulating pluripotency factors, *Oct4, Nanog*, and Sox2 expression (Anton et al., 2007; Sato et al., 2004). This observation was confirmed by down-regulation of the stem cell pluripotency genes in the *β-catenin* deficient mouse ES cells (Anton et al., 2007). Wnt signaling pathway was also associated with cell reprogramming through the telomerase reverse transcriptase (TERT) and Brahma-related gene 1 (BRG1) interaction (Barker et al., 2001) to modulate chromatin structure during reprogramming (Miki et al., 2011).

Interestingly, previous study demonstrated that Wnt3a can also stimulate human ES cell differentiation, rather than only regulate human ES cell proliferation. The canonical Wnt signaling levels are minimal in the undifferentiated human ES cells but greatly increase after Wnt3a treatment and induce differentiation (Dravid et al., 2005). Dramatic increase of reprogrammed cell numbers can be observed when ES cells, which have a low level of nuclear β-catenin, are fused with neural stem cells. This is mainly due to the low nuclear β-catenin level being able to protect fused cells from apoptosis (Lluis et al., 2010), suggests the importance of β-catenin levels in the regulation of stem cell reprogramming. This finding may help to explain the balance between the maintenance of pluripotency of stem cells and apoptosis, as excess β-catenin can induce p53 expression (Damalas et al., 1999), which was found to induce apoptosis in stem cells to maintain genome integrity. The p53 protein was reported to be a transcription regulator of the Wnt signaling and it bound on the promoter regions of some Wnt signaling members for a general stress response in the mouse ES cells (Lee et al., 2010), which may provide a feedback mechanism to control the deregulation of the β-catenin during the reprogramming process.

It should be noted that inappropriate activation of components of this signaling pathway have been observed in many human cancers and differentiated stem cells, in which the high levels of *β-catenin* signaling were usually detected (Dravid et al., 2005; Fodde and Brabletz, 2007; ten Berge et al., 2008a; Vermeulen et al., 2010). Except for *p53* described previously, some components of the Wnt pathway can be regarded as both oncogenes and TSGs. For example, *AXIN2, APC, DKK1,* and *WIF1* are negative regulators of this pathway, and are called TSGs. In summary, the detailed mechanism of Wnt signaling in the control of stemness transition and reprogramming of somatic cells needs to be further explored.

3. Possible mechanisms to regulate TSGs expression in reprogramming

It is well-accepted now that epigenetic regulations are important events to control gene expression in human cells. Promoter hypermethylation and histone modification are two major events to regulate gene expression in various human tumors. The *DNA methyltransferase*

(*DNMTs*), *histone deacetylases* (*HDACs*), *histone acetyl transferase* (*HATs*), and *histone methyltransferase* are the key regulators to controlgene expression in the genome. Epigenetic changes of gene expression were reported to be important during the iPS cell reprogramming (Han and Sidhu, 2008). The epigenetic changes can also help to maintain the pluripotency by regulating the expression of the key transcription factors, Oct4, Nanog, and Sox2 (van Vlerken et al., 2012). In previous studies, mouse ES cell genomes were found to contain less methylation than the somatic cells, while human ES cells show a distinct epigenetic profile, when compared to somatic cells (Jackson et al., 2004; Lagarkova et al., 2006; Zvetkova et al., 2005). A silenced TSG, *p16*, was found to be re-expressed during the reprogramming process (Ron-Bigger et al., 2010). On the other hand, a previous study suggested that the promoter region of INK4A/ARF was found to be hypermethylated in the iPS and ES cells. Inhibition of *DNMTs* by inhibitor and siRNA increased the *INK4A* and *p21* (*CIP1/WAF1*) expression in human umbilical cord blood-derived multipotent stem cells (So et al., 2011). However, the epigenetic regulation of TSGs during the reprogramming process is still not fully understood now. It is necessary to further explore epigenetic changes of TSGs in the reprogramming process and relevant other cellular activities.

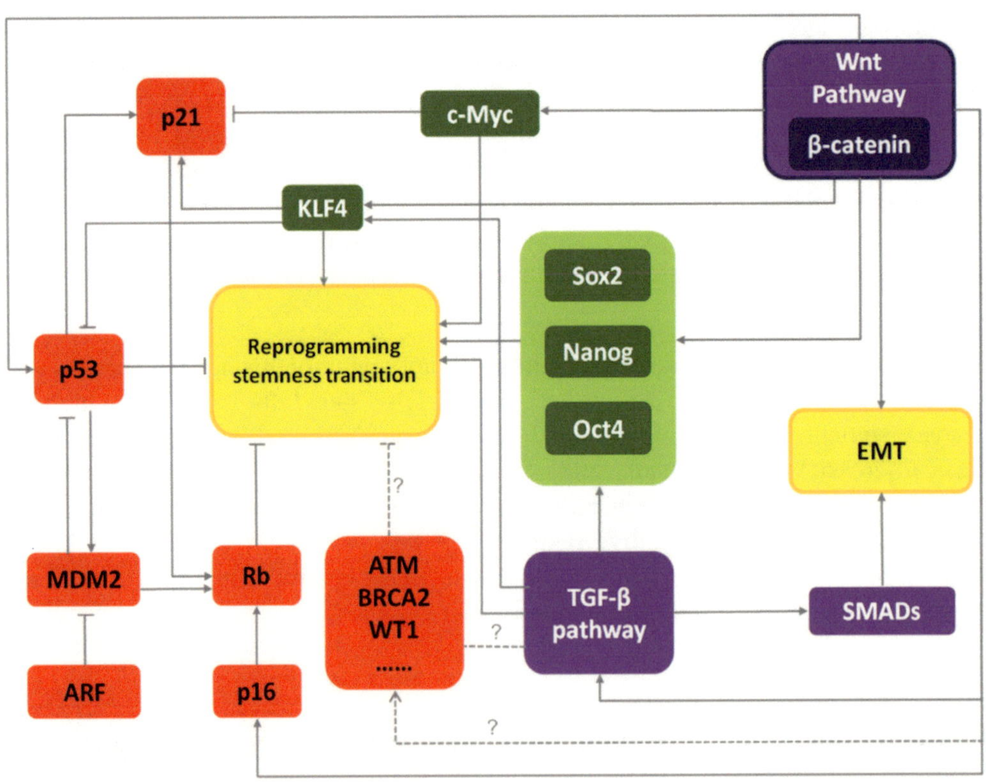

Figure 3. Hypothesis of integrated networks of TGSs, Wnt/β-catenin and TGF-β pathways in controlling reprogramming, stemness transition and EMT events. These pathways may play central roles in regulating other TSGs, transcriptional factors and other signaling pathways.

4. Conclusions

The known and unknown TSGs are the important participators in the regulation of cell reprogramming and stemness transition. These genes are components of various signaling pathways, and play different roles in maintaining cell pluripotency, regulating cell differentiation and proliferation, cell cycle control, apoptosis, and other cell fate decisions. These genes controlling cellular activities act in a time-dependent or a dosage-dependent manner in various tissues. Although detailed underlying mechanisms are not fully clear now, more and more evidence indicates that some TSG signaling activities are determinant forces in important cellular processes, including cell reprogramming. A proposed hypothesis illustrates this in Figure 3. Understanding of the delicate control of these signaling networks in living cells will provide more insight in reprogramming studies and regenerative medicine.

Acknowledgements

The University of Hong Kong Seed Funding Program for Basic Research (Project Codes: 201007159005 and 201111159142) to YC.

Author details

Arthur Kwok Leung Cheung, Yee Peng Phoon, Hong Lok Lung, Josephine Mun Yee Ko, Yue Cheng* and Maria Li Lung

*Address all correspondence to: yuecheng@hku.hk

Center for Nasopharyngeal Carcinoma Research, Center for Cancer Research, Department of Clinical Oncology, University of Hong Kong, Hong Kong (SAR), PR China

References

[1] Anton, R., Kestler, H.A., and Kuhl, M. (2007). Beta-catenin signaling contributes to stemness and regulates early differentiation in murine embryonic stem cells. FEBS Lett *581*, 5247-5254.

[2] Barker, N., Hurlstone, A., Musisi, H., Miles, A., Bienz, M., and Clevers, H. (2001). The chromatin remodelling factor Brg-1 interacts with beta-catenin to promote target gene activation. EMBO J *20*, 4935-4943.

[3] Batlle, E., Henderson, J.T., Beghtel, H., van den Born, M.M., Sancho, E., Huls, G., Meeldijk, J., Robertson, J., van de Wetering, M., Pawson, T., *et al.* (2002). Beta-catenin

and TCF mediate cell positioning in the intestinal epithelium by controlling the expression of EphB/ephrinB. Cell *111*, 251-263.

[4] Bonizzi, G., Cicalese, A., Insinga, A., and Pelicci, P.G. (2012). The emerging role of p53 in stem cells. Trends in Molecular Medicine *18*, 6-12.

[5] Brockes, J.P., and Kumar, A. (2008). Comparative aspects of animal regeneration. Annual Review of Cell & Developmental Biology *24*, 525-549.

[6] Camarda, G., Siepi, F., Pajalunga, D., Bernardini, C., Rossi, R., Montecucco, A., Meccia, E., and Crescenzi, M. (2004). A pRb-independent mechanism preserves the postmitotic state in terminally differentiated skeletal muscle cells. The Journal of Cell Biology *167*, 417-423.

[7] Chang, C.J., Chao, C.H., Xia, W., Yang, J.Y., Xiong, Y., Li, C.W., Yu, W.H., Rehman, S.K., Hsu, J.L., Lee, H.H., *et al.* (2011). p53 regulates epithelial-mesenchymal transition and stem cell properties through modulating miRNAs. Nat Cell Biol *13*, 317-323.

[8] Chen, X., Xu, H., Yuan, P., Fang, F., Huss, M., Vega, V.B., Wong, E., Orlov, Y.L., Zhang, W., Jiang, J., *et al.* (2008). Integration of external signaling pathways with the core transcriptional network in embryonic stem cells. Cell *133*, 1106-1117.

[9] Cicalese, A., Bonizzi, G., Pasi, C.E., Faretta, M., Ronzoni, S., Giulini, B., Brisken, C., Minucci, S., Di Fiore, P.P., and Pelicci, P.G. (2009). The tumor suppressor p53 regulates polarity of self-renewing divisions in mammary stem cells. Cell *138*, 1083-1095.

[10] Clarke, A.R., Maandag, E.R., van Roon, M., van der Lugt, N.M., van der Valk, M., Hooper, M.L., Berns, A., and te Riele, H. (1992). Requirement for a functional Rb-1 gene in murine development. Nature *359*, 328-330.

[11] Cole, M.F., Johnstone, S.E., Newman, J.J., Kagey, M.H., and Young, R.A. (2008). Tcf3 is an integral component of the core regulatory circuitry of embryonic stem cells. Genes Dev 22, 746-755.

[12] Cox, J.L., and Rizzino, A. (2010). Induced pluripotent stem cells: what lies beyond the paradigm shift. Experimental Biology & Medicine *235*, 148-158.

[13] Cui, W., Fowlis, D.J., Bryson, S., Duffie, E., Ireland, H., Balmain, A., and Akhurst, R.J. (1996). TGFbeta1 inhibits the formation of benign skin tumors, but enhances progression to invasive spindle carcinomas in transgenic mice. Cell *86*, 531-542.

[14] Damalas, A., Ben-Ze'ev, A., Simcha, I., Shtutman, M., Leal, J.F., Zhurinsky, J., Geiger, B., and Oren, M. (1999). Excess beta-catenin promotes accumulation of transcriptionally active p53. EMBO J *18*, 3054-3063.

[15] Dannenberg, J.H., van Rossum, A., Schuijff, L., and te Riele, H. (2000). Ablation of the retinoblastoma gene family deregulates G(1) control causing immortalization and increased cell turnover under growth-restricting conditions. Genes & Development *14*, 3051-3064.

[16] Demidov, O.N., Timofeev, O., Lwin, H.N., Kek, C., Appella, E., and Bulavin, D.V. (2007). Wip1 phosphatase regulates p53-dependent apoptosis of stem cells and tumorigenesis in the mouse intestine. Cell Stem Cell *1*, 180-190.

[17] Dravid, G., Ye, Z., Hammond, H., Chen, G., Pyle, A., Donovan, P., Yu, X., and Cheng, L. (2005). Defining the role of Wnt/beta-catenin signaling in the survival, proliferation, and self-renewal of human embryonic stem cells. Stem Cells *23*, 1489-1501.

[18] Fodde, R., and Brabletz, T. (2007). Wnt/beta-catenin signaling in cancer stemness and malignant behavior. Current Opinion in Cell Biology *19*, 150-158.

[19] Greber, B., Wu, G., Bernemann, C., Joo, J.Y., Han, D.W., Ko, K., Tapia, N., Sabour, D., Sterneckert, J., Tesar, P., *et al.* (2010). Conserved and divergent roles of FGF signaling in mouse epiblast stem cells and human embryonic stem cells. Cell Stem Cell *6*, 215-226.

[20] Gu, B., Watanabe, K., and Dai, X. (2010). Epithelial stem cells: an epigenetic and Wnt-centric perspective. Journal of Cellular Biochemistry *110*, 1279-1287.

[21] Han, J., and Sidhu, K.S. (2008). Current concepts in reprogramming somatic cells to pluripotent state. Curr Stem Cell Res Ther *3*, 66-74.

[22] Hanna, J., Saha, K., Pando, B., van Zon, J., Lengner, C.J., Creyghton, M.P., van Oudenaarden, A., and Jaenisch, R. (2009). Direct cell reprogramming is a stochastic process amenable to acceleration. Nature *462*, 595-601.

[23] Hao, J., Li, T.G., Qi, X., Zhao, D.F., and Zhao, G.Q. (2006). WNT/beta-catenin pathway up-regulates Stat3 and converges on LIF to prevent differentiation of mouse embryonic stem cells. Dev Biol *290*, 81-91.

[24] Hong, H., Takahashi, K., Ichisaka, T., Aoi, T., Kanagawa, O., Nakagawa, M., Okita, K., and Yamanaka, S. (2009). Suppression of induced pluripotent stem cell generation by the p53-p21 pathway. Nature *460*, 1132-1135.

[25] Huang, J., Chen, T., Liu, X., Jiang, J., Li, J., Li, D., Liu, X.S., Li, W., Kang, J., and Pei, G. (2009). More synergetic cooperation of Yamanaka factors in induced pluripotent stem cells than in embryonic stem cells. Cell Res *19*, 1127-1138.

[26] Huh, M.S., Parker, M.H., Scime, A., Parks, R., and Rudnicki, M.A. (2004). Rb is required for progression through myogenic differentiation but not maintenance of terminal differentiation. The Journal of cell biology *166*, 865-876.

[27] Ichida, J.K., Blanchard, J., Lam, K., Son, E.Y., Chung, J.E., Egli, D., Loh, K.M., Carter, A.C., Di Giorgio, F.P., Koszka, K., *et al.* (2009). A small-molecule inhibitor of tgf-Beta signaling replaces sox2 in reprogramming by inducing nanog. Cell Stem Cell *5*, 491-503.

[28] Jacks, T., Fazeli, A., Schmitt, E.M., Bronson, R.T., Goodell, M.A., and Weinberg, R.A. (1992). Effects of an Rb mutation in the mouse. Nature *359*, 295-300.

[29] Jackson, M., Krassowska, A., Gilbert, N., Chevassut, T., Forrester, L., Ansell, J., and Ramsahoye, B. (2004). Severe global DNA hypomethylation blocks differentiation and induces histone hyperacetylation in embryonic stem cells. Mol Cell Biol *24*, 8862-8871.

[30] James, D., Levine, A.J., Besser, D., and Hemmati-Brivanlou, A. (2005). TGFbeta/activin/nodal signaling is necessary for the maintenance of pluripotency in human embryonic stem cells. Development *132*, 1273-1282.

[31] Jiang, J., and Ng, H.H. (2008). TGFbeta and SMADs talk to NANOG in human embryonic stem cells. Cell Stem Cell *3*, 127-128.

[32] Kalani, M.Y., Cheshier, S.H., Cord, B.J., Bababeygy, S.R., Vogel, H., Weissman, I.L., Palmer, T.D., and Nusse, R. (2008). Wnt-mediated self-renewal of neural stem/progenitor cells. Proc Natl Acad Sci U S A *105*, 16970-16975.

[33] Kanatsu-Shinohara, M., Inoue, K., Lee, J., Yoshimoto, M., Ogonuki, N., Miki, H., Baba, S., Kato, T., Kazuki, Y., Toyokuni, S., *et al.* (2004). Generation of pluripotent stem cells from neonatal mouse testis. Cell *119*, 1001-1012.

[34] Kang, Y., and Massague, J. (2004). Epithelial-mesenchymal transitions: twist in development and metastasis. Cell *118*, 277-279.

[35] Katoh, M. (2007). WNT signaling pathway and stem cell signaling network. Clin Cancer Res *13*, 4042-4045.

[36] Kawamura, T., Suzuki, J., Wang, Y.V., Menendez, S., Morera, L.B., Raya, A., Wahl, G.M., and Belmonte, J.C. (2009). Linking the p53 tumour suppressor pathway to somatic cell reprogramming. Nature *460*, 1140-1144.

[37] Kikuchi, A., Yamamoto, H., and Sato, A. (2009). Selective activation mechanisms of Wnt signaling pathways. Trends in Cell Biology *19*, 119-129.

[38] Kinzler, K.W., and Vogelstein, B. (1996). Lessons from hereditary colorectal cancer. Cell *87*, 159-170.

[39] Knudson, A.G., Jr. (1971). Mutation and cancer: statistical study of retinoblastoma. Proc Natl Acad Sci U S A *68*, 820-823.

[40] Korinek, V., Barker, N., Moerer, P., van Donselaar, E., Huls, G., Peters, P.J., and Clevers, H. (1998). Depletion of epithelial stem-cell compartments in the small intestine of mice lacking Tcf-4. Nat Genet *19*, 379-383.

[41] Krizhanovsky, V., and Lowe, S.W. (2009). Stem cells: The promises and perils of p53. Nature *460*, 1085-1086.

[42] Lagarkova, M.A., Volchkov, P.Y., Lyakisheva, A.V., Philonenko, E.S., and Kiselev, S.L. (2006). Diverse epigenetic profile of novel human embryonic stem cell lines. Cell Cycle *5*, 416-420.

[43] Lam, M.Y., and Nadeau, J.H. (2003). Genetic control of susceptibility to spontaneous testicular germ cell tumors in mice. APMIS *111*, 184-190.

[44] Lane, D.P. (1992). Cancer. p53, guardian of the genome. Nature *358*, 15-16.

[45] Lang, G.A., Iwakuma, T., Suh, Y.A., Liu, G., Rao, V.A., Parant, J.M., Valentin-Vega, Y.A., Terzian, T., Caldwell, L.C., Strong, L.C., *et al.* (2004). Gain of function of a p53 hot spot mutation in a mouse model of Li-Fraumeni syndrome. Cell *119*, 861-872.

[46] Lee, E.Y., Chang, C.Y., Hu, N., Wang, Y.C., Lai, C.C., Herrup, K., Lee, W.H., and Bradley, A. (1992). Mice deficient for Rb are nonviable and show defects in neurogenesis and haematopoiesis. Nature *359*, 288-294.

[47] Lee, K.H., Li, M., Michalowski, A.M., Zhang, X., Liao, H., Chen, L., Xu, Y., Wu, X., and Huang, J. (2010). A genomewide study identifies the Wnt signaling pathway as a major target of p53 in murine embryonic stem cells. Proc Natl Acad Sci U S A *107*, 69-74.

[48] Li, H., Collado, M., Villasante, A., Strati, K., Ortega, S., Canamero, M., Blasco, M.A., and Serrano, M. (2009). The Ink4/Arf locus is a barrier for iPS cell reprogramming. Nature *460*, 1136-1139.

[49] Li, R., Liang, J., Ni, S., Zhou, T., Qing, X., Li, H., He, W., Chen, J., Li, F., Zhuang, Q., *et al.* (2010). A mesenchymal-to-epithelial transition initiates and is required for the nuclear reprogramming of mouse fibroblasts. Cell Stem Cell *7*, 51-63.

[50] Liu, C., Kelnar, K., Liu, B., Chen, X., Calhoun-Davis, T., Li, H., Patrawala, L., Yan, H., Jeter, C., Honorio, S., *et al.* (2011). The microRNA miR-34a inhibits prostate cancer stem cells and metastasis by directly repressing CD44. Nature Medicine *17*, 211-215.

[51] Liu, Y., Clem, B., Zuba-Surma, E.K., El-Naggar, S., Telang, S., Jenson, A.B., Wang, Y., Shao, H., Ratajczak, M.Z., Chesney, J., *et al.* (2009a). Mouse fibroblasts lacking RB1 function form spheres and undergo reprogramming to a cancer stem cell phenotype. Cell Stem Cell *4*, 336-347.

[52] Liu, Y., Hoya-Arias, R., and Nimer, S.D. (2009b). The role of p53 in limiting somatic cell reprogramming. Cell Research *19*, 1227-1228.

[53] Lluis, F., Pedone, E., Pepe, S., and Cosma, M.P. (2008). Periodic activation of Wnt/beta-catenin signaling enhances somatic cell reprogramming mediated by cell fusion. Cell Stem Cell *3*, 493-507.

[54] Lluis, F., Pedone, E., Pepe, S., and Cosma, M.P. (2010). The Wnt/beta-catenin signaling pathway tips the balance between apoptosis and reprograming of cell fusion hybrids. Stem Cells *28*, 1940-1949.

[55] Lowry, W.E., Blanpain, C., Nowak, J.A., Guasch, G., Lewis, L., and Fuchs, E. (2005). Defining the impact of beta-catenin/Tcf transactivation on epithelial stem cells. Genes Dev *19*, 1596-1611.

[56] Luis, T.C., Naber, B.A., Roozen, P.P., Brugman, M.H., de Haas, E.F., Ghazvini, M., Fibbe, W.E., van Dongen, J.J., Fodde, R., and Staal, F.J. (2011). Canonical wnt signaling regulates hematopoiesis in a dosage-dependent fashion. Cell Stem Cell *9*, 345-356.

[57] Macarthur, B.D., Ma'ayan, A., and Lemischka, I.R. (2009). Systems biology of stem cell fate and cellular reprogramming. Nat Rev Mol Cell Biol *10*, 672-681.

[58] Maherali, N., and Hochedlinger, K. (2009). Tgfbeta signal inhibition cooperates in the induction of iPSCs and replaces Sox2 and cMyc. Current Biology *19*, 1718-1723.

[59] Mani, S.A., Guo, W., Liao, M.J., Eaton, E.N., Ayyanan, A., Zhou, A.Y., Brooks, M., Reinhard, F., Zhang, C.C., Shipitsin, M., *et al.* (2008). The epithelial-mesenchymal transition generates cells with properties of stem cells. Cell *133*, 704-715.

[60] Marion, R.M., Strati, K., Li, H., Murga, M., Blanco, R., Ortega, S., Fernandez-Capetillo, O., Serrano, M., and Blasco, M.A. (2009). A p53-mediated DNA damage response limits reprogramming to ensure iPS cell genomic integrity. Nature *460*, 1149-1153.

[61] Marson, A., Foreman, R., Chevalier, B., Bilodeau, S., Kahn, M., Young, R.A., and Jaenisch, R. (2008). Wnt signaling promotes reprogramming of somatic cells to pluripotency. Cell Stem Cell *3*, 132-135.

[62] Maurelli, R., Zambruno, G., Guerra, L., Abbruzzese, C., Dimri, G., Gellini, M., Bondanza, S., and Dellambra, E. (2006). Inactivation of p16INK4a (inhibitor of cyclin-dependent kinase 4A) immortalizes primary human keratinocytes by maintaining cells in the stem cell compartment. FASEB Journal *20*, 1516-1518.

[63] Menendez, S., Camus, S., Herreria, A., Paramonov, I., Morera, L.B., Collado, M., Pekarik, V., Maceda, I., Edel, M., Consiglio, A., *et al.* (2012). Increased dosage of tumor suppressors limits the tumorigenicity of iPS cells without affecting their pluripotency. Aging Cell *11*, 41-50.

[64] Menendez, S., Camus, S., and Izpisua Belmonte, J.C. (2010). p53: guardian of reprogramming. Cell Cycle *9*, 3887-3891.

[65] Merrill, B.J. (2008). Develop-WNTs in somatic cell reprogramming. Cell Stem Cell *3*, 465-466.

[66] Miki, T., Yasuda, S.Y., and Kahn, M. (2011). Wnt/beta-catenin signaling in embryonic stem cell self-renewal and somatic cell reprogramming. Stem Cell Rev *7*, 836-846.

[67] Molchadsky, A., Rivlin, N., Brosh, R., Rotter, V., and Sarig, R. (2010). p53 is balancing development, differentiation and de-differentiation to assure cancer prevention. Carcinogenesis *31*, 1501-1508.

[68] Morel, A.P., Lievre, M., Thomas, C., Hinkal, G., Ansieau, S., and Puisieux, A. (2008). Generation of breast cancer stem cells through epithelial-mesenchymal transition. PLoS ONE [Electronic Resource] *3*, e2888.

[69] Natalwala, A., Spychal, R., and Tselepis, C. (2008). Epithelial-mesenchymal transition mediated tumourigenesis in the gastrointestinal tract. World J Gastroenterol *14*, 3792-3797.

[70] Ogawa, K., Saito, A., Matsui, H., Suzuki, H., Ohtsuka, S., Shimosato, D., Morishita, Y., Watabe, T., Niwa, H., and Miyazono, K. (2007). Activin-Nodal signaling is involved in propagation of mouse embryonic stem cells. J Cell Sci *120*, 55-65.

[71] Pajcini, K.V., Corbel, S.Y., Sage, J., Pomerantz, J.H., and Blau, H.M. (2010). Transient inactivation of Rb and ARF yields regenerative cells from postmitotic mammalian muscle. Cell Stem Cell *7*, 198-213.

[72] Poss, K.D., Wilson, L.G., and Keating, M.T. (2002). Heart regeneration in zebrafish. Science *298*, 2188-2190.

[73] Puri, P.L., Iezzi, S., Stiegler, P., Chen, T.T., Schiltz, R.L., Muscat, G.E., Giordano, A., Kedes, L., Wang, J.Y., and Sartorelli, V. (2001). Class I histone deacetylases sequentially interact with MyoD and pRb during skeletal myogenesis. Molecular cell *8*, 885-897.

[74] Qi, X., Li, T.G., Hao, J., Hu, J., Wang, J., Simmons, H., Miura, S., Mishina, Y., and Zhao, G.Q. (2004). BMP4 supports self-renewal of embryonic stem cells by inhibiting mitogen-activated protein kinase pathways. Proceedings of the National Academy of Sciences of the United States of America *101*, 6027-6032.

[75] Reya, T., and Clevers, H. (2005). Wnt signalling in stem cells and cancer. Nature *434*, 843-850.

[76] Reya, T., Duncan, A.W., Ailles, L., Domen, J., Scherer, D.C., Willert, K., Hintz, L., Nusse, R., and Weissman, I.L. (2003). A role for Wnt signalling in self-renewal of haematopoietic stem cells. Nature *423*, 409-414.

[77] Ron-Bigger, S., Bar-Nur, O., Isaac, S., Bocker, M., Lyko, F., and Eden, A. (2010). Aberrant epigenetic silencing of tumor suppressor genes is reversed by direct reprogramming. Stem Cells *28*, 1349-1354.

[78] Sage, J., Mulligan, G.J., Attardi, L.D., Miller, A., Chen, S., Williams, B., Theodorou, E., and Jacks, T. (2000). Targeted disruption of the three Rb-related genes leads to loss of G(1) control and immortalization. Genes & Development *14*, 3037-3050.

[79] Saha, K., and Jaenisch, R. (2009). Technical challenges in using human induced pluripotent stem cells to model disease. Cell Stem Cell *5*, 584-595.

[80] Sato, N., Meijer, L., Skaltsounis, L., Greengard, P., and Brivanlou, A.H. (2004). Maintenance of pluripotency in human and mouse embryonic stem cells through activation of Wnt signaling by a pharmacological GSK-3-specific inhibitor. Nat Med *10*, 55-63.

[81] Scheel, C., Eaton, E.N., Li, S.H., Chaffer, C.L., Reinhardt, F., Kah, K.J., Bell, G., Guo, W., Rubin, J., Richardson, A.L., *et al.* (2011). Paracrine and autocrine signals induce and maintain mesenchymal and stem cell states in the breast. Cell *145*, 926-940.

[82] Singla, D.K., Schneider, D.J., LeWinter, M.M., and Sobel, B.E. (2006). wnt3a but not wnt11 supports self-renewal of embryonic stem cells. Biochem Biophys Res Commun *345*, 789-795.

[83] Slack, R.S., Skerjanc, I.S., Lach, B., Craig, J., Jardine, K., and McBurney, M.W. (1995). Cells differentiating into neuroectoderm undergo apoptosis in the absence of functional retinoblastoma family proteins. The Journal of cell biology *129*, 779-788.

[84] So, A.Y., Jung, J.W., Lee, S., Kim, H.S., and Kang, K.S. (2011). DNA methyltransferase controls stem cell aging by regulating BMI1 and EZH2 through microRNAs. PLoS ONE [Electronic Resource] *6*, e19503.

[85] Takahashi, K. (2010). Direct reprogramming 101. Development Growth & Differentiation *52*, 319-333.

[86] Takahashi, K., Tanabe, K., Ohnuki, M., Narita, M., Ichisaka, T., Tomoda, K., and Yamanaka, S. (2007). Induction of pluripotent stem cells from adult human fibroblasts by defined factors. Cell *131*, 861-872.

[87] Takahashi, K., and Yamanaka, S. (2006). Induction of pluripotent stem cells from mouse embryonic and adult fibroblast cultures by defined factors. Cell *126*, 663-676.

[88] Takao, Y., Yokota, T., and Koide, H. (2007). Beta-catenin up-regulates Nanog expression through interaction with Oct-3/4 in embryonic stem cells. Biochem Biophys Res Commun *353*, 699-705.

[89] Tanaka, E.M., Gann, A.A., Gates, P.B., and Brockes, J.P. (1997). Newt myotubes reenter the cell cycle by phosphorylation of the retinoblastoma protein. Journal of Cell Biology *136*, 155-165.

[90] Tanaka, E.M., and Weidinger, G. (2008). Micromanaging regeneration. Genes & Development *22*, 700-705.

[91] Tapia, N., and Scholer, H.R. (2010a). p53 connects tumorigenesis and reprogramming to pluripotency. Journal of Experimental Medicine *207*, 2045-2048.

[92] Tapia, N., and Scholer, H.R. (2010b). p53 connects tumorigenesis and reprogramming to pluripotency. J Exp Med *207*, 2045-2048.

[93] ten Berge, D., Brugmann, S.A., Helms, J.A., and Nusse, R. (2008a). Wnt and FGF signals interact to coordinate growth with cell fate specification during limb development. Development *135*, 3247-3257.

[94] ten Berge, D., Koole, W., Fuerer, C., Fish, M., Eroglu, E., and Nusse, R. (2008b). Wnt signaling mediates self-organization and axis formation in embryoid bodies. Cell Stem Cell *3*, 508-518.

[95] Thiery, J.P. (2002). Epithelial-mesenchymal transitions in tumour progression. Nature Reviews Cancer *2*, 442-454.

[96] Thiery, J.P. (2003). Epithelial-mesenchymal transitions in development and pathologies. Current Opinion in Cell Biology *15*, 740-746.

[97] Utikal, J., Polo, J.M., Stadtfeld, M., Maherali, N., Kulalert, W., Walsh, R.M., Khalil, A., Rheinwald, J.G., and Hochedlinger, K. (2009). Immortalization eliminates a roadblock during cellular reprogramming into iPS cells. Nature *460*, 1145-1148.

[98] van Vlerken, L.E., Hurt, E.M., and Hollingsworth, R.E. (2012). The role of epigenetic regulation in stem cell and cancer biology. Journal of Molecular Medicine *90*, 791-801.

[99] Vermeulen, L., De Sousa, E.M.F., van der Heijden, M., Cameron, K., de Jong, J.H., Borovski, T., Tuynman, J.B., Todaro, M., Merz, C., Rodermond, H., *et al.* (2010). Wnt activity defines colon cancer stem cells and is regulated by the microenvironment. Nat Cell Biol *12*, 468-476.

[100] Wahl, B.T.S.a.G.M. (2011). p53, Stem Cells and Reprogramming: Tumor Suppression beyond Guarding the Genome. Genes & Cancer *2*, 404-419.

[101] Wellner, U., Schubert, J., Burk, U.C., Schmalhofer, O., Zhu, F., Sonntag, A., Waldvogel, B., Vannier, C., Darling, D., zur Hausen, A., *et al.* (2009). The EMT-activator ZEB1 promotes tumorigenicity by repressing stemness-inhibiting microRNAs. Nat Cell Biol *11*, 1487-1495.

[102] Wenzel, P.L., Wu, L., de Bruin, A., Chong, J.L., Chen, W.Y., Dureska, G., Sites, E., Pan, T., Sharma, A., Huang, K., *et al.* (2007). Rb is critical in a mammalian tissue stem cell population. Genes Dev *21*, 85-97.

[103] Wu, L., de Bruin, A., Saavedra, H.I., Starovic, M., Trimboli, A., Yang, Y., Opavska, J., Wilson, P., Thompson, J.C., Ostrowski, M.C., *et al.* (2003). Extra-embryonic function of Rb is essential for embryonic development and viability. Nature *421*, 942-947.

[104] Wu, Y., and Zhou, B.P. (2008). New insights of epithelial-mesenchymal transition in cancer metastasis. Acta Biochimica et Biophysica Sinica *40*, 643-650.

[105] Xu, R.H., Sampsell-Barron, T.L., Gu, F., Root, S., Peck, R.M., Pan, G., Yu, J., Antosiewicz-Bourget, J., Tian, S., Stewart, R., *et al.* (2008). NANOG is a direct target of TGFbeta/activin-mediated SMAD signaling in human ESCs. Cell Stem Cell *3*, 196-206.

[106] Ying, Q.L., Nichols, J., Chambers, I., and Smith, A. (2003). BMP induction of Id proteins suppresses differentiation and sustains embryonic stem cell self-renewal in collaboration with STAT3. Cell *115*, 281-292.

[107] Zhao, T., and Xu, Y. (2010). p53 and stem cells: new developments and new concerns. Trends in Cell Biology *20*, 170-175.

[108] Zvetkova, I., Apedaile, A., Ramsahoye, B., Mermoud, J.E., Crompton, L.A., John, R., Feil, R., and Brockdorff, N. (2005). Global hypomethylation of the genome in XX embryonic stem cells. Nat Genet *37*, 1274-1279.

Chapter 5

Modeling Tumorigenesis in *Drosophila*: Current Advances and Future Perspectives

Fani Papagiannouli and Bernard M. Mechler

Additional information is available at the end of the chapter

http://dx.doi.org/10.5772/55686

1. Introduction

1.1. Tumor suppressor genes, a historical perspective

Cancer is essentially considered as a genetic disease caused by the accumulation of multiple genetic or epigenetic lesions in tumor-suppressor genes and oncogenes [1]. Although the notion that retinoblastoma could be an inherited disease was already formulated at the end of the 19th Century a solid genetic basis was established with the discovery of both proto-oncogenes, whose gain-of function mutations or altered expression is associated with the cancerous state, and tumor suppressor genes (TSGs), whose inactivation releases the "brakes" inhibiting cell proliferation. Analysis of both proto-oncogenes and TSGs revealed also that cancer results from an alteration of the normal pathway of cell fate and differentiation. The hallmarks of cancer, as laid down by Hanahan and Weinberg to explain the complex biology of cancer, comprise six major developmental changes taking successively place in human tumors. These cancer "characteristics" include sustained proliferative signaling, evasion of growth suppressors, resistance to cell death, replicative immortality, angiogenesis as well as cell invasion and metastasis. Underlying these hallmarks are genome instability, inflammation, reprogramming of energy metabolism and evading immune destruction [2].

Cancer cells are the foundation of the cancer disease, as they initiate formation of tumors and drive tumor progression forward. Based on the sequence of events in which cells accumulate genetic lesions, tumor progression and metastasis are highly variable, even among tumors of the same type [3]. Previously, cancer cells within tumors were thought to be largely made of homogenous cells until relatively late in the course of tumor progression, when hyperproliferation combined with increased genetic instability spawn distinct clonal populations. Now we know that tumors rather than homogenous masses of proliferating cancer cells are complex tissues composed of distinct cell types participating in heterotypic interactions with each another. Reflecting such clonal heterogeneity is the finding that many human tumors display a complex histological pattern, characterized by regions exhibiting various degrees of differentiation, proliferation, vascularity, inflammation and invasiveness [2]. In addition, tumors exhibit another dimension of complexity arising from the surrounding normal cells of the "tumor microenvironment" [2] (analyzed in part 3).

Over the last decades the origin of oncogenesis has been the subject of different theoretical "fashions". In the current view taking into account the role of oncogenes and tumor suppressor genes cancer results from a failure occurring more in the control of cell differentiation than in cell proliferation [4, 5]. Nowadays, cancer is generally considered to result from a block or an error in the normal progression of differentiation. As a result, the cancer cells escape the mechanisms controlling normal growth and proliferation. Several decades ago, pioneer studies in the field of *Drosophila*, mouse somatic cells and human genetics revealed that neoplasia may result from a loss of function in regulatory genes controlling cell growth and differentiation [6-9]. In the following years, research in developmental biology has greatly contributed to cancer research. Indeed very often the initiating event in the formation of a malignant tumor is a block of a critical step in normal differentiation, usually through inactivation of a single gene, and can be accompanied by events occurring in parallel. In the case of tumor-suppressor genes, proliferation of tumor cells is suppressed by the same set of genes that suppress the proliferation of normal cells in the same cell type during the process of differentiation. Studies of the *Drosophila lethal [2] giant larvae (lgl)* gene, the first cloned tumor suppressor gene, have shown beyond doubt that tumors can be produced at a defined period of development by the impairment of a gene that normally regulates a critical step of differentiation [5, 10, 11]. At that time, such precise time delimitation in the process of development would not have been possible to achieve with mammalian cells, but the observations made thereafter in mammals were consistent with the conclusion derived from the *Drosophila* study.

1.1.1. Identification of the first tumor suppressor genes in Drosophila melanogaster

Over the past 40 years it has become increasingly evident that cancer is causally related to mutations in specific genes. These genes are instrumental to developmental processes such as cell-cell communication, signal transduction, regulation of gene expression, translation, cytoskeletal organization, protein folding and transport, and differential regulation of cell cycle [12]. The *Drosophila* field has made marked contributions in many of the mechanisms that are fundamental to cancer processes, several of which have been later validated in vertebrates. Less well known is the precursor role of *Drosophila* in the cancer field, as some of the earliest tumor suppressors were identified in flies. The first tumor-containing mutant was recognized in 1967 in a wild type laboratory stock of *Drosophila melanogaster* [6]. The mutant gene was soon identified by Ed Lewis as an allele of the already known *lgl* gene, which was discovered by Bridges in 1933 [13]. During the 1940-50s, this gene has been the subject of a number of developmental studies performed by Hadorn and his collaborators. The phenotypic studies performed by Hadorn's group on *lgl* and other *Drosophila* genes have greatly contributed to the conceptual basis of developmental genetics. In 1955 Hadorn published in German his seminal monography on "*Letalfaktoren und ihrer Bedeutung für Erbpathologie und Genphysiologie der Entwicklung*" which in 1961 was translated into English as "*Developmental Genetics and Lethal Factors*". Comparative developmental studies conducted thereafter showed that one of the pleiotropic effects of the mutation was the formation of a malignant neuroblastoma in the larval brain and the appearance of imaginal disc tumors [7, 14]. Molecular studies on the *lgl* gene were initiated in 1985 by Mechler and his co-workers by cloning the first tumor suppressor gene [10]. Subsequent analysis of *lgl* has demonstrated unequivocally that the tumorous phenotype results from the lack of *lgl* function and showed that tumorigenesis can be rescued by reintegrating a wild type copy of *lgl* into the genome of *lgl*-deficient animals [15]. Biochemical and genetic analysis of the *lgl* gene and its human homologue *hugl-1,* showed that the encoded proteins are components of the cytoskeleton and interact physically with a domain located near the carboxyl terminal of the non-muscle myosin II heavy chain [11, 16-20]. Further studies also revealed that the Lgl protein can interact with the Nucleosome Assembly Protein-1 (NAP-1), a component of the cyclinB-p34^{cdc2} kinase complex [21, 22] and NAP-1 is intimately associated with the cytoskeletal matrix during interphase, accumulates in the nucleus during prophase where it becomes associated with the spindle apparatus [21].

Recent contributions show that the Lgl protein may directly contribute to genetic regulation in association with the heavy chain of nonmuscle myosin II, or nmMHC [23, 24]. In particular mutations in *lgl* or heteroallelic combinations between *lgl* and *zipper*, encoding nmMHC, were found to block the disintegration of the salivary glands by blocking the program induced by the molting hormone ecdysone [23]. An interaction between both proteins was found to be required for the binding of specific nuclear remodeling proteins to chromatin [24]. Defect in this interaction may result in a block of genetic cascade initiated by the ecdysone hormone and lead to a transcriptionally frozen genome. The outcome of these analyses shed light on the key roles that tumor suppressor genes may play not only in the mechanism of cell shape and tissue organization but also in the regulation of developmental programs.

Subsequently to these initial studies, Gateff isolated a series of other recessive mutations in distinct genetic loci, which gave rise to four specific types of tumors. These tumors affected either the developing larval brain, the imaginal discs, the hematopoietic organs, or the germ cells [25, 26]. Shortly after *lgl* was reported [6], a genetic screen assaying imaginal disc morphology identified a mutation in the *discs large* (*dlg*) gene, coding for a septate junction tumor suppressor gene [27] with a second mutation identified few years later [28]. Twenty years later, a third mutation with the strikingly similar phenotypes, called *scribble* (*scrib*), was independently isolated in two different labs. Initially *scrib* was found in a screen for regulators of epithelial architecture [29]. Parallel to this investigation, a P-element mutagenesis screen led to the identification of the recessive *scrib*^{vartul} mutation causing late larval lethality, imaginal disc overgrowth and brain tumors with a complex syndrome reminiscent of that observed in mutations of the other tumor suppressors [30]. Therefore, already at the very early days of tumor genetics, *Drosophila* has been an extremely favorable object of study. Since then a great number of tumor suppressor genes have been identified and *Drosophila* has largely contributed to our understanding of the basic biology and cellular mechanisms of tumorigenesis including cell growth and proliferation, apoptosis, maintenance of cell polarity and architecture.

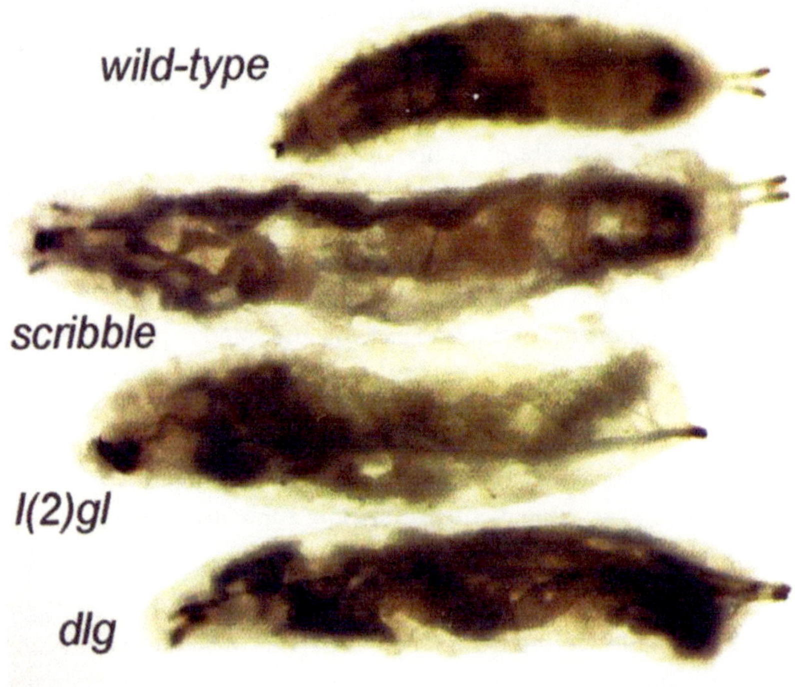

Figure 1. Size comparison between wild type 3rd instar larvae and *scrib*, *l2gl* and *dlg* giant larvae.

2. *Drosophila* as a unique model system to study tumor suppression

In order for an animal species to serve as a model of human biology it must fulfill two key criteria. The model system should be amenable to a broad set of experimental approaches and to be similar enough to humans so that findings from the model system can be exported to higher organisms and facilitate research in humans. *Drosophila*, being at the forefront of genetic research for the past one hundred years, together with the high degree of conservation with humans at the gene and cellular level, has proved itself as an essential partner in discoveries related to genetics, cell biology, human disease and cancer metastasis.

There are also many technical advantages in using *Drosophila* over vertebrate models. Flies are easy and inexpensive to culture in laboratory conditions, have a much shorter life cycle, produce large number of offsprings with feature-rich morphology and large numbers of externally laid embryos that can be genetically modified in numerous ways. Each female flies lay up to approximately 50 to 100 eggs per day for up to 20 days. It takes approximately 10 days at 25°C for an embryo to develop into a fertile adult fly. Thus, it is easy to generate large numbers of embryos or flies for experimental studies and genome-wide screens. Moreover, there are generally few limited restrictions on their use in the laboratory as there are minimal ethical and safety issues [31, 32].

2.1. *Drosophila* in a century of "tool-building" research

The first documented use of *Drosophila* in the laboratory was in William Castle's group at Harvard in 1901, but the "father" of *Drosophila* research is without doubt Thomas Hunt Morgan [33]. It was almost 100 years ago that Thomas Hunt Morgan reported the identification of the *white* gene in *Drosophila* [34]. Morgan greatly refined the theory of inheritance first proposed by Gregor Mendel, by using *Drosophila* to define genes and establish that they are found within chromosomes, long before it was even established that the DNA is the genetic material [31, 35]. Morgan won the Nobel Prize in Physiology or Medicine in 1933 "for his discoveries concerning the role played by the chromosome in heredity". One of Morgan's students, Hermann Muller, won the Nobel Prize in Physiology or Medicine in 1945 "for the discovery of the production of mutations by means of x-ray irradiation". Using *Drosophila* in 1920s, Muller discovered that x-rays cause massive increase in the rate of mutations and that the mutations can be passed from one generation to the other [36]. The genetics approaches used in the first 50 years of research in *Drosophila* (1910-1960), led to the development of important concepts and tools e.g. balancer chromosomes, that allowed the study of many other biological processes in the years to come [35, 37].

Interestingly, researches realized in the early fifties that genetic approaches could be used to study problems other than heredity. The continuous development of research tools between the years 1960-2010 has driven numerous new discoveries in fruit flies. In the mid-seventies, the available genetic tools in *Drosophila* offered the opportunity to address how embryonic pattern formation is controlled and to identify the genes involved in these processes [37, 38]. By carrying out a systematic chemical mutagenesis screen on the different fly chromosomes and analyzing the larval cuticle patterns, Nüsslein-Volhard and Wieschaus identified 139 genes that affect the development of the fly larva [39-41]. This analysis led to the identification of numerous genes participating in the Hedgehog, Wingless, Decapentaplegic (the *Drosophila* Tumour growth factor-β; TGF-β), and Notch pathways. In 1995, Christiane Nüsslein-Volhard, Eric Wieschaus and Ed Lewis won the Nobel Prize in Physiology or Medicine "for their discoveries concerning the control of early embryonic development". These findings have clearly demonstrated the power of forward genetics in solving complex developmental questions. Further genetic screens shed light on factors involved in various developmental aspects such as neuronal migration and growth cone guidance, circadian rhythms, learning and memory [37]. All these studies proved that despite the great morphological differences between flies and humans, many of the underlying building blocks and cellular processes are strikingly similar and conserved through evolution [31].

The range of genetic tools that have become available for *Drosophila* over the past century surpass by far those of any other multicellular organism [31]. Two experimental key features, namely the successful and efficient removal or addition of single genes or gene products, are important for any model organism to be successfully used in the laboratory. In *Drosophila*, genes can be inactivated in a random fashion using chemical mutagenesis followed by screening for specific phenotypes. Current tools allow very rapid mapping of chemically induced mutations that have robust phenotypes, permitting the isolation of null alleles, hypomorphs, neomorphs as well as conditional alleles, making possible the functional analysis of single genes. These screens, combined with duplications and deletions covering almost all chromosomes, have greatly facilitated gene mapping. The recent improvements in whole genome sequencing techniques and single-cell profiling will enhance even more the speed of accessibility to genomic information [37, 42]. Apart from the imprecise excision of transposable elements [43], one can create mutations by selective removal or replacement of sequences using the "targeted-knockout" technology [44], as well as by using RNA interference (RNAi) to reduce expression on any gene in a tightly regulated temporal and spatial pattern [45, 46].

The use of P-element-mediated transformation, available since 1982 [47] has allowed the insertion of single genes and any DNA fragment of interest in the fly genome, and has opened the field to even more sophisticated genetic manipulations [37]. This technique was significantly improved with the P[acman] technology that allows the insertion DNA fragments in specific docking sites spread throughout the *Drosophila* genome [48, 49]. Efficient transformation has been achieved by using the Flipase-Flipase Recognition Target (FLP-FRT) recombination system, which enables the creation of mutant patches of tissues or cells in an otherwise heterozygous background [50, 51]. This system led to the development of a highly efficient "mitotic recombination system" that knockouts defined genetic function in specific cells, tissues and organs. The yeast site-specific recombinase FLP, coupled with centromere-linked insertions of the FRT target site on all major chromosome arms of *Drosophila* [52], allows the generation of genetic mosaics, within an otherwise wild type organism, by removing almost every fly gene function. Generation of genetic mosaics is particularly useful for elucidating the function of genes which, when mutated, would otherwise kill the organism and subsequently laid the carpet in understanding how cells within tumors can interact with their surrounding microenvironment. A considerable improvement of the FLP-FRT system for analyzing mosaic tissue was the development of the MARCM (Mosaic Analysis with Repressible Cell Marker) technique [53]. Prior to the introduction of MARCM, homozygous mutant cells were identified by the absence of a visible marker such as GFP or *lacZ* in comparison to the surrounding heterozygous environment and the wild type "twin clone". By using the MARCM technique the homozygous clones can be positively marked using e.g. GFP or RFP, which can be of particular importance for the analysis of single cells in a disease model.

Another use of the P-element-mediated transformation facilitated the development of the *UAS-GAL4* system in order to ectopically express or overexpress a gene of interest in almost any tissue or cell [54]. The binary *UAS-GAL4* system allows a gene DNA sequence fused to the UAS (upstream activating sequence) to be ectopically expressed via the enhancer/ promoter of a second gene that drives synthesis of the UAS-binding GAL4 transcription factor. Thousands of *UAS* and *GAL4* fly lines are now available and their use can either modify or even abrogate gene expression in selected cell populations, in a specific temporal and spatial pattern [37, 55, 56]. Moreover, the "*Vienna Drosophila RNAi* Center" and "Bloomington Stock Center" house a collection of transgenic fly lines carrying inducible *UAS-RNAi* constructs against single protein coding genes. Currently they accommodate over 22.000 different transgenic lines, which provide knockdowns for over 90% of *Drosophila* genes. Further development of the *UAS-GAL4* system led to the TARGET (temporal and regional gene expression targeting) system, which utilizes a temperature-sensitive form of GAL80 repressor that binds GAL4 and blocks its transcription activity at the restrictive temperature, while a shift to a permissive temperature results in GAL80 losing the ability to bind GAL4 [57, 58].

Finally, P-element technology also allowed the tagging of most genes *in vivo* e.g. with GFP, RFP or YFP, permitting sophisticated manipulations in a genomic context [59]. Transgenic flies containing enhancer-trap or protein-trap versions of individual genes allowing the analysis of the gene expression pattern and protein localization are available at "FlyTrap" (http://flytrap.med.yale.edu/index.html) [60-63] and "FlyPROT" (http://www.flyprot.org/index.php) [64].

2.2. *Drosophila* is a model system relevant to human biology

While *Drosophila* has long served as a model for basic biological research, more recently its potential as a model for unraveling molecular mechanisms of human diseases has become widely appreciated, and numerous publications and conferences illustrate the use of *Drosophila* to unravel the mechanisms of human diseases [65-69]. Release of the first sequence of the *Drosophila* genome in March 2000 (11 months ahead of the human genome release) allowed the actual comparison to the human genome. This comparison has greatly consolidated *Drosophila's* legitimacy as a model organism for medical research [31, 70]. The sequence and annotation are freely available in "Flybase", an outstanding online database combining all current knowledge on single *Drosophila* genes including sequence and expression data, mutations, interactions and up to date scientific references. Comparative studies have shown the molecular mechanisms underlying the development of *Drosophila* and humans are highly conserved and that the *Drosophila* genome contains functional homologues of nearly 75% of the human disease genes, we can understand why this aspect of *Drosophila* research continuously expands. Moreover, over 85% of human genes that have been associated with a disease, have a *Drosophila* counterpart. These findings constitute a strong basis for the continuous expansion of *Drosophila* research in relation to human diseases.

What makes *Drosophila* also practical in the analysis of gene function is the nearly absence of genetic redundancy. The duplication of the vertebrate genome during evolution has resulted in the occurrence of multiple paralogs, e.g. Hox genes that control the body plan along the anterior-posterior axis [71], with their subsequent evolution that has generated gene expansion ad diversity in protein function [72, 73]. Genetic expansion means that when knocking down a gene, other genes or homologues can compensate for its function. Yet, the extra genes rarely represent novel functions as they simply allow more complex and subtle regulation of core molecular mechanisms. In this respect, the absence of gene duplication in *Drosophila* provides the advantage of elucidating more readily the fundamental function of single genes, e.g. in tumor development, and then apply it back to vertebrates and humans as the mechanism is very likely conserved [74]. Indeed, lack of redundancy can expose the physiologically relevant phenotype of gene homologues.

The similarities between flies and humans are further supported by the fact that components of signal transduction pathways and the molecular mechanisms involved in specification, development, cell cycle regulation and human diseases were first identified in flies. The genes, which have been characterized in flies, were subsequently studied in mice and humans, and their names were adopted or adapted from their *Drosophila* homologues. For example, the mammalian Notch 1-4 named after the *Drosophila* Notch (fly wings having a large notch on the wing), "*sonic hedgehog*" named after the *Drosophila* "*hedgehog*" (round larvae with extra bristles) and *Wnt* from the *Drosophila* "*wingless* and *INT*-related" gene [31].

3. Recent advances in modeling tumor progression and metastasis in *Drosophila melanogaster*

Since the discovery of oncogenes and tumor suppressor genes, intense research in many laboratories all over the world has brought us to the point where we are starting to understand the main principles underlying molecular changes in the course of tumor progression [3, 75]. The development of new technologies revealed the complex molecular nature of tumorigenesis in which tumor progression can be envisaged as a network of simultaneous events within both tumor cells and the stroma. Because cancer is an age-associated disease in humans, using *Drosophila* to model cancer development, progression and metastasis was debatable [76, 77]. However, over the last decades the study of *Drosophila* has significantly contributed to the understanding of key cancer events, including the loss of cell polarity, the competition between tumor and normal cells, as well as metastasis [78].

In addition to the importance of tumor cells themselves, their neighboring cells and the surrounding stroma are now recognized as important regulators of cancer progression [79]. Cell competition is a type of short-range cell-cell interaction in which cells expressing different levels of a particular protein are able to discriminate between their relative levels so that the one cell, the "loser", disappears from the tissue whereas the other, the "winner", survives and proliferates to cover the space left by the disappearing cell. Some tumor-promoting mutations in *Drosophila* are able to induce cell competition between the cancer cells and their micro- and macroenvironment [80-86]. Metastasis is the latest phase of cancer progression during which cells detach from their original niche to invade distant tissues [87]. For several decades, our understanding of the molecular processes leading to metastasis was largely derived from studies of cancer cell lines *in vitro*, xenografts of human tumors and a limited number of transgenic or knockout mouse models [1, 88, 89]. However, understanding the individual steps leading to tumorigenesis or analyzing multiple genetic interactions in mice is difficult. Existing models are currently being re-evaluated given the increasing awareness of tumor complexity. Therefore, using a model system that allows the efficient analysis of the combinations of genetic events that trigger tumor initiation and metastasis during cancer development is the major challenge in cancer research at the moment. *Drosophila melanogaster* provides a model of choice for cancer analysis as the collection of sophisticated genetic manipulation techniques have been invaluable for dissecting signaling pathways that affect cell specification, differentiation and growth [90-93]. Indeed, *Drosophila* cancer models are very helpful in unraveling the chronological sequence of events leading to human cancer. For example, in metastasized human tumors elucidating the identity of the initial mutations is often tedious as oncologists are in most cases looking at the end point of the disease progression. Finally, *Drosophila* genetics is very powerful as we can dissect the triggering events initiating cancer and the subsequent steps leading to the progression of the disease [94].

3.1. Modeling cell competition and metastasis

With these added complexities in mind, the analysis of cancer-disposing mutations in only a subset of cells or in clones within the context of a wild type surrounding tissue is gaining more interest because it offers a reasonable approximation of the clonal nature of human cancers, compared to the analysis of the multi-step model of tumor progression in the context of a whole organism. A great number of very interesting publications provided us with information about new and unexpected findings on the role of the polarity genes *scrib, lgl* and *dlg* in cancer initiation, progression and metastasis. Nowadays, it becomes obvious that they play a broader role than initially thought, through the cooperation with individual partners and signaling pathways and have helped us to understand the role of cell competition and of the tumor microenvironment during tumor survival and progression.

Analysis of *scrib*⁻ mutant clones in the *Drosophila* eye imaginal discs has shown that tumor development is suppressed by the JNK-mediated apoptotic pathway activated by the surrounding wild type cells, whereas the neoplastic and metastatic potential is regained through the synergistic effect of a simultaneous up-regulation of Ras signalling within the same clones [84, 95, 96]. These results underline the effect of the surrounding normal cells on the transformed *scrib*⁻ clonal cells, which leads to a cell competition similar to the one observed in the mammalian cancers [1, 96-100]. In a model for *scrib* tumorigenesis, the analysis of the downstream pathways in *scrib*⁻ epithelial clones revealed that excessive cell proliferation is restrained by JNK-mediated apoptosis. Upon simultaneous activation of either Ras or Notch, JNK-mediated apoptosis is blocked, and Ras/Notch together with JNK cooperatively promote tumor growth and invasion [96]. In other words, whereas JNK activity normally promotes the apoptosis of *scrib*-deficient cells, it becomes a driver of cellular overgrowth, tumorigenesis and invasion in the presence of oncogenically activated Ras or Notch signalling [76, 101]. These tumors present similar characteristics to human cancers that lack Scrib, including basement membrane and extracellular matrix degradation, loss of E-cad expression, combined with migration, invasion and secondary tumor formation [101] Another report provided a molecular link between loss of polarity and tumorigenesis, since *scrib*⁻, *dlg*⁻ and *lgl*⁻ clonal cells in a wild type surrounding become metastatic only in combination with *Ras*^{v12} activation, resulting in JNK activation and E-cad inactivation [102]. The analysis of the JNK-mediated tumorigenesis, which in *Drosophila* cells reveals a cooperation with Ras similar to that taking place in mammalian breast epithelial cells, indicates sthat the knowledge gained from analysis in *Drosophila* can help us elucidate tumor formation in the mammalian system.

Mutations inactivating the Salvator-Warts-Hippo (Sav-Wts-Hpo) pathway can also cause super-competition, contributing to the overgrowth of cells expressing these mutations in the presence of wild type cells [80, 103]. Since the first discovery of the Sav-Wts-Hpo pathway in *Drosophila,* the role of these genes in restricting cell growth and proliferation, and inducing apoptosis has triggered a great interest in its study. The components of this pathway act as important tumor suppressors that regulate tissue growth by promoting cell cycle exit and apoptosis [80, 104-112]. Recent data from mammals and *Drosophila* show the occurrence of a very conserved pathway that links the pathway of cell polarity to the regulation of tissue growth.

In the model of Scrib tumorigenesis, induction of apoptosis in *scrib*-clones could not explain how *scrib*⁻ cells are prevented from overproliferating. This was answered by the finding that cell competition between *scrib* and wild type cells prevents overproliferation by suppressing Yorkie (Yki; a transcription factor, which is suppressed by the Sav/Wts/Hpo pathway) activity in *scrib*⁻ cells [113, 114]. Suppressing Yki activation is critical for *scrib*⁻ clone elimination by cell competition. Cell competition leads to activation of JNK in *scrib*⁻ cells and JNK antagonises Yki activity, which leads to elimination of the clone. Experimental Yki elevation is sufficient to promote neoplastic growth in *scrib*⁻ cells [114]. Along the same line, when *lgl* is mutated in a mosaic tissue, the *lgl*⁻ clonal cells become the "losers" in cell competition. However, simultaneous overexpression of the Ras signalling pathway or of *yki* in *lgl*⁻ clones, causes overgrowths and JNK-mediated apoptosis at the periphery of the transformed clones [115-120]. Moreover, JNK-mediated elimination of *lgl*⁻ clonal cells is relieved and the overgrowth potential is re-established by upregulation of c-Myc, demonstrating the the death of *lgl*⁻ clones is essentially driven mainly by c-Myc-induced cell competition [121]. Simultaneous downregulation of the *lgl* and the JNK pathway in the whole-animal system results in a phenotypic reversion of tumor growth, absence of the giant larvae and recurrence of pupariation, thereby showing that JNK activity is essential for overgrowth and invasion of *lgl* tumorous discs [122]. Moreover, in the developing *Drosophila* eye and imaginal disc epithelia Lgl, αPKC and Crumbs proteins regulate cell proliferation and survival by controlling the activity of the Sav-Wts-Hpo pathway [115, 123-125].

Among the wide palette of cellular events leading to JNK activation is the dTNF (tumor necrosis factor)/Eiger. Eiger is the only *Drosophila* member of the TNF superfamily and its deregulated expression in imaginal disc cells results in JNK-mediated apoptosis [76, 126]. JNK-dependent cell death in *scrib* and *dlg* clones requires dTNF, acting as a "tumor death factor" [127]. On the other hand, in tumors deficient for *scrib* and *dlg* that also express *Ras*, the TNF signal is converted into a signal, which promotes tumor growth and invasion [126]. More precisely, upon dTNF downregulation, cell death in *dlg* and *scrib* clones was blocked and *in situ* outgrowths appeared, probably by TNF-mediated extra-cellular matrix (ECM) remodelling [76, 126]. When generated in an *eiger*⁻ mutant background, Ras^{v12}*scrib*⁻ clones displayed non-invasive *in situ* overgrowth. Similarly, in whole Ras^{v12}*scrib*⁻*dTNF*⁻ animals, development proceeded up to pupal stages, overcoming the "giant larvae" phenotype [76, 126]. These recent results suggest that several of the critical overgrowth phenotypes of *scrib, dlg* and *lgl* in the clonal and whole-tissue context are mediated by dTNF and that dTNF pro-tumor function depends partially on JNK activation in tumor cells, which provides a switch from *in situ* to invasive growth. Immunostaining experiments that detected dTNF in a punctuated, intracellular vesicle pattern at the periphery of hemocytes in association with the *dlg*⁻ clones, indicating that dTNF expression in hemocytes is sufficient for dTNF/JNK pathway activation within the *dlg*⁻ clones, and mark the importance of hemolymph and non-cell autonomous immune response in tumor progression [76, 126].

Until very recently, the mechanism by which surrounding normal tissue exerts antitumor effects against *dlg*, *scrib* or *lgl* clones remained elucive. New results from clonal analysis in *Drosophila* imaginal discs have shown that JNK activation from the wild type surrounding leads to upregulation of PVR, the *Drosophila* PDGF/VEGF receptor, which subsequently activates the ELMO/Mbc phagocytic pathway, which in turn eliminates the oncogenic clonal cells by engulfment [128]. From an evolutionary point of view, the development of such mechanism, which senses and eliminates "neoplastic" tumor-suppressor mutant cells such as those of *scrib*$^-$ and *dlg*$^-$ but not "hyperplastic" ones (in which despite of overproliferation, cells are normally shaped and retain a differentiated epithelial monolayer, such as those of the Hippo pathway and PTEN) [128], shows the necessity to specifically eliminate the high-risk malignant neoplastic cells before they confer any harm to the organism. Taken together, these studies demonstrate that hemocytes together with the tumor microenvironment act as regulators of epithelial delamination required for tumor invasion. Due to the ease of genetic manipulations, *Drosophila* research can bring meaningful insights to our understanding of the mechanisms of communication between cancerous and normal cells as well as between tumor tissue and the immune system [87].

3.2. *Drosophila* provides critical insights on how conserved mechanisms contribute in cancer and tumorous development

Drosophila research has also contributed to cancer analysis by identifying genes and signaling pathways later found to be critical for tumorigenesis in mammalian systems. In several cases *Drosophila* has been used to establish specific model systems in order to understand processes that seem to be more complex in vertebrates and mammals [81, 101, 129, 130]. One of the most extensively studied *Drosophila* models of tumor biology is the asymmetric cell division of neuroblasts, the *Drosophila* neuronal stem cells. In *Drosophila* neuroblasts (NB), the PAR3-PAR6-atypical protein kinase C (aPKC) complex segregates apically and recruits the adaptor protein Inscutable (Insc), which connects this complex to the partner of Insc (Pins), guanine nucleotide associated protein-α_1 (Gα_1), mushroom body defect (MUD), and p150^{glued} to the crescent directing the orientation of the mitotic spindle during asymmetric cell divisions. aPKC promotes the exclusion of partner of numb (PON), Lgl and Numb, which along with Miranda (Mira), Brain Tumor (Brat) and Prospero (Pros), localize to the basal crescents [101, 130, 131]. Interestingly, most of these genes have functional mammalian homologues and a very recent study points out the role of the Par3 in mammalian skin tumorigenesis [132]. In mouse skin tumorigenesis, Par3 deficiency results in reduced papilloma formation and growth. Furthermore, Par3 expression is reduced in both mouse and human keratoacanthomas, indicating the tumor-suppressive properties of Par3. More insights into tumor physiology came from a very interesting study in *Drosophila* NBs that indicates the critical role of starvation in promoting the overgrowth *pins* larval brains. Energy stress, mediated by loss of TOR and PI3K components, in combination with loss of *pins* results in loss of asymmetric NB division and brain overgrowth (Rossi, 2012). Since the PI3K and TOR signaling pathways are vital to the growth and survival of mammalian cancers [133, 134] using *Drosophila* in order to dissect the cross talk of these pathways to preexisting tumor susceptibility defects e.g. polarity defects, in a simple animal model is of great importance.

The usefulness of *Drosophila* is further illustrated in the development of a *Drosophila* model for human brain cancer. Glioblastoma (GMB) is the most frequent and malignant form of high-grade glioma that infiltrate the brain, grow rapidly and are refractory to current therapies [135]. One key to development of new and effective therapies against these tumors is to understand the fundamental genetic, cellular and molecular logic underlying gliomagenesis. Signature genetic lesions in glioblastomas include mutation of the epidermal growth factor receptor tyrosine kinase (EGFR) and mutations activating components of the PI3K pathway. *Drosophila* studies using lineage analysis combined with cell-type specific markers demonstrated that EGFR-Ras and PI3K can induce fly glial neoplasia through activation of a combinatorial genetic network composed in part of other genetic pathways also commonly mutated in human glioblastomas [135]. In the future, large-scale forward genetic screens with this model may reveal new insights into the origins of glioblastoma and may also provide new therapeutic strategy for the treatment of this form of human tumor.

Drosophila has been also used to elucidate the molecular mechanisms of human hereditary diffuse gastric cancer (HDGC) [136]. Gastric cancer, as several human cancers, originates from epithelial cells. Mutations in the CDH1 gene, which encodes the cell adhesion molecule E-cadherin (E-cad), are associated with HDGC in humans. In order to understand the role of E-cad in this disease, a *Drosophila* model has been developed in which mutated forms of E-cad can be studied *in vivo* [136, 137]. Moreover, genetic and molecular studies of *Drosophila* hematopoiesis can also contribute to our knowledge of the hematopoietic niche and hematopoietic malignancies in humans. Vertebrate hematopoietic stem cells give rise to an hierarchically organized set of progenitors and deregulation of the hematopoietic differentiation program can lead to numerous pathologies including leukemias. With the discovery that many transcriptional regulators and signaling pathways controlling blood development are conserved between humans and flies, *Drosophila* is particularly suitable model for investigating the mechanisms underlying the generation of blood cell lineages and blood cell homeostasis [138].

Interesting results using a *Drosophila* cancer model demonstrated that apoptosis activation in differentiation-compromised cells constitutes a mechanism for early prevention of cancer [139]. Apoptosis is a highly conserved cellular function to remove excessive or unstable cells in diverse developmental processes and disease-responses. An important example is the elimination of cells unable to differentiate, which have the potential to generate tumors. Using cell-type specification in *Drosophila* Ingrid Lohmann and her colleagues identified a conserved regulatory mechanism that underlies cell-type specific removal of uncommitted cells by apoptosis as a cancer prevention mechanism [139]. Under normal conditions the transcription factor Cut activates differentiation, while it simultaneously represses cell death via the direct regulation of a pro-apoptotic gene. However, loss of Cut and subsequent release of apoptosis leads to overproliferation of the mispecified cells that can acquire metastatic potential in a sensitized background. Importantly, this regulatory wiring is also found in vertebrates in which other cell-type specification factors may similarly be employed to suppress tumor formation. Thus, coupling of differentiation and apoptosis by individual transcription factors

1	*Drosophila* flies are easy and inexpensive to culture in laboratory conditions, have relatively low set-up and maintenance costs
2	Short life cycle
3	High fecundity (produce large number of off-springs with feature-rich morphology)
4	The *Drosophila* community is open and generous in sharing reagents within the community.
5	No ethical issues and regulatory considerations.
6	Genetic advantages • flies have only 4 pairs of chromosomes • males lack genetic recombination, making genetic crosses easier • flies tend to lack genetic redundancy
7	Signaling pathways controlling growth, differentation and development, which are involved in tumor formation in the fly are largely conserved between *Drosophila* and humans
8	Availability of numerous genetic tools & reagents for generating mutants and analysis of gene expression by using methods producing over- & ectopic- expression. • The use of "balancer chromosomes" with multiple DNA inversions prevent female recombination and allows the perdurability of mutations on the original carrier chromosome • Wide variety of gene targeting strategies, e.g. UAS-GAL4 system combined with RNAi knock-down allow the tissue-specific analysis of tumor suppressor gene and oncogene function • Mosaic analysis of animals containing mutant clones next to wild type tissue, using FLP-FRT and MARCM recombination systems, allows the analysis of tumor microenvironment in invasion, metastasis & inflammation.
9	Possibility to perform genome-wide screens using chemical mutagenesis, tissue-specific RNAi knockdown, effectors and modifier screens to identify genes involved in specific developmental pathways and assign and validate new gene functions.
10	Use of *Drosophila* tumor models for pharmacological screening and development of new therapeutic strategies for cancer.

Table 1. Advantages of *Drosophila melanogaster* for the analysis of cancer.

is a widely used and evolutionarily conserved cancer prevention module, which is hard-wired into the developmental program [139].

Furthermore, genetic analysis of border cell migration in the *Drosophila melanogaster* ovary provides clues that improve our understanding of ovarian cancer metastasis at the molecular level that might also lead to therapeutic targets [140]. The border cells of the *Drosophila* ovary provide a particularly simple example of cell migration in which cells derived from an epithelium invade a neighboring tissue. The large numbers of genes that control border cell migration identified in genetic screens emphasized also the requirement of multiple extracellular signals for border cell motility [141]. Interestingly, the motile and invasive properties of the border cells seem to share common characteristics with mammalian ovarian carcinoma cells. Based on work done in *Drosophila*, the function of some mammalian proteins such as myosin VI, have been tested for their ability to regulate motility of ovarian carcinoma cells.

Another example can be found in the EGFR pathway, which functions redundantly to PVR to stimulate border cell migration [142]. Overexpression of EGFR has been reported in up to 70% of ovarian carcinomas [143]. The role of E-cad is also here critical. Normal cells of ovarian surface epithelium express little or no E-cad. However, many primary ovarian carcinomas, similar to border cells, express E-cad at the cell surface and in the cytoplasm [144, 145]. Although cell-surface expression of E-cad is reduced in many metastatic carcinomas, these tumors frequently still have detectable intracellular E-cad [146], indicating that these cells, similar to border cells, have acquired the ability to downregulate E-cad activity at a post-transcriptional level [140].

Numerous molecules identified in *Drosophila* genetic screens have proven to be important to human cancers [140, 147]. For example, the Hedgehog gene and the Wnt homologue Wingless were originally identified in genetic screens for mutations that disrupt embryonic patterning. Subsequent studies of signaling proteins such as Hh, Wnt [140, 148-150], Notch [151, 152] and RhoGTPases [153] as well as of integrin-related adaptor proteins [154, 155] and Hox genes [156, 157], revealed crucial functions not only in normal mammalian development but also in various cancers. It is therefore well accepted that genetic approaches to the study of normal cellular behavior in simple model organisms can yield fundamental insights into the molecular underpinnings of cancer.

4. New perspectives in modeling tumorigenesis in *Drosophila melanogaster*

Drosophila is emerging as a valuable system for use in clinical drug discovery and therapeutic process [52, 129, 158-160]. So far, *Drosophila* was not a favorable model in drug discovery. The main reason was that "Drosophilists" were mainly concerned with answering fundamental development and cell biology questions, and elucidating principles of basic mechanisms and not practical issues of therapeutics [129]. This view is slowly changing as interest in therapeutics and pressure for practical outcomes increases and combined with the development of powerful tools allows *Drosophila* to catch-up in the field.

The remarkable degree in conservation of biochemical pathways that control processes such as cell proliferation, differentiation and migration as well as nervous system function in behavior and cognition, sustains perfectly the use of invertebrate model genetic organisms as tools for drug discovery and validation [52]. *Drosophila's* genetic and genomic tools can be adapted to build sophisticated disease models for studying cancer and metastasis, and for therapeutic development. While testing with mammalian models is essential prior to approval for human trials, the use of invertebrate animal models that are amenable to molecular genetic manipulations, provides experimental and biological advantages that can streamline drug discovery and testing process. Among the benefits of a genetics-based approach is the ability to screen for proteins that may be novel drug targets, and in genetic backgrounds that could more accurately reflect a specific disease state [52]. New drugs can be tested in *Drosophila* much faster than in mammalian models and can even be used for high throughput screening processes as an alternative to cell culture [31].

Drosophila may constitute an appropriate model system that can be used for screening a "whole animal setting" as an alternative to cell-cultured based methods. In most pharmaceutical industries the discovery of new compounds with potential positive pharmacological effect relies on the screening of small molecule libraries for interactions with purified proteins, or for the ability to induce a desired physiological response in cultured cells [52]. One of the main problems is that complex processes such as tumor metastasis cannot be recapitulated in a cell or an organ culture. Moreover, cells and organs are physiologically connected and their interplay could be critical in the development of some diseases. Furthermore, the time component of the disease progression is not cannot be easily recapitulated *in vitro* [161]. The second problem is that after the initial screen, the next phase of drug testing, which requires the use of intact mammalian animal models to assay the effectiveness of candidate compounds, usually fails. If in this step of drug discovery process, the compounds isolated *in vitro* or in cell culture, are invalid in mouse, the result is an enormous waste of funds and efforts [158]. At the same time, whole animal vertebrate models are not particularly suitable for high-throughput methodologies and if then only at an extremely high cost. To by pass these limitations, efforts are now being made to screen chemical libraries on whole-animals like *Drosophila* with genetic amenability, low cost and culture conditions compatible with large-scale high-throughput screening.

Furthermore, performing drug screening in the *Drosophila* "whole animal system" does not dependent on the prior identification of a target and permits the selection of compounds with an improved safety profile. A screen based on a phenotypic observation has the advantage of being independent of the specific molecular target involved. Then, depending on the readout used to assay the effectiveness of the candidate drug compound, a large variety of bio-active molecules may be detected in the same screen [158]. Finally, similar to the established "enhancer" or "suppressor" screens, *Drosophila* gives the possibility to test chemical libraries in the genetic background of a disease for their ability to reverse the abnormal phenotype to wild type or partially rescue the disease phenotype [158, 161].

Several groups today develop the associated technology to use *Drosophila* in the first phase of screening for drug compounds, and subsequently test them in more expensive mammalian models. Moreover, the fact that it is usually easier and more straightforward to manipulate the genetic background of *Drosophila* and mimic a disease state, opens also new possibilities for efficient drug testing in a disease-related content. For example, the development of genetically modified animals with fluorescently tagged proteins would allow the use of standard plate-reader spectrofluorometer for whole-animal screens [161].

When the development of mosaic tissues is essential for the analyses of a disease model, the use of MARCM provides notable advantages for effective drug discovery. One is the ability to follow the morphology of mutant cells and tissues which could be useful for assessing the efficacy of a therapeutic compound [52]. When a mutation in a given gene causally produces a disease, it is possible that this mutation elicits a change in expression of other genes and in the function of proteins. These alterations may contribute to the pathologies associated with the disease. The characterization of these changes constitutes then the first step needed to develop rational therapeutic strategies. Finally, the MARCM methodology should provide ways to identify mutant cells or tissues for a given gene, isolate and subject them to proteomic and genomic analyses which would determine modifications in gene expression and protein interaction profiles [52].

The phenotypes of complex trait diseases such as obesity, heart disease and cancer are the result of modifications occurring in multiple biochemical pathways. The disease phenotype can be caused by improper activation or inhibition of a protein that acts in any of the contributing pathways. Restoration of the normal phenotype would be expected if the output from the primary biochemical pathway affected is rescued via drug-based therapy [52]. However, if multiple pathways contribute to a phenotype, it stands to reason that modifying the activity of a parallel pathway could also elicit a positive therapeutic effect. The use of genetic model organisms has long been a successful means for elucidation of biochemical and physiological pathways, and one of the most powerful strategies available for uncovering genes that act together in producing a phenotype is a search for genetic interaction or a modifier screen. Modifier screens work by generating animals with a mutation in a gene of interest that elicits a sensitized phenotype, and then screening for mutations in progeny that enhance or suppress (i.e., modify) the primary phenotype [52]. *Drosophila* disease models are currently used in drug screening for treatment of diseases such as Alzheimer's and Huntingtons's disease, Fragile X Syndrome and muscular dystrophy [160, 161]. The use of drug discovery especially for muscle diseases is of particular importance as the muscles are difficult to reconstitute *in vitro* and elucidation of the physiology of muscle related diseases and relevant treatment is still poorly understood However, the similarity in architecture, composition and function between *Drosophila* and vertebrates may trigger studies in the fruit fly and provide these diseases with some valuable therapeutic answers [161].

Drug discovery has also proved very effective for the identification of cancer treatments such as the multiple endocrine neoplasia type-2 (MEN2) [129, 162]. MEN2 is a "one-hit" solid tumor syndrome, characterized by mutations in the Ret protein, a transmembrane receptor tyrosine kinase. Patients with mutated oncogenic isoforms of Ret, develop medullary thyroid carcinomas (MTC) that lead to metastasis, which seem to be resistant to traditional chemotherapies. To develop a whole-animal transgenic model, various oncogenic Ret isoforms were targeted to the developing fly eye epithelium. The fly "rough-eye" phenotype is characterized by eye overgrowth, switch in cell fate and other aspects, proving the effectiveness of fly as model and useful readout for screening. The screening for clinical relevant compounds led in the tumorous developing flies by Cagan and his group resulted in the identification of Vandetanib, a broad kinase inhibitor, which Cagan called "the worst kinase inhibitor". Although not very specific and effective, this kinase inhibitor was indeed effective in rescuing the fly phenotype because it regulated the activity of other kinases such as Ras, Src and PI3K (all of which are involved in cancer). Other compounds were more capable of rescuing the rough eye phenotype but reduced animal viability. Yet, Vandetanib displayed little toxicity to the animal as a whole, indicating that tumors might display a lower tolerance threshold for drugs than the entire animal. Obviously the "off-target" effects of Vandetanib, by suppressing kinases other than

Ret, are important for its effectiveness. Classical drug screenings would reject Vandetanib as too inefficient to the target and too low in its specificity [129, 162]. Cell-culture and subsequent fly work proved to be extremely valuable, as Vandetanib was shown to be efficient in Phase II clinical trials, Currently Vandetanib is in Phase III of clinical assays. This further proves the power of *Drosophila* not only for clinical relevant drug discovery but also for shaping how we should approach drug discovery for treating diseases.

A new *in vivo* drug screening in *Drosophila* has been performed to target cancer stem cells (CSCs) in the group of Norbert Perrimon [163]. Cancer cells represent a small number of cells within tumors with a self-renewing capacity that can regenerate tumor cells types through their stem cell-like renewal capacity. Their resistance to chemotherapy is the main reason why chemotherapeutic treatments are ineffective and the disease often relapses. In order to cope with the challenge of finding drugs that specifically target the CSCs, the Perrimon laboratory uses the *Drosophila* gut as the stem cell system to develop novel methods and approaches to screen for anti-cancer drugs that target CSCs *in vivo* in the gut microenvironment. By directing the expression of oncogenes in *Drosophila* transgenic fly models combined with a screening method that involves monitoring of tumor size by luciferase reporter activity, they identified 25 compounds that reduced tumor size. Further confirmation was validated with dissection of the gut, histochemical-imaging of the gut specific cells and determination of the specificity of the drugs. For example, some drugs were targeting only the CSCs whereas others were targeting CSCs but at the same time promoted growth of the wild type stem cells and in some cases also affected stress pathways in the daughter cells.

5. Limitations in using *Drosophila* as a model system: how far can we go?

Drosophila melanogaster, as a model system for studying tumorigenesis and human disease has certain limitations, especially in regard to the biological processes that evolved only within the vertebrate lineage [164]. The main limitation of *Drosophila* arises from the fact that some diseases cannot be modeled because the corresponding genes and organs present in humans are missing in flies [161]. For example, *Drosophila* has a single cardiac chamber that functions as a heart in the context of an open circulatory system. Moreover, the *Drosophila* myocardium receives oxygenation through diffusion and does not have coronary arteries [165]. A second limitation arises from differences in cellular and molecular processes of *Drosophila* in comparison to humans. For example, there are cases in which one or several key molecules mediating a human disease-specific pathway are missing in flies [160]. Ultimately, some areas such as learning, endocrine function and mammary gland development, may prove difficult to study in a simple invertebrate like *Drosophila* and so the study of these particular disorders may not benefit substantially from *Drosophila* genetics [166]. Another example is modeling tumor metastasis. In mammals malignant cells undergoing metastasis enter the local blood or lymph vessel before colonizing a distant tissue and forming metastatic tumors. This is very difficult to model in *Drosophila* as flies have a rudimentary hematopoietic system and a dramatic different lymphatic system compared to mammals [78]. However, one should point out that despite these differences the "master regulators" of heart, eye, kidney etc. have proved to be remarkably conserved [162]. This means that in the case of e.g. spinal malformation, flies could be used to model bone formation *per se*, but as Notch signaling has a pivotal role in regulating this process, any knowledge obtained about interactions between components of this pathway in *Drosophila* could be relevant to processes that these genes control during spinal formation in humans [164].

Are there limitations in using *Drosophila* for treatment-relevant drug discovery? The limitations in this case results from the anatomical and molecular differences of small model organisms in comparison to humans, as this may cause the elimination of a significant fraction of the hits generated The use of *Drosophila* models should be viewed as complementary alternatives to cellular assays and *in vitro* screenings made in mammalian cells, rather than the absolute shortcut to screen drugs for human treatments [158]. Their added value for drug discovery varies from disease to disease, and mainly depends on the availability of other options. Indeed, assays in *Drosophila* are complementary to *in vitro* and cellular systems, and in comparison to rodent model systems *Drosophila's* small size and culture conditions fulfill the requirements for large-scale screens [158]. Furthermore, another limitation also results from the dose-dependence of the drug treatment. In rodent model systems the drug dosage may differ according to the mode of penetration and the nature of the drug. In *Drosophila* dose-response experiments are easily feasible but the compounds are essentially delivered to the fruit flies through the media [167]. Thus it is important that the results obtained with *Drosophila* are compared with data obtained on laboratory rodents and, when possible, in humans. Furthermore, in numerous cases the results will be more qualitative than quantitative. Although the conclusions derived from *Drosophila* studies may remain too uncertain for pharmacologists, the data obtained from invertebrate-based screening could lead to important breakthrough, particularly for those diseases in which the pathophysiology is poorly understood [158, 161].

Often model systems are used to understand life and basic biological and cellular mechanisms. A better understanding of a specific human disease often comes as a consequence of the better overall understanding of biological processes [159]. Within this context, *Drosophila* is a valuable tool to categorize putative candidate genes for further downstream functional analysis in vertebrates. It can be used to dissect the likelihood that individual genes in a gene-cluster contribute to disease susceptibility, identify the relevance of a gene to a disease-pathway and get insights on gene specific functions manifested at the level of a tissue or involving cell-cell communication. Using *Drosophila* models, preliminary experiments with other model systems (such as mice) may be reduced and experiments in higher organisms can be better focused. Using *Drosophila* in a systematic approach together with other models and tools, seems promising in order to significantly reduce the turnout time from genetic results into biologically meaningful data [168]. Conclusively, although *Drosophila* will probably not always serve as a perfect model for human disease modeling, the common underlying molecular interactions and signaling pathways between flies and humans will continue to allow researchers to use *Drosophila* in order to answer existing questions, pose new hypotheses, get entry points in elucidating cancer and personalized cancer therapy, and complement studies from other model systems and vertebrate organisms [164].

6. The expanding role of *Drosophila* in cancer research: Bridging past, present and future

Undeniably the study of *Drosophila* has brought major contributions to the understanding to the fundamentals of cancer and has further given strong impulses not only in basic but also in applied research. The unrivalled advantages and tools offered by *Drosophila* will ensure that it will remain a premier research organism for many years to come. The advantages of *Drosophila* as a model system includes the availability of genetic tools developed in a century of "tool-building" research, its short life cycle and ability to unravel the basic function of genes in a straightforward way. The use of visible mutations and chromosome mapping coupled to currently available complex genetic databases including genomic and proteomic sequences, together with help of systematic gene disruption (RNAi libraries), microarray analysis, protein interaction maps and Flybase, lay the carpet for a renewed new age of research in *Drosophila,* and will allow scientists to address new questions on biological processes which were previously inaccessible. In turn the new discoveries will foster new research and answer to more precise questions about signaling pathways and behavior of individual cells in cancer. The *Drosophila* research will continuously contribute to a better understanding of tumor formation and progression, and may thus improve therapies in treating cancer.

Could *Drosophila* still be a valid model for understanding tumor formation and could it still provide a lead for curing cancer? Although the fruit fly does not appear to be suited for studying vertebrate-specific tissues, such as brain and heart development or neural crest migration [37], *Drosophila* may still help to identify critical key-genes, discover new biological pathways, define new research approaches, and therefore pioneer numerous fields in the understanding of cancer, including vertebrate biology. *Drosophila* can also be used to unravel the sequence of events leading to tumor formation and to trace the initial stages of tumor formation. One of the main outcomes of the genome analysis has been the finding of a high degree of conservation among genes. Subsequent analyses revealed that 75% of the genes involved in human diseases have homologues in *Drosophila.* There is also a high degree of functional conservation between the signal transduction pathways and a high degree of structural and functional conservation between cell adhesion proteins of *Drosophila* and humans showing that the fundamental biological processes have a common origin and has been relatively conserved during the 600 million years of evolutionary divergence between invertebrates and vertebrates. New emerging challenges in the study of tumorigenic inflammation, in *in vivo* screening for drug acting on cancer stem cells, in the therapeutic drug design and discovery will provide us with new insights into a "multi-target" approach for treating cancer. Finally, innovative technologies such as microarrays and nanotechnology, combined with novel methods in computation and bioinformatics, could be used in combination with genome-wide analysis in *Drosophila,* functional maps of the chromatin landscape [169-171], cis-regulatory map of the *Drosophila* genome and pattern of co-binding partners in transcription [172], as well as high-resolution of transcriptome dynamics throughout development [173, 174] to define more accurately the network of genes and pathways that would permit initial cancer cells to build expanding tumors. These new directions highlight not only the value of basic research but also the intrinsic advantages of *Drosophila* as a model organism for studying the complexity of cancer.

Acknowledgements

We apologize to all whose work has not been sited due to space limitations. This work was supported by the "Zentraler Forschungspool" funding of the University of Heidelberg to F.P. and the Czech Grant Foundation P302/11/1640, the Charles University Center UNCE204022 and the First Faculty of Medicine Prvouk/1LF/1 to B.M.M.

Author details

Fani Papagiannouli[1] and Bernard M. Mechler[2,3,4]

1 Centre for Organismal Studies Heidelberg (COS), University of Heidelberg, Heidelberg, Germany

2 Institute of Cellular Biology and Pathology.First Faculty of Medicine, Charles University in Prague, Prague, Czech Republic

3 Deutsches Krebsforschungszentrum, Heidelberg, Germany

4 Vellore Institute of Technology University, Vellore, Tamil Nadu, India

References

[1] Kango-Singh M, Halder G. Drosophila as an emerging model to study metastasis. Genome Biol. 2004;5(4):216.

[2] Hanahan D, Weinberg RA. Hallmarks of cancer: the next generation. Cell. 2011 Mar 4;144(5):646-74.

[3] Hanahan D, Weinberg RA. The hallmarks of cancer. Cell. 2000 Jan 7;100(1):57-70.

[4] Bhatt AM, Zhang Q, Harris SA, White-Cooper H, Dickinson H. Gene structure and molecular analysis of Arabidopsis thaliana ALWAYS EARLY homologs. Gene. 2004 Jul 21;336(2):219-29.

[5] Harris H. A long view of fashions in cancer research. Bioessays. 2005 Aug;27(8): 833-8.

[6] Gateff E, Schneiderman HA. Developmental studies of a new mutant of Drosophila melanogaster lethal malignant brain tumor (l(2)gl4). Am Zool. 1967;7:760.

[7] Gateff E, Schneiderman HA. Neoplasms in mutant and cultured wild-tupe tissues of Drosophila. Natl Cancer Inst Monogr. 1969 Jul;31:365-97.

[8] Harris H, Miller OJ, Klein G, Worst P, Tachibana T. Suppression of malignancy by cell fusion. Nature. 1969 Jul 26;223(5204):363-8.

[9] Knudson AG, Jr. Mutation and cancer: statistical study of retinoblastoma. Proc Natl Acad Sci U S A. 1971 Apr;68(4):820-3.

[10] Mechler BM, McGinnis W, Gehring WJ. Molecular cloning of lethal(2)giant larvae, a recessive oncogene of Drosophila melanogaster. EMBO J. 1985 Jun;4(6):1551-7.

[11] Mechler BM, Torok I, Schmidt M, Opper M, Kuhn A, Merz R, et al. Molecular basis for the regulation of cell fate by the lethal (2) giant larvae tumour suppressor gene of Drosophila melanogaster. Ciba Found Symp. 1989;142:166-78; discussion 78-80.

[12] Gateff E. Tumor suppressor and overgrowth suppressor genes of Drosophila melanogaster: developmental aspects. Int J Dev Biol. 1994 Dec;38(4):565-90.

[13] Bridges CB, Brehme KS. The mutants of Drosophila melanogaster: Carnegie Institution; 1944.

[14] Gateff E SH. Developmental capacities of benign and malignant neoplasms of Drosophila. Wilhelm Roux' Archiv. 1974;176(23-65).

[15] Opper M, Schuler G, Mechler BM. Hereditary suppression of lethal (2) giant larvae malignant tumor development in Drosophila by gene transfer. Oncogene. 1987 May; 1(2):91-6.

[16] Merz R, Schmidt M, Torok I, Protin U, Schuler G, Walther HP, et al. Molecular action of the l(2)gl tumor suppressor gene of Drosophila melanogaster. Environ Health Perspect. 1990 Aug;88:163-7.

[17] Torok I, Hartenstein K, Kalmes A, Schmitt R, Strand D, Mechler BM. The l(2)gl homologue of Drosophila pseudoobscura suppresses tumorigenicity in transgenic Drosophila melanogaster. Oncogene. 1993 Jun;8(6):1537-49.

[18] Strand D, Jakobs R, Merdes G, Neumann B, Kalmes A, Heid HW, et al. The Drosophila lethal(2)giant larvae tumor suppressor protein forms homo-oligomers and is associated with nonmuscle myosin II heavy chain. J Cell Biol. 1994 Dec;127(5): 1361-73.

[19] Strand D, Raska I, Mechler BM. The Drosophila lethal(2)giant larvae tumor suppressor protein is a component of the cytoskeleton. J Cell Biol. 1994 Dec;127(5):1345-60.

[20] Kalmes A, Merdes G, Neumann B, Strand D, Mechler BM. A serine-kinase associated with the p127-l(2)gl tumour suppressor of Drosophila may regulate the binding of

p127 to nonmuscle myosin II heavy chain and the attachment of p127 to the plasma membrane. J Cell Sci. 1996 Jun;109 (Pt 6):1359-68.

[21] Strand D. The Tumor Suppressor l(2)gl: A Myosin-Binding Protein Family. Kohama HMaK, editor: R. G. Landes Company; 1998.

[22] Li M, Strand D, Krehan A, Pyerin W, Heid H, Neumann B, et al. Casein kinase 2 binds and phosphorylates the nucleosome assembly protein-1 (NAP1) in Drosophila melanogaster. J Mol Biol. 1999 Nov 12;293(5):1067-84.

[23] Farkas R, Mechler BM. The timing of drosophila salivary gland apoptosis displays an l(2)gl-dose response. Cell Death Differ. 2000 Jan;7(1):89-101.

[24] Farkas R, Kucharova-Mahmood S, Mentelova L, Juda P, Raska I, Mechler B. Cytoskeletal proteins regulate chromatic access of BR-C transcription factor and Rpd3-Sin3A histone deacetylase complex in Drosophila salivary glands. Nucleus. 2011:(in print).

[25] Gateff E. Malignant neoplasms of genetic origin in Drosophila melanogaster. Science. 1978 Jun 30;200(4349):1448-59.

[26] Gateff E. Cancer, genes, and development: the Drosophila case. Adv Cancer Res. 1982;37:33-74.

[27] Stewart M, Murphy C, Fristrom JW. The recovery and preliminary characterization of X chromosome mutants affecting imaginal discs of Drosophila melanogaster. Dev Biol. 1972 Jan;27(1):71-83.

[28] Gateff E. Malignant neoplasms of the hematopoietic system in three mutants of Drosophila melanogaster. Ann Parasitol Hum Comp. 1977 Jan-Feb;52(1):81-3.

[29] Bilder D, Perrimon N. Localization of apical epithelial determinants by the basolateral PDZ protein Scribble. Nature. 2000 Feb 10;403(6770):676-80.

[30] Li M, Marhold J, Gatos A, Torok I, Mechler BM. Differential expression of two scribble isoforms during Drosophila embryogenesis. Mech Dev. 2001 Oct;108(1-2):185-90.

[31] Jennings BH. Drosophila-a versatile model in biology & medicine. materials today. 2011;14(5):190-5.

[32] Stocker H, Gallant P. Getting started : an overview on raising and handling Drosophila. Methods Mol Biol. 2008;420:27-44.

[33] Kohler RE. Lords of the fly: Drosophila genetics and the experimental life. Chicago: University of Chicago Press; 1994.

[34] Morgan TH. Sex Limited Inheritance in Drosophila. Science. 1910 Jul 22;32(812): 120-2.

[35] Green MM. 2010: A century of Drosophila genetics through the prism of the white gene. Genetics. 2010 Jan;184(1):3-7.

[36] Muller HJ. The Measurement of Gene Mutation Rate in Drosophila, Its High Variability, and Its Dependence upon Temperature. Genetics. 1928 May;13(4):279-357.

[37] Bellen HJ, Tong C, Tsuda H. 100 years of Drosophila research and its impact on vertebrate neuroscience: a history lesson for the future. Nat Rev Neurosci. 2010 Jul;11(7): 514-22.

[38] Nusslein-Volhard C, Wieschaus E. Mutations affecting segment number and polarity in Drosophila. Nature. 1980 Oct 30;287(5785):795-801.

[39] Wieschaus E, Nusslein-Volhard C, Kluding H. Kruppel, a gene whose activity is required early in the zygotic genome for normal embryonic segmentation. Dev Biol. 1984 Jul;104(1):172-86.

[40] Jurgens G, Wieschaus E, Nusslein-Volhard C, Kluding H. Mutations affecting the pattern of the larval cuticle in Drosophila melanogaster. Rouxs Arch Dev Biol. 1984;193:283-95.

[41] Lewis EB, Bacher F. Methods of feeding ethyl methane sulfonate (EMS) to Drosophila males. Inf Serv. 1968;43(193):193.

[42] Hobert O. The impact of whole genome sequencing on model system genetics: get ready for the ride. Genetics. 2010 Feb;184(2):317-9.

[43] Bellen HJ, Levis RW, Liao G, He Y, Carlson JW, Tsang G, et al. The BDGP gene disruption project: single transposon insertions associated with 40% of Drosophila genes. Genetics. 2004 Jun;167(2):761-81.

[44] Rong YS, Titen SW, Xie HB, Golic MM, Bastiani M, Bandyopadhyay P, et al. Targeted mutagenesis by homologous recombination in D. melanogaster. Genes Dev. 2002 Jun 15;16(12):1568-81.

[45] Dietzl G, Chen D, Schnorrer F, Su KC, Barinova Y, Fellner M, et al. A genome-wide transgenic RNAi library for conditional gene inactivation in Drosophila. Nature. 2007 Jul 12;448(7150):151-6.

[46] Ni JQ, Liu LP, Binari R, Hardy R, Shim HS, Cavallaro A, et al. A Drosophila resource of transgenic RNAi lines for neurogenetics. Genetics. 2009 Aug;182(4):1089-100.

[47] Rubin GM, Spradling AC. Genetic transformation of Drosophila with transposable element vectors. Science. 1982 Oct 22;218(4570):348-53.

[48] Venken KJ, He Y, Hoskins RA, Bellen HJ. P[acman]: a BAC transgenic platform for targeted insertion of large DNA fragments in D. melanogaster. Science. 2006 Dec 15;314(5806):1747-51.

[49] Groth AC, Fish M, Nusse R, Calos MP. Construction of transgenic Drosophila by using the site-specific integrase from phage phiC31. Genetics. 2004 Apr;166(4):1775-82.

[50] Golic KG, Lindquist S. The FLP recombinase of yeast catalyzes site-specific recombination in the Drosophila genome. Cell. 1989 Nov 3;59(3):499-509.

[51] Golic KG. Site-specific recombination between homologous chromosomes in Drosophila. Science. 1991 May 17;252(5008):958-61.

[52] Bell AJ, McBride SM, Dockendorff TC. Flies as the ointment: Drosophila modeling to enhance drug discovery. Fly (Austin). 2009 Jan-Mar;3(1):39-49.

[53] Lee T, Luo L. Mosaic analysis with a repressible cell marker (MARCM) for Drosophila neural development. Trends Neurosci. 2001 May;24(5):251-4.

[54] Brand AH, Perrimon N. Targeted gene expression as a means of altering cell fates and generating dominant phenotypes. Development. 1993 Jun;118(2):401-15.

[55] Pfeiffer BD, Jenett A, Hammonds AS, Ngo TT, Misra S, Murphy C, et al. Tools for neuroanatomy and neurogenetics in Drosophila. Proc Natl Acad Sci U S A. 2008 Jul 15;105(28):9715-20.

[56] Margadant C, Raymond K, Kreft M, Sachs N, Janssen H, Sonnenberg A. Integrin alpha3beta1 inhibits directional migration and wound re-epithelialization in the skin. J Cell Sci. 2009 Jan 15;122(Pt 2):278-88.

[57] McGuire SE, Le PT, Osborn AJ, Matsumoto K, Davis RL. Spatiotemporal rescue of memory dysfunction in Drosophila. Science. 2003 Dec 5;302(5651):1765-8.

[58] McGuire SE, Mao Z, Davis RL. Spatiotemporal gene expression targeting with the TARGET and gene-switch systems in Drosophila. Sci STKE. 2004 Feb 17;2004(220):pl6.

[59] Venken KJ, Bellen HJ. Transgenesis upgrades for Drosophila melanogaster. Development. 2007 Oct;134(20):3571-84.

[60] Morin X, Daneman R, Zavortink M, Chia W. A protein trap strategy to detect GFP-tagged proteins expressed from their endogenous loci in Drosophila. Proc Natl Acad Sci U S A. 2001 Dec 18;98(26):15050-5.

[61] Kelso RJ, Buszczak M, Quinones AT, Castiblanco C, Mazzalupo S, Cooley L. Flytrap, a database documenting a GFP protein-trap insertion screen in Drosophila melanogaster. Nucleic Acids Res. 2004 Jan 1;32(Database issue):D418-20.

[62] Buszczak M, Paterno S, Lighthouse D, Bachman J, Planck J, Owen S, et al. The carnegie protein trap library: a versatile tool for Drosophila developmental studies. Genetics. 2007 Mar;175(3):1505-31.

[63] Quinones-Coello AT, Petrella LN, Ayers K, Melillo A, Mazzalupo S, Hudson AM, et al. Exploring strategies for protein trapping in Drosophila. Genetics. 2007 Mar;175(3):1089-104.

[64] Miles A, Zhao J, Klyne G, White-Cooper H, Shotton D. OpenFlyData: an exemplar data web integrating gene expression data on the fruit fly Drosophila melanogaster. J Biomed Inform. 2010 Oct;43(5):752-61.

[65] Singh A, Irvine KD. Drosophila as a model for understanding development and disease. Dev Dyn. 2012 Jan;241(1):1-2.

[66] Gilbert LI. Drosophila is an inclusive model for human diseases, growth and development. Mol Cell Endocrinol. 2008 Oct 10;293(1-2):25-31.

[67] Schneider D. Using Drosophila as a model insect. Nat Rev Genet. 2000 Dec;1(3): 218-26.

[68] Botas J. Drosophila researchers focus on human disease. Nat Genet. 2007 May;39(5): 589-91.

[69] Pfleger CM, Reiter LT. Recent efforts to model human diseases in vivo in Drosophila. Fly (Austin). 2008 May-Jun;2(3):129-32.

[70] Reiter LT, Potocki L, Chien S, Gribskov M, Bier E. A systematic analysis of human disease-associated gene sequences in Drosophila melanogaster. Genome Res. 2001 Jun;11(6):1114-25.

[71] McGinnis W, Krumlauf R. Homeobox genes and axial patterning. Cell. 1992 Jan 24;68(2):283-302.

[72] Miklos GL, Rubin GM. The role of the genome project in determining gene function: insights from model organisms. Cell. 1996 Aug 23;86(4):521-9.

[73] Venken KJ, Carlson JW, Schulze KL, Pan H, He Y, Spokony R, et al. Versatile P[acman] BAC libraries for transgenesis studies in Drosophila melanogaster. Nat Methods. 2009 Jun;6(6):431-4.

[74] Tenenbaum D. What's All the Buzz? Fruit Flies Provide Unique Model for Cancer Research. J Natl Cancer Inst. 2003 Dec 3;95(23):1742-4.

[75] Crnic I, Christofori G. Novel technologies and recent advances in metastasis research. Int J Dev Biol. 2004;48(5-6):573-81.

[76] Vidal M. The dark side of fly TNF: an ancient developmental proof reading mechanism turned into tumor promoter. Cell Cycle. 2010 Oct 1;9(19):3851-6.

[77] Junttila MR, Evan GI. p53--a Jack of all trades but master of none. Nat Rev Cancer. 2009 Nov;9(11):821-9.

[78] Miles WO, Dyson NJ, Walker JA. Modeling tumor invasion and metastasis in Drosophila. Dis Model Mech. 2011 Nov;4(6):753-61.

[79] Bissell MJ, Radisky D. Putting tumours in context. Nat Rev Cancer. 2001 Oct;1(1): 46-54.

[80] Moreno E. Is cell competition relevant to cancer? Nat Rev Cancer. 2008 Feb;8(2): 141-7.

[81] Brumby AM, Richardson HE. Using Drosophila melanogaster to map human cancer pathways. Nat Rev Cancer. 2005 Aug;5(8):626-39.

[82] Humbert PO, Grzeschik NA, Brumby AM, Galea R, Elsum I, Richardson HE. Control of tumourigenesis by the Scribble/Dlg/Lgl polarity module. Oncogene. 2008 Nov 24;27(55):6888-907.

[83] Mohamet L, Hawkins K, Ward CM. Loss of function of e-cadherin in embryonic stem cells and the relevance to models of tumorigenesis. J Oncol. 2011;2011:352616.

[84] Pagliarini RA, Xu T. A genetic screen in Drosophila for metastatic behavior. Science. 2003 Nov 14;302(5648):1227-31.

[85] Schmeichel KL. A fly's eye view of tumor progression and metastasis. Breast Cancer Res. 2004;6(2):82-3.

[86] Woodhouse EC, Liotta LA. Drosophila invasive tumors: a model for understanding metastasis. Cell Cycle. 2004 Jan;3(1):38-40.

[87] Parisi F, Vidal M. Epithelial delamination and migration: lessons from Drosophila. Cell Adh Migr. 2011 Jul-Aug;5(4):366-72.

[88] Van Dyke T, Jacks T. Cancer modeling in the modern era: progress and challenges. Cell. 2002 Jan 25;108(2):135-44.

[89] Balmain A. Cancer as a complex genetic trait: tumor susceptibility in humans and mouse models. Cell. 2002 Jan 25;108(2):145-52.

[90] St Johnston D. The art and design of genetic screens: Drosophila melanogaster. Nat Rev Genet. 2002 Mar;3(3):176-88.

[91] Blair SS. Genetic mosaic techniques for studying Drosophila development. Development. 2003 Nov;130(21):5065-72.

[92] Kornberg TB, Krasnow MA. The Drosophila genome sequence: implications for biology and medicine. Science. 2000 Mar 24;287(5461):2218-20.

[93] Rubin GM, Lewis EB. A brief history of Drosophila's contributions to genome research. Science. 2000 Mar 24;287(5461):2216-8.

[94] Tanenbaum DM. What's All the Buzz? Fruit Flies Provide Unique Model for Cancer Research. Journal of the National Cancer Institute. 2003;95(23):1742-4.

[95] Brumby AM, Richardson HE. scribble mutants cooperate with oncogenic Ras or Notch to cause neoplastic overgrowth in Drosophila. EMBO J. 2003 Nov 3;22(21): 5769-79.

[96] Leong GR, Goulding KR, Amin N, Richardson HE, Brumby AM. Scribble mutants promote aPKC and JNK-dependent epithelial neoplasia independently of Crumbs. BMC Biol. 2009;7:62.

[97] Etienne-Manneville S. Scribble at the crossroads. J Biol. 2009;8(12):104.

[98] Tapon N. Modeling transformation and metastasis in Drosophila. Cancer Cell. 2003 Nov;4(5):333-5.

[99] Vidal M, Salavaggione L, Ylagan L, Wilkins M, Watson M, Weilbaecher K, et al. A role for the epithelial microenvironment at tumor boundaries: evidence from Drosophila and human squamous cell carcinomas. Am J Pathol. 2010 Jun;176(6):3007-14.

[100] Wu M, Pastor-Pareja JC, Xu T. Interaction between Ras(V12) and scribbled clones induces tumour growth and invasion. Nature. 2010 Jan 28;463(7280):545-8.

[101] Martin-Belmonte F, Perez-Moreno M. Epithelial cell polarity, stem cells and cancer. Nat Rev Cancer. 2012 Jan;12(1):23-38.

[102] Igaki T, Pagliarini RA, Xu T. Loss of cell polarity drives tumor growth and invasion through JNK activation in Drosophila. Curr Biol. 2006 Jun 6;16(11):1139-46.

[103] Tyler DM, Li W, Zhuo N, Pellock B, Baker NE. Genes affecting cell competition in Drosophila. Genetics. 2007 Feb;175(2):643-57.

[104] Hariharan IK, Bilder D. Regulation of imaginal disc growth by tumor-suppressor genes in Drosophila. Annu Rev Genet. 2006;40:335-61.

[105] Harvey K, Tapon N. The Salvador-Warts-Hippo pathway - an emerging tumour-suppressor network. Nat Rev Cancer. 2007 Mar;7(3):182-91.

[106] Saucedo LJ, Edgar BA. Filling out the Hippo pathway. Nat Rev Mol Cell Biol. 2007 Aug;8(8):613-21.

[107] Bennett FC, Harvey KF. Fat cadherin modulates organ size in Drosophila via the Salvador/Warts/Hippo signaling pathway. Curr Biol. 2006 Nov 7;16(21):2101-10.

[108] Silva EA, Lee BJ, Caceres LS, Renouf D, Vilay BR, Yu O, et al. A novel strategy for identifying mutations that sensitize Drosophila eye development to caffeine and hydroxyurea. Genome. 2006 Nov;49(11):1416-27.

[109] Willecke M, Hamaratoglu F, Kango-Singh M, Udan R, Chen CL, Tao C, et al. The fat cadherin acts through the hippo tumor-suppressor pathway to regulate tissue size. Curr Biol. 2006 Nov 7;16(21):2090-100.

[110] Yu J, Zheng Y, Dong J, Klusza S, Deng WM, Pan D. Kibra functions as a tumor suppressor protein that regulates Hippo signaling in conjunction with Merlin and Expanded. Dev Cell. 2010 Feb 16;18(2):288-99.

[111] Genevet A, Wehr MC, Brain R, Thompson BJ, Tapon N. Kibra is a regulator of the Salvador/Warts/Hippo signaling network. Dev Cell. 2010 Feb 16;18(2):300-8.

[112] Baumgartner R, Poernbacher I, Buser N, Hafen E, Stocker H. The WW domain protein Kibra acts upstream of Hippo in Drosophila. Dev Cell. 2010 Feb 16;18(2):309-16.

[113] Doggett K, Grusche FA, Richardson HE, Brumby AM. Loss of the Drosophila cell polarity regulator Scribbled promotes epithelial tissue overgrowth and cooperation with oncogenic Ras-Raf through impaired Hippo pathway signaling. BMC Dev Biol. 2011;11:57.

[114] Chen CL, Schroeder MC, Kango-Singh M, Tao C, Halder G. Tumor suppression by cell competition through regulation of the Hippo pathway. Proc Natl Acad Sci U S A. 2012 Jan 10;109(2):484-9.

[115] Grzeschik NA, Parsons LM, Allott ML, Harvey KF, Richardson HE. Lgl, aPKC, and Crumbs regulate the Salvador/Warts/Hippo pathway through two distinct mechanisms. Curr Biol. 2010 Apr 13;20(7):573-81.

[116] Grzeschik NA, Parsons LM, Richardson HE. Lgl, the SWH pathway and tumorigenesis: It's a matter of context & competition! Cell Cycle. 2010 Aug 15;9(16):3202-12.

[117] Tamori Y, Bialucha CU, Tian AG, Kajita M, Huang YC, Norman M, et al. Involvement of Lgl and Mahjong/VprBP in cell competition. PLoS Biol. 2010;8(7):e1000422.

[118] Mair W. How normal cells can win the battle for survival against cancer cells. PLoS Biol. 2010;8(7):e1000423.

[119] Alderton GK. Tumorigenesis: To the death! Nat Rev Cancer. 2010 Sep;10(9):598.

[120] Menendez J, Perez-Garijo A, Calleja M, Morata G. A tumor-suppressing mechanism in Drosophila involving cell competition and the Hippo pathway. Proc Natl Acad Sci U S A. 2010 Aug 17;107(33):14651-6.

[121] Froldi F, Ziosi M, Garoia F, Pession A, Grzeschik NA, Bellosta P, et al. The lethal giant larvae tumour suppressor mutation requires dMyc oncoprotein to promote clonal malignancy. BMC Biol. 2010;8:33.

[122] Zhu M, Xin T, Weng S, Gao Y, Zhang Y, Li Q, et al. Activation of JNK signaling links lgl mutations to disruption of the cell polarity and epithelial organization in Drosophila imaginal discs. Cell Res. 2010 Feb;20(2):242-5.

[123] Robinson BS, Huang J, Hong Y, Moberg KH. Crumbs regulates Salvador/Warts/Hippo signaling in Drosophila via the FERM-domain protein Expanded. Curr Biol. 2010 Apr 13;20(7):582-90.

[124] Ling C, Zheng Y, Yin F, Yu J, Huang J, Hong Y, et al. The apical transmembrane protein Crumbs functions as a tumor suppressor that regulates Hippo signaling by binding to Expanded. Proc Natl Acad Sci U S A. 2010 Jun 8;107(23):10532-7.

[125] Chen CL, Gajewski KM, Hamaratoglu F, Bossuyt W, Sansores-Garcia L, Tao C, et al. The apical-basal cell polarity determinant Crumbs regulates Hippo signaling in Drosophila. Proc Natl Acad Sci U S A. 2010 Sep 7;107(36):15810-5.

[126] Cordero JB, Macagno JP, Stefanatos RK, Strathdee KE, Cagan RL, Vidal M. Oncogenic Ras diverts a host TNF tumor suppressor activity into tumor promoter. Dev Cell. 2010 Jun 15;18(6):999-1011.

[127] Igaki T. Correcting developmental errors by apoptosis: lessons from Drosophila JNK signaling. Apoptosis. 2009 Aug;14(8):1021-8.

[128] Ohsawa S, Sugimura K, Takino K, Xu T, Miyawaki A, Igaki T. Elimination of oncogenic neighbors by JNK-mediated engulfment in Drosophila. Dev Cell. 2011 Mar 15;20(3):315-28.

[129] Rudrapatna VA, Cagan RL, Das TK. Drosophila cancer models. Dev Dyn. 2012 Jan; 241(1):107-18.

[130] Chang KC, Wang C, Wang H. Balancing self-renewal and differentiation by asymmetric division: insights from brain tumor suppressors in Drosophila neural stem cells. Bioessays. 2012 Apr;34(4):301-10.

[131] Gonczy P. Mechanisms of asymmetric cell division: flies and worms pave the way. Nat Rev Mol Cell Biol. 2008 May;9(5):355-66.

[132] Iden S, van Riel WE, Schafer R, Song JY, Hirose T, Ohno S, et al. Tumor type-dependent function of the par3 polarity protein in skin tumorigenesis. Cancer Cell. 2012 Sep 11;22(3):389-403.

[133] Engelman JA. Targeting PI3K signalling in cancer: opportunities, challenges and limitations. Nat Rev Cancer. 2009 Aug;9(8):550-62.

[134] Busaidy NL, Farooki A, Dowlati A, Perentesis JP, Dancey JE, Doyle LA, et al. Management of metabolic effects associated with anticancer agents targeting the PI3K-Akt-mTOR pathway. J Clin Oncol. 2012 Aug 10;30(23):2919-28.

[135] Read RD. Drosophila melanogaster as a model system for human brain cancers. Glia. 2011 Sep;59(9):1364-76.

[136] Caldeira J, Pereira PS, Suriano G, Casares F. Using fruitflies to help understand the molecular mechanisms of human hereditary diffuse gastric cancer. Int J Dev Biol. 2009;53(8-10):1557-61.

[137] Pereira PS, Teixeira A, Pinho S, Ferreira P, Fernandes J, Oliveira C, et al. E-cadherin missense mutations, associated with hereditary diffuse gastric cancer (HDGC) syndrome, display distinct invasive behaviors and genetic interactions with the Wnt and Notch pathways in Drosophila epithelia. Hum Mol Genet. 2006 May 15;15(10): 1704-12.

[138] Crozatier M, Vincent A. Drosophila: a model for studying genetic and molecular aspects of haematopoiesis and associated leukaemias. Dis Model Mech. 2011 Jul;4(4): 439-45.

[139] Zhai Z, Ha N, Papagiannouli F, Hamacher-Brady A, Brady N, Sorge S, et al. Antagonistic regulation of apoptosis and differentiation by the Cut transcription factor rep-

resents a tumor-suppressing mechanism in Drosophila. PLoS Genet. 2012 Mar; 8(3):e1002582.

[140] Naora H, Montell DJ. Ovarian cancer metastasis: integrating insights from disparate model organisms. Nat Rev Cancer. 2005 May;5(5):355-66.

[141] Naora H. Developmental patterning in the wrong context: the paradox of epithelial ovarian cancers. Cell Cycle. 2005 Aug;4(8):1033-5.

[142] Duchek P, Somogyi K, Jekely G, Beccari S, Rorth P. Guidance of cell migration by the Drosophila PDGF/VEGF receptor. Cell. 2001 Oct 5;107(1):17-26.

[143] Bartlett JM, Langdon SP, Simpson BJ, Stewart M, Katsaros D, Sismondi P, et al. The prognostic value of epidermal growth factor receptor mRNA expression in primary ovarian cancer. Br J Cancer. 1996 Feb;73(3):301-6.

[144] Sundfeldt K, Piontkewitz Y, Ivarsson K, Nilsson O, Hellberg P, Brannstrom M, et al. E-cadherin expression in human epithelial ovarian cancer and normal ovary. Int J Cancer. 1997 Jun 20;74(3):275-80.

[145] Sundfeldt K. Cell-cell adhesion in the normal ovary and ovarian tumors of epithelial origin; an exception to the rule. Mol Cell Endocrinol. 2003 Apr 28;202(1-2):89-96.

[146] Marques FR, Fonsechi-Carvasan GA, De Angelo Andrade LA, Bottcher-Luiz F. Immunohistochemical patterns for alpha- and beta-catenin, E- and N-cadherin expression in ovarian epithelial tumors. Gynecol Oncol. 2004 Jul;94(1):16-24.

[147] Edwards PA. The impact of developmental biology on cancer research: an overview. Cancer Metastasis Rev. 1999;18(2):175-80.

[148] Ng JM, Curran T. The Hedgehog's tale: developing strategies for targeting cancer. Nat Rev Cancer. 2011 Jul;11(7):493-501.

[149] Cordero JB, Sansom OJ. Wnt signalling and its role in stem cell-driven intestinal regeneration and hyperplasia. Acta Physiol (Oxf). 2012 Jan;204(1):137-43.

[150] Jessen S, Gu B, Dai X. Pygopus and the Wnt signaling pathway: a diverse set of connections. Bioessays. 2008 May;30(5):448-56.

[151] Radtke F, Raj K. The role of Notch in tumorigenesis: oncogene or tumour suppressor? Nat Rev Cancer. 2003 Oct;3(10):756-67.

[152] Bossuyt W, De Geest N, Aerts S, Leenaerts I, Marynen P, Hassan BA. The atonal proneural transcription factor links differentiation and tumor formation in Drosophila. PLoS Biol. 2009 Feb 24;7(2):e40.

[153] Berthold J, Schenkova K, Rivero F. Rho GTPases of the RhoBTB subfamily and tumorigenesis. Acta Pharmacol Sin. 2008 Mar;29(3):285-95.

[154] Hannigan G, Troussard AA, Dedhar S. Integrin-linked kinase: a cancer therapeutic target unique among its ILK. Nat Rev Cancer. 2005 Jan;5(1):51-63.

[155] Butcher DT, Alliston T, Weaver VM. A tense situation: forcing tumour progression. Nat Rev Cancer. 2009 Feb;9(2):108-22.

[156] Shah N, Sukumar S. The Hox genes and their roles in oncogenesis. Nat Rev Cancer. 2010 May;10(5):361-71.

[157] Grier DG, Thompson A, Kwasniewska A, McGonigle GJ, Halliday HL, Lappin TR. The pathophysiology of HOX genes and their role in cancer. J Pathol. 2005 Jan;205(2): 154-71.

[158] Giacomotto J, Segalat L. High-throughput screening and small animal models, where are we? Br J Pharmacol. 2010 May;160(2):204-16.

[159] Aitman TJ, Boone C, Churchill GA, Hengartner MO, Mackay TF, Stemple DL. The future of model organisms in human disease research. Nat Rev Genet. 2011 Aug;12(8): 575-82.

[160] Pandey UB, Nichols CD. Human disease models in Drosophila melanogaster and the role of the fly in therapeutic drug discovery. Pharmacol Rev. 2011 Jun;63(2):411-36.

[161] Segalat L. Invertebrate animal models of diseases as screening tools in drug discovery. ACS Chem Biol. 2007 Apr 24;2(4):231-6.

[162] Kasai Y, Cagan R. Drosophila as a tool for personalized medicine: a primer. Per Med. 2010 Nov;7(6):621-32.

[163] Perrimon N, Friedman A, Mathey-Prevot B, Eggert US. Drug-target identification in Drosophila cells: combining high-throughout RNAi and small-molecule screens. Drug Discov Today. 2007 Jan;12(1-2):28-33.

[164] Bier E. Drosophila, the golden bug, emerges as a tool for human genetics. Nat Rev Genet. 2005 Jan;6(1):9-23.

[165] Wolf MJ, Rockman HA. Drosophila melanogaster as a model system for genetics of postnatal cardiac function. Drug Discov Today Dis Models. 2008 Oct 1;5(3):117-23.

[166] Chien S, Reiter LT, Bier E, Gribskov M. Homophila: human disease gene cognates in Drosophila. Nucleic Acids Res. 2002 Jan 1;30(1):149-51.

[167] Kaletta T, Hengartner MO. Finding function in novel targets: C. elegans as a model organism. Nat Rev Drug Discov. 2006 May;5(5):387-98.

[168] Roeder T, Isermann K, Kabesch M. Drosophila in asthma research. Am J Respir Crit Care Med. 2009 Jun 1;179(11):979-83.

[169] Kharchenko PV, Alekseyenko AA, Schwartz YB, Minoda A, Riddle NC, Ernst J, et al. Comprehensive analysis of the chromatin landscape in Drosophila melanogaster. Nature. 2011 Mar 24;471(7339):480-5.

[170] Furlong EE. Molecular biology: A fly in the face of genomics. Nature. 2011 Mar 24;471(7339):458-9.

[171] Beller M, Oliver B. One hundred years of high-throughput Drosophila research. Chromosome Res. 2006;14(4):349-62.

[172] Negre N, Brown CD, Ma L, Bristow CA, Miller SW, Wagner U, et al. A cis-regulatory map of the Drosophila genome. Nature. 2011 Mar 24;471(7339):527-31.

[173] Graveley BR, Brooks AN, Carlson JW, Duff MO, Landolin JM, Yang L, et al. The developmental transcriptome of Drosophila melanogaster. Nature. 2011 Mar 24;471(7339):473-9.

[174] Chintapalli VR, Wang J, Dow JA. Using FlyAtlas to identify better Drosophila melanogaster models of human disease. Nat Genet. 2007 Jun;39(6):715-20.

Chapter 6

Polyphenolic Compounds Targeting p53-Family Tumor Suppressors: Current Progress and Challenges

Nelly Etienne-Selloum, Israa Dandache,
Tanveer Sharif, Cyril Auger and
Valérie B. Schini-Kerth

Additional information is available at the end of the chapter

http://dx.doi.org/10.5772/56102

1. Introduction

The chemotherapeutic properties of polyphenols have recently received an increasing interest since it has been established that these compounds can modulate each step of the cancer progression process (initiation, proliferation, survival, migration, angiogenesis, and metastasis). Polyphenols are believed to be multi-targets drugs and in the present chapter we will give an overview of recent investigations concerning apoptosis induction by three major compounds, resveratrol, curcumin and epigallocatechin-3-gallate (EGCG) mainly through the regulation of the p53 tumor suppressor pathway. The potential regulation by polyphenols of p53 expression at the transcriptional and post-translational levels has been extensively described. Interestingly, polyphenolic compounds are also able to trigger apoptosis of numerous cancer cells, independently of the p53 status (wild-type, mutated or deficient). Moreover alternative mechanisms supported by recent studies highlight the role of p73, a p53 related tumor suppressor, as another key target for polyphenols. Then the molecular mechanisms involved in tumor suppressors (mainly p53 and p73) expression by polyphenols will be discussed with a specific focus on the role of oxidative stress which is believed to be a key element in polyphenols-induced cancer cells death.

2. Anticancer properties of polyphenols: Chemoprevention and chemotherapy

Polyphenols are natural compounds characterized by a structure containing at least one benzene ring substituted by at least one hydroxyl group. Beside this chemical hallmark, phenolic products currently constitute a large and still expanding complex and heterogeneous family of molecules (more than 8000 phenolic structures currently known) with a great diversity of structure and size ranging from the low molecular weight simple phenols up to the high molecular weight tannins [1-3]. Polyphenols are also one of the largest and most widespread classes of constituents present in plant kingdom and more particularly in plant-derived foods and beverages giving them their color and taste properties. Polyphenols can be structurally divided into two main families: flavonoids and non-flavonoids. Flavonoids are especially abundant in fruits, vegetables, seeds, spices, herbs, tea, cocoa, and wine. The six major subclasses of flavonoids are anthocyanidins (e.g., cyanidin, delphinidin; primary sources: red berries, red cabbages, cherries, grapes, and onions), flavan-3-ols (e.g., catechin, epicatechin, EGCG; primary sources: tea, grapes, cocoa, apples, and red wine), flavanones (e.g., hesperitin, naringenin; primary sources: oranges, lemons, and grapefruits), flavones (e.g., apigenin, luteolin; primary sources: celery, parsley, and thyme), flavonols (e.g., kaempferol, myricetin, quercetin; primary sources: apples, beans, broccoli, and onions), and isoflavonoids (e.g., daidzein, genistein; primary sources: legumes and soy products). Phenolic acids represent a large subclass of non-flavonoid polyphenolic compounds which can be further divided into two main types: benzoic acids (e.g., gallic acid, ellagic acid, vanillic acid; primary sources: tea, red wine, berries, nuts, and herbs) and cinnamic acids (e.g., caffeic acid, chlorogenic acid; primary sources: coffee, berries, plum, and apple). Other important classes of non flavonoids with healthy properties are stilbenes, such as resveratrol (primary sources: red wine, berries and nuts) and curcuminoids such as curcumin the main component of dried turmeric and curry powder [4]. Polyphenols are considered as secondary plant metabolites and have been associated with several functions in plants such as resistance against microbial pathogens and insects, protection against DNA-damaging UV light, reproduction, nutrition and growth [3]. In parallel to their protective properties in plants, polyphenols have long been regarded as a pool of bioactive natural products with potential benefits for human health. Plant extracts, herbs and spice containing these compounds have been used for thousands of years in traditional medicines. Nowadays, plant polyphenols enjoy an ever-increasing recognition not only by scientific community but also, and most remarkably, by the general public because of their presence and abundance in fruits, seeds, vegetables and derived foodstuffs and beverages, whose regular consumption has been claimed to be beneficial for human health [3, 5]. Indeed, epidemiological and experimental studies have shown the potential of polyphenols or polyphenolic nutritional sources in reducing the risk of chronic diseases such as cardiovascular diseases [6-10] and cancers [10-14], as well as the risk of degenerative diseases [10, 15, 16]. Altogether these observations led to the current nutritional recommendations to eat five servings of fruits and vegetables per days in order to keep healthy.

A wealth of data, including epidemiological and animal studies, has described the chemopreventive and anticancer properties of polyphenolic compounds, such as resveratrol, curcumin or tea catechins, or polyphenol-rich nutritional sources [13, 14, 17-19]. Nonetheless, recent investigations have highlighted additional mechanisms responsible for direct anti-proliferative and chemotherapeutic properties of polyphenols. Indeed, these compounds can interfere with the initiation, as well as the progression of cancer through the modulation of different cellular events, such as cell cycle arrest by decreasing cyclins or apoptosis induction through cytochrome c release, activation of caspases and down- or up-regulation of Bcl-2 family members, and inhibition of survival/proliferation signals (AKT, MAPK, NF-κB, etc.). Furthermore, they play an important role in inflammation (COX-2, TNF secretion, etc.), as well as in suppression of key proteins involved in angiogenesis and metastasis [13]. Importantly, it has been established that tumor suppressors like p53 and its analogs are key molecular targets of polyphenols responsible for their pro-apoptotic effect in human and animal cancer models. Here we provide an overview of the molecular mechanisms involved in p53 family proteins modulation by three major and well characterized polyphenolic compounds, resveratrol, curcumin and EGCG.

3. p53 family proteins are chemotherapeutic targets of polyphenols

Since the discovery of p53 in 1979 [20-22] numerous studies have been conducted related to its functions in response to stress and its regulatory mechanisms. p53 is a sequence-specific nuclear transcription factor that binds to defined consensus sites within DNA as a tetramer and represses transcription of a set of genes involved in cell growth stimulation, while activating a different set of genes involved in cell cycle control, like p21. It causes growth arrest providing a window for DNA repair or elimination of cells with severely damaged DNA strands. In some conditions, p53 activation triggers the transcription of pro-apoptotic genes such as Bax or PUMA, as well as the repression of anti-apoptotic genes like survivin [23]. Moreover, p53 can induce transcription-independent apoptosis. This mechanism involves early p53 translocation to mitochondria where it binds to Bcl-2 family proteins, such as Bax, Bak and Bcl-XL, activating cytochrome c release and caspases cascade [24]. Undoubtedly p53 exerts major anti-neoplastic effects and is considered actually as the "guardian of the genome" [25]. Tumor suppressive capabilities of p53 are related to a coordinated regulatory circuit that monitors and responds to a variety of stress signals, including DNA damage, abnormal oncogenic events, telomere erosion and hypoxia [26]. Importantly, in unstressed cells, p53 is latent and is maintained at low level by targeted ubiquitin-mediated degradation related to its interaction with ubiquitine ligases, mainly MDM2 [27]. Regarding the "guardian" functions of p53, mutations of p53 gene or disruptions of p53 coordination such as post-translational inactivation, can disturb the normal physiological balance, and lead to cancer if genome disarrangement reachs a critical value [28]. Indeed, low level of functionnal p53 is a common characteristic of cancer from several localizations including lung, colon, rectum, breast, brain, bladder, stomac, prostate, ovary, liver or lymphoid organs [29]. Somatic p53 missense inactivating mutations are found in approximately 50% of human cancers [30] and this inactivating mutations render the mutant p53 protein unable to carry out its normal function, that is, transcriptional transactivation of downstream target genes that regulate cell cycle and apoptosis [31-33]. On the other hand, p53 pathway can be also inactivated in wild-type (WT) p53-carrying tumors via indirect mechanisms such as MDM2 amplification leading to p53 destabilization [34, 35].

Recently, cDNAs with strong homologies to p53 have been identified and their products were termed p63 and p73 [36-38]. Both proteins are structurally similar and functionally related to p53, and consequently the entire p53 family may be regarded as a unique signalling network controlling cell proliferation, differentiation and death. Interestingly, in contrary to p53, the role of the other two p53-related proteins in tumor suppression is less obvious, since they are rarely deleted or mutated in cancer, and the respective knockout mice die tumor-free from developmental defects [39-41]. However, increasing number of evidences suggest that both p63 and p73 have a role in tumor suppression. Indeed, different studies indicated that TAp73 and TAp63, the transcriptionally active isoforms, can induce cell cycle arrest, senescence, DNA repair, and apoptosis in response to chemotherapeutic drugs, independently of p53 [42-45]. In addition, even if not mutated, p63 and p73 can be aberrantly expressed in cancer. More particularly, the dominant negative and transcriptionally inactive isoforms ΔNp63 and ΔNp73 are frequently overexpressed in a wide range of tumors, in which they are associated with poor prognosis [46]. Actually, the imbalance in the TAp73/ΔNp73 may be more critical for tumorigenesis and response to chemotherapy than mutations [47]. In summary, despite their differences, the three members of the p53 family may be considered as therapeutic targets for cancer management.

Many *in vitro* studies as well as few *in vivo* studies have shown that resveratrol, curcumin and EGCG, as well as nutritional sources of polyphenols induce overexpression of wild-type p53 (Table 1-4). The p53-related anticancer properties of these three isolated molecules have been extensively evaluated but other polyphenolic compounds such as genistein, luteolin, quercetin, and wogonin have been shown also to upregulate wild-type p53 protein in several cancer cell lines [48-51]. The polyphenol-induced stabilization and expression of wild-type p53 is often associated with a G1 or G2/M phase cell cycle arrest together with transcriptional regulation of target genes such as p21, Bax, PUMA and apoptosis induction [52-56]. The key role of p53 in polyphenol-induced anticancer properties is supported by studies indicating that p53 downregulation counteracts apoptosis triggered by natural products. Indeed, p53 silencing by siRNA abrogate the cytotoxic effect of curcumin in chondrosarcoma cells [57] and genetic invalidation of p53 by shRNA leads to inhibition of EGCG plus luteonin-induced apoptosis of lung cancer cells [58]. In addition, EGCG fails to induce significant cytotoxic effect in p53-null PC-3 prostate cancer cells, but forced expression of p53 in such cell line leads to sensitization to the polyphenolic compound [53]. Indeed, in the later study EGCG induces p53 phosphorylation on Serine 15 and upregulation of p53 and p21 expression together with cell cycle arrest and apoptosis. However the key role of p53 in the anticancer properties of polyphenols is still controversial, especially for curcumin, since many studies have shown its anti-proliferative properties in several p53-mutated or p53-null cancer cell lines (Table 2). For instance, curcumin has significant anti-proliferative effects in two p53-mutated human glioblastoma cell lines, indicating alternative and p53-independent pathway involved in such anticancer properties [59]. Similarly, curcumin reduces glioblastoma cells viability irrespective of p53 mutational status [60]. In this study, curcumin-induced cancer cell death was associated with caspase-3 activity in p53-wild-type cells, but not in p53-mutated cells, indicating that polyphenols can trigger p53- and caspases-independent cell death. p53-independent anticancer properties of polyphenols have been also described in many other cancer cells [61-67]. Interestingly, curcumin reduces the expression of the mutated form of p53 in MDA-MB-231 breast cancer cells together with cell cycle arrest [68], suggesting that a polyphenol-dependent regulatory process can also modulate the expression of a non-functional tumor suppressor. However, despite potential apoptosis induction by polyphenols in absence of functional p53 protein, its wild-type expression makes cancer cells more sensitive to pro-apoptotic effects of polyphenols. Recently, Ferraz da Costa et al. have demonstrated that transient transfection of wild-type p53 in human non-small lung carcinoma cell line H1299 (p53 negative) dramatically increased susceptibility to resveratrol-induced apoptosis [69]. Altogether these data indicate that p53 participates to the cytotoxic effect of polyphenols but also that alternative pathways might be involve in their anticancer properties.

One of this alternative pathway might involve Egr-1, an immediate early-response gene induced by stress, injury, mitogens, and differentiation [70]. Egr-1 regulates the expression of genes involved in the control of growth and apoptosis by transactivating many proteins including p21. One study has shown that transcription of the p21 gene is activated by Egr-1 independently of p53 but under the control of MAPKs in response to curcumin treatment in U-87MG human glioblastoma cells [71]. In addition, the apoptotic effect of resveratrol in colorectal cancer cells as well as EGCG-mediated cytotoxicity in pulmonary cancer cells are also associated with Egr-1 upregulation [72, 73].

Alternatively to p53, its functionally related proteins p63 and p73 might represent targets for polyphenols. Nevertheless only few data are available concerning a potential regulatory effect of polyphenolic compounds on p63 and p73 (Table1-4). Different flavones (luteolin, apigenin, chrysin) and flavonols (quercetin, kaempferol, myricetin) are able to induce cytotoxicity in p53-mutated oesophageal squamous carcinoma cells together with upregulation of p63 and p73 [74]. Similarly, EGCG induces selective apoptosis in multiple myeloma cells with overexpression of p63 and p73 without any change in the p53 expression level [75], as well as overexpression of p73 in p53-mutated T-lymphocyte leukemic cells [76]. As previously mentioned, different isoforms of p73 have been described and quercetin has been shown to control the subcellular localization of the dominant negative isoform ΔNp73 in melanoma cells expressing wild-type p53. In this model, quercetin caused redistribution of ΔNp73 into the cytoplasm and nucleus, which has been associated with increased p53 transcriptional activity and apoptosis [47, 77]. Beside isolated compounds, more complex sources of polyphenol such as red wine polyphenolic extract or berries-derived product can also modulate p53 and/or p73 expression level, *in vitro* and *in vivo* (Table 4) [78-81]. Interestingly, a synthetic analogue of curcumin increases p73 expression level in two distinct p53-wild-type pancreatic cancer cell lines, BxPC-3 and Colo-357 together with upregulation of pro-apoptotic effector Bax and simultaneous downregulation of the anti-apoptotic protein Bcl-2 [82]. Curcumin itself has been shown to stimulate p53 and also p73 expression in p53-mutated C33A cervical cancer cells [83]. Moreover, EGCG upregulates transcriptional target of p53, in a p53-independent but p73-dependent manner in mouse embryonic fibroblasts [84]. These data suggest that independently of the p53 status (wild-type, mutated or deleted), p73 seems to be involde in the anticancer effect of polyphenolic compounds. Many others studies have shown the potential of polyphenols to induce apoptosis of cancer cells in a p53-dependent but also a p53-independent manner (Table 1-4). In summary data concerning the role of tumor suppressors in the polyphenol-induced anticancer effects are inconsistent, probably dependent on the cell type, and consequently remain controversial. Moreover, the molecular mecanisms responsible for p53-family tumor suppressors regulation by polyphenols are only partially elucidated. However some evidences indicate that polyphenols might modulate p53 or p73 expression as well as their stabilization which are under the control of phosphorylation and acetylation levels.

Resveratrol

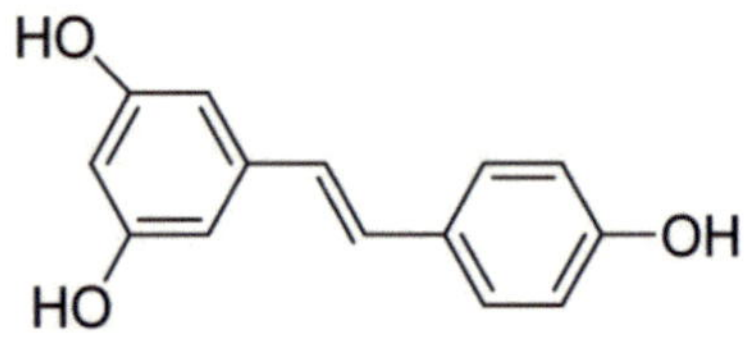

Cancer model	Described effects on TSG (p53/p73)	References
Prostate cancer cells (LNCaP, DU145, p53-mutated CWR22Rv1, p53-null PC-3)	- No change in p53 mRNA, increased expression of p53-p(ser15) and/or p53-ac(lys382) and total p53 protein - p53 translocation to mitochondria - cell cycle alteration and apoptosis induction maintained in p53-mutated cancer cells - potentiation of radiation–induced p53 expression in p53-mutated cancer cells	[135, 158, 164, 165, 175, 176]
Ovarian carcinoma cells (OVCAR-3)	- nuclear accumulation of p53-p(Ser15)	[110]
Breast cancer cells (MCF-7, p53-mutated MDA-MB-231, p53-mutated MDA-MB-435)	- increased expression of p53-p(ser15), p53-p(ser20), p53-p(ser392) and p53-ac(lys382/lys373)	[177-180]
	- no change in total p53 protein expression, p53-independent apoptosis	[178, 181]
	- increased expression of p53 mRNA and total protein	[182-184]
	- no change in p53 mRNA	[185]
	- p53-independent cytochrome c release	[181]
	- increased p53-dependent transcriptional activity	[50]
Colon cancer cells (HCT116, p53-null HCT116)	- increased p53-p(ser15) expression - resveratrol-induced senescence is p53 dependent	[103]
Pancreatic cancer cells (capan-2, colo357)	- upregulation and nuclear accumulation of p53 in both cell line (restoration of wild-type expression)	[186]

Resveratrol

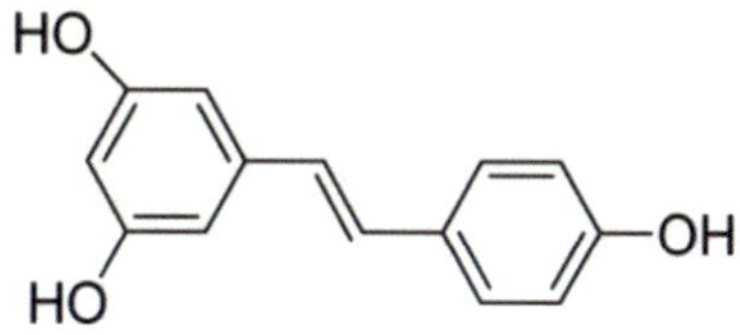

Cancer model	Described effects on TSG (p53/p73)	References
Glioblastoma cells (A172, p53-mutated T98G)	- no change in p53 mRNA	[187]
Hepatocellular carcinoma cells (HepG2)	- no change in p53 mRNA	[185]
Osteosarcoma cells (U-2 OS)	- increased p53-p(ser15) and p53-p(Ser37) expression	[188]
Lung adenocarcinoma cells (A549)	- increased p53-p(ser15) and p53-p(Ser37) expression	[188]
Head and neck squamous cancer cells (UMSCC-22B)	- increased p53-p(ser15) and total p53 expression	[109]
Cervical cancer cells (HeLa)	- increased p53-ac(lys373) and total p53 expression	[185]
Hodgkin lymphoma cells (L-428)	- increased p53-p(ser15) expression	[129]
Follicular lymphoma cells (LY8)	- increased p53-p(ser15) and total p53 expression	[190]
Acute lymphoblastic leukemia cells (MOLT-4)	- increased p53-p(ser15) expression	[189]
Neuroblastoma cells (B65, NUB-7)	- increased p53-p(ser15) and total p53 expression - nuclear translocation of p53	[94, 132]
DMBA-TPA-induced mouse skin tumor ; DEN-induced rat hepatocellular carcinoma	- increased p53-p(ser15) and total p53 expression	[18, 191, 192]
	- increased wild-type p53 and decreased mutated-p53 expression	[193]

Cancer cell lines express wild-type p53 except where otherwise stated; ac(lys.)=acetylated lysine, p(ser.)=phosphorylated serine

Table 1. p53 family-related anticancer properties of resveratrol

Curcumin

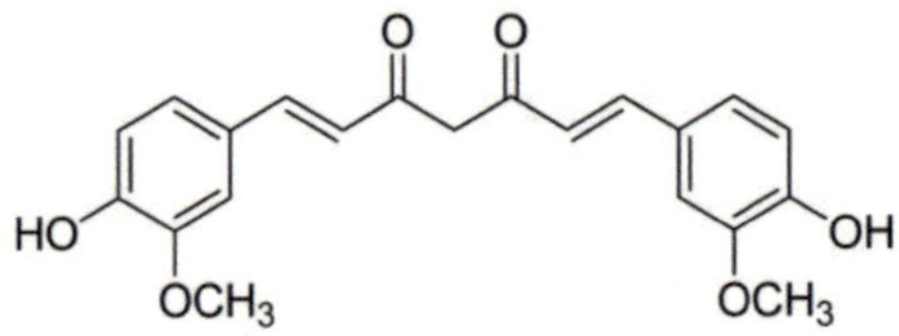

Cancer model	Described effects on TSG (p53/p73)	References
Breast cancer cells (MCF-7, p53-mutated MDA-MB-231, p53-mutated SkBr3)	- increased expression of p53-p(ser15), no change or increased expression of total p53 -decreased expression of mutated p53	[68, 178, 194-197]
Cervical cancer cells (p53-mutated C33A, Caski)	-increased expression of p53 and p73	[83, 198]
Ovarian cancer cells (HEY, OVCA429, p53-mutated OCC1, p53-null SKOV3, CaOV3, Ho-8910)	-p53-independent cell death	[106]
	-increased expression of p53-p(ser15) -increased expression of p53	[107, 199, 200]
Prostate cancer cells (LNCaP, p53-null PC3)	-increased expression of p53-p(ser15),p53-ac(lys) and total p53 protein -p53-independent cell death -p53 translocation to mitochondria	[95, 100, 201]
Bladder cancer cells (p53 mutated-T-24 and AY-27)	-no change or increased expression of p53	[202, 203]
Erhlich Ascite carcinoma cells	-increased expression of p53-ac(lys373) and total p53	[120]
Colorectal cancer cells (LoVo, HCT116, p53-null HCT116, p53-mutated HT29, p53-mutated Colo205)	-increased expression of total p53	[52, 204]
	-increased expression of p53-p(ser15), p53-p(ser33) and total p53	[64]
	-no change in p53 expression	[173]
	-increased expression of p53-p(ser15) -decreased or unchanged expression of mutated p53	[156, 205]

Curcumin

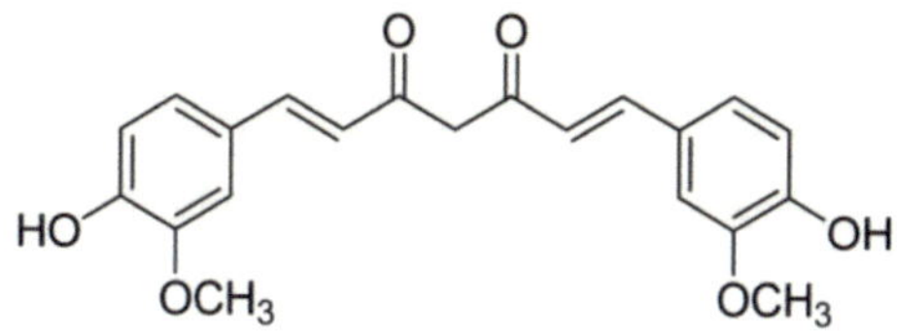

Cancer model	Described effects on TSG (p53/p73)	References
	-cell cycle arrest, senescence and autophagy independent of p53 expression -cytochrome c release independent of p53 expression -increased expression of total p53	[206, 207]
Colitis-associated colorectal cancer in mice	-no change in p53 expression	[17]
Acute lymphoblastic leukemia cells (B6p210, T315I)	-increased expression of total p53	[208]
Chondrosarcoma cells and xenograft in nude mice (JJ012)	-increased expression of total p53 in vitro and in vivo -p53-dependent apoptosis	[57]
Melanoma cells (MMRU, p53-mutated PMWK, B16BL6)	-no change in p53 expression	[209, 210]
Glioblastoma cells (C6, U-87MG, p53-mutated U138MG and U251, DBTRG, T98G, T67)	-p53-independent cell death -unchanged or increased expression of p53	[60, 71, 163, 211, 212]
Neuroblastoma cells (SK-N-AS, NUB-7, p53-mutated SK-N-BE(2))	-p53-independent cell death -nuclear translocation of p53	[65, 94]

Cancer cell lines express wild-type p53 except where otherwise stated; ac(lys.)=acetylated lysine, p(ser.)=phosphorylated serine

Table 2. p53 family-related anticancer properties of curcumin

Epigallocatechin-3-gallate (EGCG)

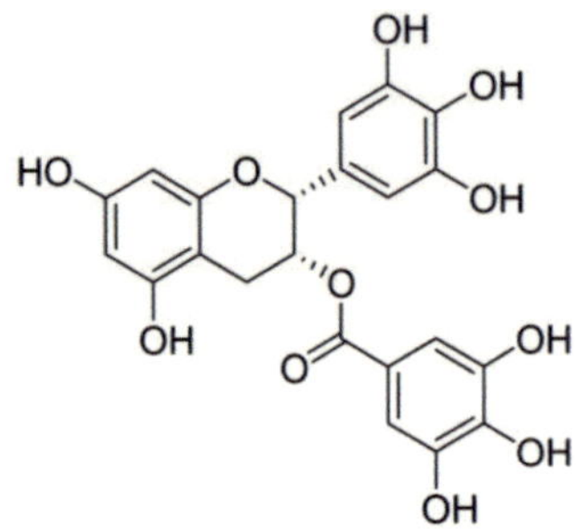

Cancer model	Described effects on TSG (p53/p73)	References
Breast Cancer cells (MCF7, p53-mutated MDA-MB-468)	-increased expression of p53-p(ser15) and total p53 -p53-independent cell death	[213, 214]
Prostate cancer cells (LNCaP, p53-null PC-3, p53-expressing PC-3, PC3-ML, p53-mutated DU-145)	-increased expression of p53-p(ser6), p53-p(ser15), p53-p(ser20), p53-p(ser37), p53-p(ser392) , p53-ac(lys373), p53-ac(lys382) and total p53 -p53-dependent and independent cell death -increased expression of p73	[53, 66, 89, 139, 140, 170, 215]
PC3-ML cells (prostate cancer) xenograft in mice	-increased expression of p53 and p73 (synergistic effect with paclitaxel and docetaxel)	[170]
Cervical cancer cells (Hela)	-increased expression of p53	[216]
Ovarian cancer cells (PA-1, p53-null SKOV3, p53-mutated OVCAR-3)	-p53-independent cell death	[217]
Hepatocellular carcinoma cells (HepG2, p53-null Hep3B)	-increased expression of p53 -p53-independent cytotoxicity	[218, 219]
Colorectal cancer cells (HCT116, p53-mutated HT-29)	-increased expression of p53	[55, 155, 157]
Head and neck squamous carcinoma cells (KB, Hep2, Tu686,	-increased expression of p53-p(ser15) and p53-p(ser37), -decreased expression of p53-p(ser6), p53-p(ser392) -unchanged or increased expression of p53	[101, 172, 220]

Epigallocatechin-3-gallate (EGCG)

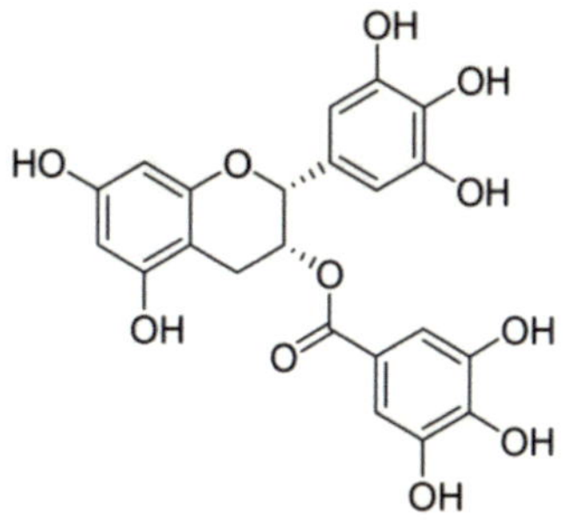

Cancer model	Described effects on TSG (p53/p73)	References
686NL, Tu212, Tu177, p53-null M4e)	-p53-dependent cytotoxicity	
Head and neck squamous cell carcinoma syngenic mouse model (SCC VII/SF cells xenograft)	-increased *in vivo* expression of p53-p(ser15)	[101]
Lung cancer cells (A549)	-increased expression of p53-p(ser15) and total p53 -absence of p73 expression -p53-dependent activation of caspases 3/7	[221]
Fibrosarcoma cells (HT-1080)	-increased expression of p53	[222]
Sarcoma xenograft (S180)	-increased *in vivo* expression of p53	[90]
Multiple myeloma cells (INA6)	-increased expression of p63 and 73, unchanged expression of p73	[75]
T lymphocyte leukemic cells (p53-mutated Jurkat, HuT-102, C91-PL, p53-mutated CEM)	-increased expression of p73 -increased expression of p53	[76, 223]

Cancer cell lines express wild-type p53 except where otherwise stated; ac(lys.)=acetylated lysine, p(ser.)=phosphorylated serine

Table 3. p53 family-related anticancer properties of EGCG

Polyphenolic source	Cancer model	p53- and p73-related effects	References
Grape-derived products (red wine, grape seed extract)	C26 colorectal cancer cells xenograft in mice	-increased expression of p53 and p73	[78]
	Human colorectal cancer cells (LoVo, p53-mutated HT29, P53-mutated SW480)	-p53-independent apoptosis	[224]
	Oral squamous carcinoma cells (SCC-25, p53-mutated OEC-MI)	-p53-independent cytotoxicity	[225]
	Leukemia cells (p53-mutated Jurkat)	-increased expression of p73	[80]
	Teratocarcinoma cells (P19)	-increased expression of p53	[174]
	Prostate cancer cells (LNCaP, p53-mutated DU145)	-increased expression of p53-p(ser15) and total p53	[226]
Black and green tea	DMBA-induced mammary tumor in rat	-increased expression of wild-type p53 and decreased expression of mutated-p53	[227]
	3,4-benzopyrene-induced lung carcinoma in rat	-increased expression of p53	[228]
	Ehrlich's ascites carcinoma cell xenograft in mice	-increased expression of p53	[229]
	Oral cells from smoker and non-smoker subjects	-increased expression of p53	[230]
	Patients with high-risk oral premalignant lesions	-no association between p53 expression and clinical response	[231]
	Prostate cancer cells (LNCaP, p53-null PC3)	-increased expression of p53-ac(lys373), p53-ac(lys382) and total p53 -p53-independent apoptosis	[139, 140]
	Colorectal cancer cells (LoVo, p53-mutated HT29)	-p53-independent cytotoxicity	[232]
Berry-derived products (aronia juice, strawberry extract)	Leukemia cells (p53-mutated Jurkat)	-increased expression of p73	[79]
	Breast cancer cells (p53-mutated T47D)	-increased expression of p73	[81]

Cancer cell lines express wild-type p53 except where otherwise stated; ac(lys.)=acetylated lysine, p(ser.)=phosphorylated serine

Table 4. p53 family-related anticancer properties of polyphenolic sources

4. Polyphenols as regulator of p53 expression and localization

Under physiological conditions, the transcriptional activity of p53 is downregulated by three different ways: i) ubiquitin-mediated proteasomal degradation mainly through the action of mouse double minute protein (MDM2), ii) nuclear export leading to a decrease in nuclear level, or iii) transcriptional repression of chromatin. MDM2 is an ubiquitin E3 ligase considered as an oncoprotein because of its activity in promoting p53 ubiquitination and proteasomal degradation. Moreover MDM2 binds to the NH2 terminus of p53 and blocks its transactivational activities [27]. Interestingly, MDM2 promotes also cell cycle progression independently of p53 for instance by modulating the activity of p21 [85]. Then MDM2 itself represents a potential target for new drug with chemotherapeutic properties including polyphenolic compounds [86]. Indeed, curcumin has been identified as an inhibitor of MDM2 expression (Figure 1) *in vitro* and *in vivo* in p53-null and p53-wild-type human prostate cancer cells and this inhibitory effect seems to be related to the inhibition of the PI3K/mTOR pathway [87]. In addition, the curcumin analog EF24, which displays higher potency, increases phosphorylation of p53 together with downregulation of MDM2, which likely leads to p53 overexpression and cytotoxicity in hepatocellular carcinoma cells [88]. Similarly, EGCG reduces MDM2 expression in prostate cancer cells [89], but not in sarcoma cells [90]. Data concerning the effect of resveratrol on MDM2 expression are more controversial since upregulation or downregulation have been observed in different cancer models [91, 92].

As mentioned previously, p53 activity depends upon its expression level but also its subcellular localization. Indeed, p53 displays direct pro-apoptotic effects related to mitochondrial translocation and this pathway works in synergy with transcriptional activation function of p53 dependent upon its nuclear translocation [24, 93]. Therefore, the control of p53 subcellular localization might interfere with p53-mediated cell death. For instance, treatment of neuroblasma cells by either curcumin or resveratrol transiently upregulated p53 expression and induced nuclear translocation of p53, followed by induction of p21 and Bax expression associated with apoptosis [94]. In addition curcumin increases p53 and Bax expression in mitochondrial fraction under the control of the PI3K/Akt pathway in prostate cancer cells followed by caspase-dependent apoptosis [95]. Altogether, these data indicate that polyphenols are able to control not only p53 expression, but also its localization and therefore its pro-apoptotic activity in cancer cells. However, other post-translational regulatory effects of polyphenols have been also described and related to phosphorylation and acetylation of the tumor suppressor.

5. Polyphenols as regulator of p53 phosphorylation

Phosphorylation of serine/threonine residues are essential for stabilization and activation of p53, the most extensively studied being serine 15 (Ser 15). These phosphorylation sites are mainly concentrated in the N-terminal transactivation domain and in the C-terminal regulatory domain [96]. Recent data about p53 phosphorylation induced by resveratrol,

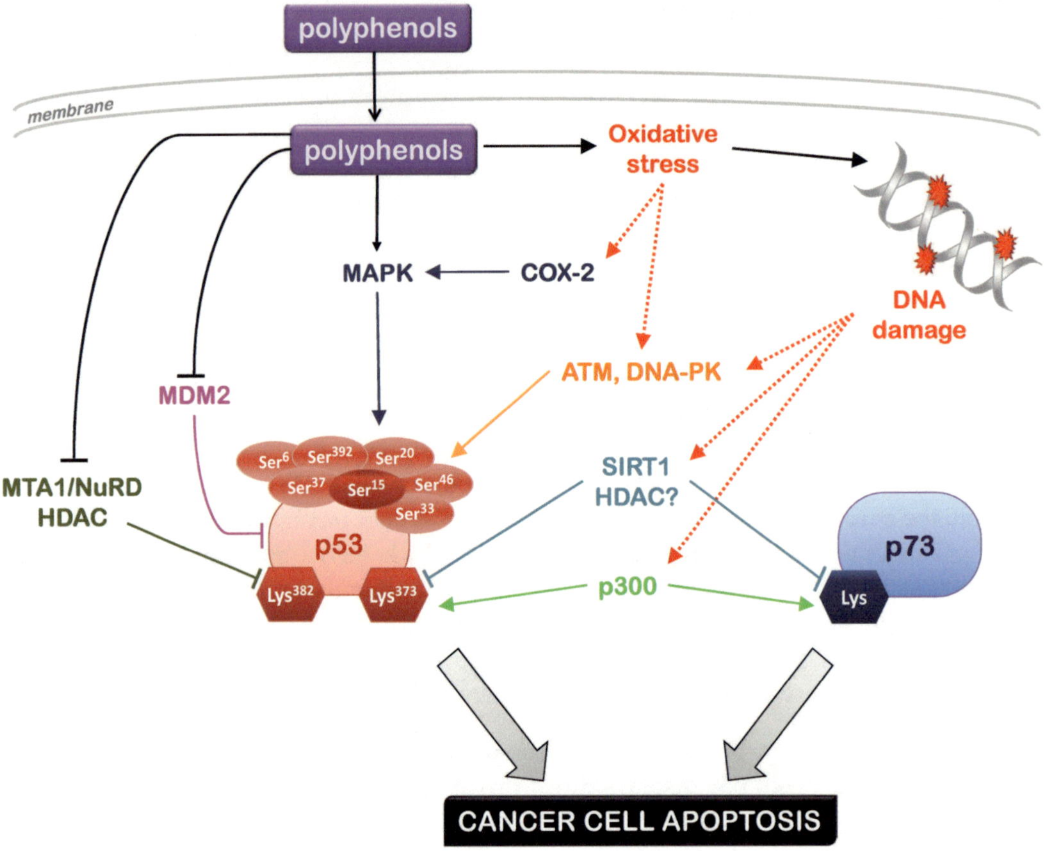

Figure 1. Overview of p53- or p73-mediated pro-apoptotic effects of polyphenols in cancer cells. Polyphenols likely induce intracellular oxidative stress and DNA damage with subsequent activation of kinases (MAPK, ATM, DNA-PK) responsible for p53 phosphorylation. Simultaneously, and also in response to DNA damage, acetylation of p53 or p73 have been described due to enhanced acetylase activity from p300 and/or to reduced deacetylase activity from SIRT1 or HDAC. In addition, p53 expression has been shown to be under the control of MDM2 as well as MTA1/NuRD, both factors being downregulated by polyphenols. Phosphorylation and acetylation, together with MDM2 inhibition result in p53, and to a less extend, p73 stabilization and sustained expression which activate cell death cascade in cancer cells.

curcumin or EGCG, *in vitro* and *in vivo*, are summarized in Tables 1, 2 and 3 as well as in Figure 1. The DNA damage is one of the main signals relayed to p53 subsequently activated by phosphorylation at serine residues that are the target of ataxia-telangiectasia mutated (ATM), ataxia telangiectasia and Rad3 related (ATR) and DNA-dependent protein kinase (DNA-PK) [97]. The DNA damage response could be activated by chemotherapeutic drugs, UV or oxidative stress [98, 99], but activation of this pathway by polyphenols remains controversial. Watson et al. investigated the pro-apoptotic effect of curcumin which is similar in p53+/+ (wild-type) and p53-/- (knockout) HCT116 colorectal cancer cells. Moreover, they demonstrated the ability of this polyphenol to induce up regulation of p53-p(Ser 15) and total p53 without any change in the expression level of ATM, ATR or DNA-PK. In contrast, curcumin enhances p38, JNK and ERK1/2 phosphorylation in both p53+/+ and p53-/- HCT116 cell lines; this suggests that the cytotoxic effects of curcumin are independent of the DNA-damage/ATM/ATR/DNA-PK pathway but associated with Mitogen-Activated Protein Kinases (MAPKs) activities [64]. On the other hand, treatment of LNCaP prostate cancer cells or HCT116 colorectal cancer cells with curcumin induces the phosphorylation of ATM, histone H2AX (a marker of DNA damage) and p53 at Ser 15 together with increased expression of p53, suggesting p53 activation through the DNA damage/ATM pathway [52, 100]. The importance of ATM in polyphenols-induced cytotoxicity is also supported by recent data showing that EGCG lose the ability to trigger p53 phosphorylation at Ser 15 in absence of ATM [101]. In addition, genistein induced p53 phosphorylation at Ser 6, 9, 15, 20, 46 and 392 in the ATM-proficient human lymphoblastic cell lines, but not in ATM-deficient cell lines, indicating a key role of ATM kinase activity for polyphenol-induced p53 activation [102]. Moreover, stimulation of the ATM/p53 pathway by polyphenols like resveratrol has been shown to also participate in senescence of cancer cells [103]. On the other hand, quercetin strongly induced DNA-PK expression, p53 phosphorylation and apoptosis in melanoma cells, suggesting that other kinases might be activated by polyphenols [77].

Alternatively, MAPKs such as ERK1/2, p38 or JNK have been involved in p53 activation and phosphorylation [104, 105]. Therefore, the potential MAPKs/p53-dependent activation of apoptosis by polyphenols has been investigated. As previously mentioned, curcumin induces phospho-p38, phospho-ERK1/2 and phospho-JNK in colorectal as well as ovarian cancer cells [64, 106, 107]. In addition, it has been shown that resveratrol- or luteolin-induced apoptosis depends on the activities of ERK1/2, JNK and p38 kinase which target p53 phosphorylation at Ser 15 [49, 108]. An alternative pathway which implicates cyclooxygenase (COX)-2 activity and expression has been decribed by Lin et al. They have shown that resveratrol-induced apoptosis of human head and neck squamous cancer cells or human ovarian carcinoma cells is associated with p53 phosphorylation at Ser 15 and that both processes are downregulated by pERK1/2 and COX-2 specific inhibitors [109, 110]. Recent investigations indicate that ERK and p53 regulate each other and that ATM controls their interaction [104]. Therefore polyphenols might likely trigger p53 activation through ATM and MAPKs complementary pathways.

6. Polyphenols as regulators of p53 and p73 acetylation

Functions of p53 and p73 are also regulated by acetylation on different lysine (Lys) residues. These posttranslational covalent modifications occur in response to DNA damage in the close vicinity of the oligomerization domain. The main Histone Acetyl Transferases (HATs) responsible for these modifications include p300, CREB-Binding Protein (CBP), P300/CBP-Associated Factor (PCAF) and Tat-Interactive Protein of 60 kDa (TIP60) [38]. As a consequence of its acetylation, p53 is stabilized by excluding ubiquitination on the same site and acetylation also promotes p53 transcriptional activity [96]. In comparison to p53, only few data are available concerning p73 interaction with HAT and acetylation. It has been established that p300 can acetylate p73 in response to DNA damage, but p300 can also behave as a co-activator of p73 independently of its HAT activity [111]. Importantly, the level of p53 or p73 acetylation seems to be a major way of regulation for the tumor suppressors function since deacetylated p53 and p73 are compromised in their ability to induce cell cycle arrest and apoptosis [91, 112].

The transcriptional coactivator p300 is a large multidomain protein that possesses histone acetyl-transferase ability [113]. Together with its homolog CBP, p300 mediate transcription through binding to transcriptional activators such as JUN, E1A, NF-κB, as well as to the p53 family members and they have been involved in human diseases including cancers [114]. Recent studies indicate that the transcriptional activity of p53 and p73 in response to genotoxic stress is regulated by its interaction with p300 [115-117]. Indeed, it has been established that interaction between p73 and p300 acetyl-transferase promotes first p73 stability and then its transcriptional activity.

Narayanan et al., have suggested that resveratrol-induced apoptosis of prostate cancer cells is mediated by transcriptional activation of p300 which subsequently acetylates and stabilizes p53 [118]. Similarly in breast cancer cells, resveratrol enhanced p300 expression and interaction with the phosphorylated form of p53 by a MAPK-dependent mechanism [119]. Interestingly, p53-p300 interaction fails to occur in doxorubicin-resistant cells, but curcumin pre-treatment could restore this interaction. Consistently, curcumin also restored drug-induced p53 acetylation (lysine 373) and p53-dependent transcription of Bax, PUMA, and Noxa in resistant cells leading to their apoptosis [120]. Therefore, polyphenols-induced acetylation of p53 by p300 might represent a key molecular mechanism for the cytotoxic properties of these natural compounds in cancer cells including chemoresistant cells (Figure 1).

Acetylation level of tumor suppressors is dependent upon the balance between acetylation and deacetylation reactions. Indeed, deacetylation of p53 or p73 by SIRT1 (silent information regulator 1), a member of the sirtuin Histone DeACetylase (HDAC) class III family, prevents p53-mediated transactivation of cell cycle inhibitor p21 and pro-apoptotic factor Bax, allowing promotion of cell survival after DNA damage and ultimately tumorigenesis [121]. Members of the Silent Information Regulator family (SIRT or sirtuins) are evolutionary conserved NAD-dependent protein deacetylases and adenosine diphosphate (ADP)-ribosylases. There are seven identified isoforms (SIRT1-7) that differ in their subcellular localization (cytoplasmic, mitochondrial or nuclear), substrate specificities and functions [122]. The founding member of this class of deacetylases, SIRT1 (homolog of yeast silent information regulator, Sir2), is the most widely studied sirtuins. SIRT1 has been associated with aging processes as well as a variety of human diseases such as metabolic syndrome, inflammation, neurodegeneration and more recently cancer [123, 124]. SIRT1 can deacetylate a variety of histones as well as a number of non-histone substrates, the first identified of these non-histone substrates being p53 (Lys 382-p53). The SIRT1 activity on p53 results in repression of p53-dependent apoptosis in response to DNA damage and oxidative stress [125, 126]. SIRT1 deacetylates also other tumor suppressors such as p73 [91]. Then SIRT1 has been considered as an oncogenic protein because of its role in inactivating tumor suppressors such as p53, p73 but also PTEN [127], and/or activating other oncogenic proteins like N-Myc [128]. Nevertheless, the oncogenic potential of SIRT1 has been controversial and, depending on the context, SIRT1 might also act as a tumor suppressor [122]. However, inhibition of the oncogenic potential of SIRT1 is likely able to induce apoptosis by counteracting the deacetylation of p53 or p73 and other key factors such as FOXO3a [91, 125, 129].

In 2003, resveratrol was the top hit in a screen designed to identify activators of sirtuin enzymes [130] and was subsequently shown to extend lifespan in yeast. However following experiments led to confusing data suggesting that resveratrol might not be a direct activator of SIRT1 [131]. Regardless of the controversy about its mode of action, resveratrol has been confirmed to have numerous health benefits, including anticancer properties. Nevertheless, the role of SIRT1 in the anti-proliferative and pro-apoptotic effects of resveratrol on cancer cells is still unclear. Indeed, in neuroblastoma cells, resveratrol-induced apoptosis was associated with a reduced expression of SIRT1 as well as up-regulation of the acetylated and active form of p53, but the pre-treatment of cancer cells with SIRT1 enzymatic inhibitors such as sirtinol or nicotinamide has no cytotoxic effect suggesting that resveratrol-induced apoptosis is independent of SIRT1 activity [132]. In the opposite, siRNA-mediated downregulation of SIRT1 in lymphoma cells decreased the resveratrol-induced apoptosis, indicating in this case a critical role of SIRT1 in polyphenol-mediated cancer cell death [133]. Interestingly, Frazzi et al., have recently described anti-proliferative effect of resveratrol associated with downregulation of SIRT1 expression and activity together with upregulation of acetylated-Lys 373-p53, the active form of p53, and total p53 overexpression [129]. All together, these data suggest that, in the context of cancer cells, resveratrol might be an inhibitor, instead of an activator, of SIRT1 functions (Figure 1). However, only few data are available concerning potential regulation of SIRT1/p53 pathway by other polyphenolic compounds [134]. Therefore additional investigations are needed to further understand the role of SIRT1 in polyphenols-mediated anticancer effects.

On the other hand, resveratrol has been shown to enhance p53 acetylation and apoptosis in prostate cancer cells through alternative pathways. Indeed, resveratrol caused down-regulation of MTA1 protein, leading to destabilization of MTA1/NuRD complex thus allowing acetylation/activation of p53 [135]. Metastasis-associated protein 1 (MTA1) is part of the nucleosome remodelling deacetylation (NuRD) complex involved in global and gene-specific histone deacetylation, alteration of chromatin structure and transcriptional repression [136, 137]. This complex, which also contains Histone DeACetylase (HDAC)1 and HDAC2, plays an essential role in governing deacetylation of histones but also non histone proteins, such as p53 [138]. In addition, green tea polyphenols have been shown to behave as HDAC class I inhibitors which results in p21 and Bax expression irrespective of p53 status in prostate cancer cells [139, 140]. Moreover HDAC inhibition by EGCG is associated with p53 acetylation in p53-wild-type LNCAP prostate cancer cells suggesting an increase of p53 halftime and binding to p21 and Bax promoters as previously described [141]. The mechanism by which HDAC inhibition could induce apoptosis in absence of functional p53 in p53-null PC3 prostate cancer cells might be related to interaction with p73 pathway as previously suggested [142, 143] or direct regulation of p21 promoter activation [144]. Similar HDAC inhibition by curcumin has been also seen in prostate cancer cells [145]. However the exact role of HDAC inhibition in polyphenol-induced apoptosis of cancer cells remains to be elucidated especially *in vivo*.

7. Role of oxidative stress and DNA damage in p53/p73 regulation by polyphenols

Polyphenolic compounds have been extensively described as anti-oxidant molecules with the capability to scavenge reactive oxygen species (ROS), which include radical oxygen and nitrogen species such as $O2^-$ (superoxide anion), HO· (hydroxyl radical), NO· (nitric oxide radical), $ONOO^{\cdot-}$ (peroxinitrite anion) and H_2O_2 (hydrogen peroxide), as well as oxidatively generated free radicals RO· and ROO· from biomolecules like lipids, proteins or nucleic acids (DNA and RNA) [3, 146]. Polyphenols are not only able to quench the ROS but also to regulate directly the oxidative stress-mediated enzyme activity, therefore reducing the formation of ROS. These anti-oxidant properties have been linked to the polyphenol-mediated reduction of chronic disease risk including cancer chemoprevention [13, 147]. Indeed, redox changes are often reported as important inducer of neoplastic transformation as well as chemoresistance. Cerutti et al. identified for the first time in 1985 the close relationship between pro-oxidant conditions and cancer development [148]. More than twenty years later, accumulated evidences indicate that the non-physiological alterations of the intracellular redox state could be considered as a hallmark of tumor biology. Indeed, redox changes have been involved in several key events of carcinogenesis such as self-sufficiency in growth signals [149, 150], resistance to apoptosis [151, 152], sustained angiogenesis [153, 154], autophagy and invasiveness. However, recent findings also suggest that this redox changes might be exploited as therapeutic strategy to selectively kill tumor cells.

Recently and unexpectedly, it has been established that various and structurally different (flavonoids or non-flavonoids) polyphenols are able to induce ROS (mainly superoxide anions or hydrogen peroxide) formation in cancer cells and for some of them to activate the DNA-damage response pathway [51, 79, 80, 95, 102, 103, 155-158]. Heiss et al. have also shown that the resveratrol-induced senescence in colon cancer cells is dependent upon an increased formation of ROS and the subsequent phosphorylation of p53 on the Ser15, suggesting a relationship between polyphenol-induced oxidative stress and p53 activation [103]. On the other hand, curcumin and wogonin induce ROS production and cause cytotoxicity in p53+/+ and p53-/- cancer cells [56, 64], indicating that ROS formation is an event independent of p53 and might be an earlier step in the cell death pathway. This hypothesis is supported by the study showing ROS in cancer cells as earlier as 20 minutes after the beginning of wogonin treatment. Moreover, in the same study, the subsequent up-regulation of p53 (maximal activation at 16 hours) is significantly inhibited by anti-oxidants such as N-Acetyl-Cysteine. Importantly, most of the studies did not investigate the possible alternative role of p73 in p53-mutated cells, but we and others have shown that in p53-mutated or p53-deficient cells, a polyphenolic compound (EGCG) or source (red wine polyphenolic extract, polyphenol-rich aronia juice) strongly induced oxidative stress-mediated up-regulation of p73 and apoptosis [79, 80, 84]. Further reports also supported that p53-family tumor suppressors regulation might be related to oxidative stress [159].

Interestingly, the cytotoxic effects induced by curcumin or its analogue HO-3867 were reduced in non-cancerous cells as well as the ROS formation in comparison to human ovarian cancer cells. This suggests that the specific pro-oxidant activity of polyphenols in cancer cells might explain the selective anticancer properties of these compounds, sparing healthy normal cells [107, 160]. Similarly, EGCG increased preferentially ROS formation, p53 and p21 expression and cytotoxicity in colorectal cancer cells but not in human embryonic kidney cells and normal human lung cell line [157].

Oxidative stress is one of the major conditions that damages DNA, acting as a mediator of environmental stressors such as UV- and X-rays irradiation, drugs, and of metabolic imbalance [161]. Since p53 might be regulated by the redox environment [162], especially by the ROS/DNA damage pathway, it has been proposed that polyphenol-mediated anticancer effects are related to a ROS/DNA damage/p53 pathway (Figure 1). Indeed polyphenol-induced DNA damage and apoptosis have been demonstrated with various compounds such as curcumin in glioblastoma and prostate cancer cells [100, 163], resveratrol in prostate cancer cells [164, 165], EGCG in lung cancer cells and xenograft in mice [166], wogonin in glioblastoma and prostate cancer cells [51, 56], and luteolin in lung and head and neck cancer cells [58]. Therefore the current molecular mechanism of the anticancer properties of polyphenols might involve selective ROS formation together with DNA damage in cancer cells. Thus, this process might lead to the regulation of several pathways (ATM/DNA-PK, MAPKs, p300, SIRT1, HDAC, see Figure 1), and ultimately to the expression and stabilization of p53-family tumor suppressors triggering programmed cell death.

8. Therapeutic perspectives

Recent investigations have demonstrated additional or synergistic effects when polyphenols are combined with chemo- or radiotherapy. Indeed, resveratrol induces synergistic apoptosis with 5-fluorouracile [167]. Similar observations have been made with curcumin associated with doxorubicin, cisplatin, gemcitabine or radiation for cell death induction of glioblastoma cells and prostate cancer cells [60, 87]. More importantly, curcumin and its analogue, HO-3867 sensitized doxorubicin-resistant ascite carcinoma cells and breast cancer cells as well as cisplatin-resistant ovarian carcinoma cells together with enhanced p53 expression [107, 120, 168, 169]. Similarly, EGCG displayed synergistic upregulation of p53 and p73 as well as anticancer properties with taxanes (paclitaxel and docetaxel) *in vitro* but also *in vivo* in prostate cancer models [170, 171]. Because all of the previously mentioned drugs demonstrated the ability to induce DNA damage, it is likely that polyphenols might amplify these damages leading therefore to synergistic effects. Surprisingly, EGCG has also synergistic effects with targeted therapy such as erlotinib (inhibitor of epidermal growth factor receptor) to induce p53 phosphorylation on Ser15 and expression together with apoptosis [172].

Interestingly, curcumin also ameliorated oxaliplatin-induced chemoresistance in colorectal cancer cells without significant effect on p53 expression [173]. Similarly, curcumin and EGCG sensitized glioma cells *in vitro* and *in vivo* to chemotherapeutic drugs and also to radiation in a p53-independent manner [163]. These data suggest that polyphenols can effectively circumvent resistance of cancer cells to chemotherapy, but likely through a p53-independent pathway.

Numerous *in vitro* studies have demonstrated the cytotoxic effects of polyphenols by using micromolar concentrations which are much higher than current chemotherapeutic drugs under development. However, polyphenols still keep their potential as chemotherapeutic drug, firstly because of their activity on chemo- or radioresistant cancer cells and secondly because of their very low toxicity on healthy tissues giving them a large therapeutic index. Indeed, many recent *in vitro* studies have highlighted the selective pro-apoptotic properties of polyphenols or analogues with no or low cytotoxic effect on non-cancer healthy cells, such as endothelial cells, cardiomyocytes, lymphocytes, chondrocytes, ovarian cells, prostate and mammary epithelial cells, astrocytes, or neurons [54, 57, 60, 79, 95, 107, 158, 163, 169, 174]. Interestingly, the selective pro-apoptotic effect of curcumin in breast cancer cells is associated with an increased expression of p53, whereas p53 is only slightly upregulated in normal mammary epithelial cells, suggesting a selective activation of p53 pathway in cancer cells sparing normal cells [54]. Moreover, *in vivo* treatment with polyphenolic compounds or products in tumor model such as cancer cells xenografts induced a significant inhibition of tumor growth together with a very good tolerance for healthy tissues, including heart, liver, kidney, lung and haematopoietic tissue [60, 78]. However, more animal studies and human clinical trials are now necessary to clearly determine whether polyphenols or their natural nutritional sources are safe and efficient to treat cancer.

9. Concluding remarks

The present literature review has summarized the results of recent studies focusing mainly on the p53-related anticancer properties of three major polyphenolic compounds (resveratrol, curcumin and EGCG). Despite highly active research in this area, the data are still controversial concerning the possible key role of the tumor suppressor p53 in the polyphenol-mediated apoptosis of tumor cells. However, according to the emerging evidences suggesting that polyphenols might alternatively regulate also the structurally- and functionally-related tumor suppressors such as p73, these natural compounds might be considered as general executors of the p53 family-mediated programmed cell death in cancer cells. Importantly, the selective anticancer properties of polyphenols are maintained even when p53 is mutated or absent, as well as when cells are resistant to current therapies. However, further investigations are still mandatory to better understand the underlying molecular mechanism *in vitro* as well as *in vivo* before a potential clinical development.

Author details

Nelly Etienne-Selloum*, Israa Dandache, Tanveer Sharif, Cyril Auger and Valérie B. Schini-Kerth

UMR CNRS 7213, Laboratoire de Biophotonique et Pharmacologie, Université de Strasbourg, Faculté de Pharmacie, Illkirch, France

References

[1] Es-Safi N. Plant Polyphenols: Extraction, Structural Characterization, Hemisynthesis and Antioxidant Properties. In: Rao DV, editor. Phytochemicals as Nutraceuticals - Global Approaches to Their Role in Nutrition and Health: INTECH; 2012. p. 181-206.

[2] Cheynier V. Polyphenols in foods are more complex than often thought. The American journal of clinical nutrition. 2005;81(1 Suppl):223S-9S.

[3] Quideau S, Deffieux D, Douat-Casassus C, Pouysegu L. Plant polyphenols: chemical properties, biological activities, and synthesis. Angewandte Chemie. 2011;50(3): 586-621.

[4] Neveu V, Perez-Jimenez J, Vos F, Crespy V, du Chaffaut L, Mennen L, et al. Phenol-Explorer: an online comprehensive database on polyphenol contents in foods. Database (Oxford). 2010;doi: 10.1093/database/bap024. http://www.phenol-explorer.eu/ (accessed 10 January 2013).

[5] Kris-Etherton PM, Hecker KD, Bonanome A, Coval SM, Binkoski AE, Hilpert KF, et al. Bioactive compounds in foods: their role in the prevention of cardiovascular disease and cancer. The American journal of medicine. 2002;113 Suppl 9B:71S-88S.

[6] Andriantsitohaina R, Auger C, Chataigneau T, Etienne-Selloum N, Li H, Martinez MC, et al. Molecular mechanisms of the cardiovascular protective effects of polyphenols. The British journal of nutrition. 2012:1-18.

[7] Schini-Kerth VB, Etienne-Selloum N, Chataigneau T, Auger C. Vascular protection by natural product-derived polyphenols: in vitro and in vivo evidence. Planta medica. 2011;77(11):1161-7.

[8] Di Castelnuovo A, di Giuseppe R, Iacoviello L, de Gaetano G. Consumption of cocoa, tea and coffee and risk of cardiovascular disease. European journal of internal medicine. 2012;23(1):15-25.

[9] Dauchet L, Amouyel P, Hercberg S, Dallongeville J. Fruit and vegetable consumption and risk of coronary heart disease: a meta-analysis of cohort studies. The Journal of nutrition. 2006;136(10):2588-93.

[10] Sofi F, Abbate R, Gensini GF, Casini A. Accruing evidence on benefits of adherence to the Mediterranean diet on health: an updated systematic review and meta-analysis. The American journal of clinical nutrition. 2010;92(5):1189-96.

[11] Aune D, Lau R, Chan DS, Vieira R, Greenwood DC, Kampman E, et al. Nonlinear reduction in risk for colorectal cancer by fruit and vegetable intake based on meta-analysis of prospective studies. Gastroenterology. 2011;141(1):106-18.

[12] Riboli E, Norat T. Epidemiologic evidence of the protective effect of fruit and vegetables on cancer risk. The American journal of clinical nutrition. 2003;78(3 Suppl): 559S-69S.

[13] Ramos S. Cancer chemoprevention and chemotherapy: dietary polyphenols and signalling pathways. Molecular nutrition & food research. 2008;52(5):507-26.

[14] Surh YJ. Cancer chemoprevention with dietary phytochemicals. Nature reviews Cancer. 2003;3(10):768-80.

[15] Dai Q, Borenstein AR, Wu Y, Jackson JC, Larson EB. Fruit and vegetable juices and Alzheimer's disease: the Kame Project. The American journal of medicine. 2006;119(9):751-9.

[16] Bastianetto S, Krantic S, Chabot JG, Quirion R. Possible involvement of programmed cell death pathways in the neuroprotective action of polyphenols. Current Alzheimer research. 2011;8(5):445-51.

[17] Villegas I, Sanchez-Fidalgo S, de la Lastra CA. Chemopreventive effect of dietary curcumin on inflammation-induced colorectal carcinogenesis in mice. Molecular nutrition & food research. 2011;55(2):259-67.

[18] Roy P, Kalra N, Prasad S, George J, Shukla Y. Chemopreventive potential of resveratrol in mouse skin tumors through regulation of mitochondrial and PI3K/AKT signaling pathways. Pharm Res. 2009;26(1):211-7.

[19] Yang CS, Wang H. Mechanistic issues concerning cancer prevention by tea catechins. Molecular nutrition & food research. 2011;55(6):819-31.

[20] Linzer DI, Levine AJ. Characterization of a 54K dalton cellular SV40 tumor antigen present in SV40-transformed cells and uninfected embryonal carcinoma cells. Cell. 1979;17(1):43-52.

[21] DeLeo AB, Jay G, Appella E, Dubois GC, Law LW, Old LJ. Detection of a transformation-related antigen in chemically induced sarcomas and other transformed cells of the mouse. Proceedings of the National Academy of Sciences of the United States of America. 1979;76(5):2420-4.

[22] Lane DP, Crawford LV. T antigen is bound to a host protein in SV40-transformed cells. Nature. 1979;278(5701):261-3.

[23] Laptenko O, Prives C. Transcriptional regulation by p53: one protein, many possibilities. Cell death and differentiation. 2006;13(6):951-61.

[24] Speidel D. Transcription-independent p53 apoptosis: an alternative route to death. Trends Cell Biol. 2010;20(1):14-24.

[25] Lane DP. Cancer. p53, guardian of the genome. Nature. 1992;358(6381):15-6.

[26] Horn HF, Vousden KH. Coping with stress: multiple ways to activate p53. Oncogene. 2007;26(9):1306-16.

[27] Haupt Y, Maya R, Kazaz A, Oren M. Mdm2 promotes the rapid degradation of p53. Nature. 1997;387(6630):296-9.

[28] Soussi T, Wiman KG. Shaping genetic alterations in human cancer: the p53 mutation paradigm. Cancer cell. 2007;12(4):303-12.

[29] Edlund K, Larsson O, Ameur A, Bunikis I, Gyllensten U, Leroy B, et al. Data-driven unbiased curation of the TP53 tumor suppressor gene mutation database and validation by ultradeep sequencing of human tumors. Proceedings of the National Academy of Sciences of the United States of America. 2012;109(24):9551-6.

[30] Soussi T, Ishioka C, Claustres M, Beroud C. Locus-specific mutation databases: pitfalls and good practice based on the p53 experience. Nature reviews Cancer. 2006;6(1):83-90.

[31] Zhang XP, Liu F, Wang W. Coordination between cell cycle progression and cell fate decision by the p53 and E2F1 pathways in response to DNA damage. The Journal of biological chemistry. 2010;285(41):31571-80.

[32] Avery-Kiejda KA, Bowden NA, Croft AJ, Scurr LL, Kairupan CF, Ashton KA, et al. P53 in human melanoma fails to regulate target genes associated with apoptosis and the cell cycle and may contribute to proliferation. BMC cancer. 2011;11:203.

[33] Kuribayashi K, El-Deiry WS. Regulation of programmed cell death by the p53 pathway. Advances in experimental medicine and biology. 2008;615:201-21.

[34] Brooks CL, Gu W. p53 ubiquitination: Mdm2 and beyond. Molecular cell. 2006;21(3):307-15.

[35] Wade M, Wahl GM. Targeting Mdm2 and Mdmx in cancer therapy: better living through medicinal chemistry? Mol Cancer Res. 2009;7(1):1-11.

[36] Kaghad M, Bonnet H, Yang A, Creancier L, Biscan JC, Valent A, et al. Monoallelically expressed gene related to p53 at 1p36, a region frequently deleted in neuroblastoma and other human cancers. Cell. 1997;90(4):809-19.

[37] Trink B, Okami K, Wu L, Sriuranpong V, Jen J, Sidransky D. A new human p53 homologue. Nat Med. 1998;4(7):747-8.

[38] Collavin L, Lunardi A, Del Sal G. p53-family proteins and their regulators: hubs and spokes in tumor suppression. Cell death and differentiation. 2010;17(6):901-11.

[39] Mills AA, Zheng B, Wang XJ, Vogel H, Roop DR, Bradley A. p63 is a p53 homologue required for limb and epidermal morphogenesis. Nature. 1999;398(6729):708-13.

[40] Yang A, Walker N, Bronson R, Kaghad M, Oosterwegel M, Bonnin J, et al. p73-deficient mice have neurological, pheromonal and inflammatory defects but lack spontaneous tumours. Nature. 2000;404(6773):99-103.

[41] Donehower LA, Harvey M, Slagle BL, McArthur MJ, Montgomery CA, Jr., Butel JS, et al. Mice deficient for p53 are developmentally normal but susceptible to spontaneous tumours. Nature. 1992;356(6366):215-21.

[42] Guo X, Keyes WM, Papazoglu C, Zuber J, Li W, Lowe SW, et al. TAp63 induces senescence and suppresses tumorigenesis in vivo. Nature cell biology. 2009;11(12): 1451-7.

[43] Grande L, Bretones G, Rosa-Garrido M, Garrido-Martin EM, Hernandez T, Fraile S, et al. Transcription factors Sp1 and p73 control the expression of the proapoptotic protein NOXA in the response of testicular embryonal carcinoma cells to cisplatin. The Journal of biological chemistry. 2012;287(32):26495-505.

[44] Leong CO, Vidnovic N, DeYoung MP, Sgroi D, Ellisen LW. The p63/p73 network mediates chemosensitivity to cisplatin in a biologically defined subset of primary breast cancers. J Clin Invest. 2007;117(5):1370-80.

[45] Lin YL, Sengupta S, Gurdziel K, Bell GW, Jacks T, Flores ER. p63 and p73 transcriptionally regulate genes involved in DNA repair. PLoS genetics. 2009;5(10):e1000680.

[46] Deyoung MP, Ellisen LW. p63 and p73 in human cancer: defining the network. Oncogene. 2007;26(36):5169-83.

[47] Zaika AI, Slade N, Erster SH, Sansome C, Joseph TW, Pearl M, et al. DeltaNp73, a dominant-negative inhibitor of wild-type p53 and TAp73, is up-regulated in human tumors. J Exp Med. 2002;196(6):765-80.

[48] Tanigawa S, Fujii M, Hou DX. Stabilization of p53 is involved in quercetin-induced cell cycle arrest and apoptosis in HepG2 cells. Bioscience, biotechnology, and biochemistry. 2008;72(3):797-804.

[49] Shi R, Huang Q, Zhu X, Ong YB, Zhao B, Lu J, et al. Luteolin sensitizes the anticancer effect of cisplatin via c-Jun NH2-terminal kinase-mediated p53 phosphorylation and stabilization. Molecular cancer therapeutics. 2007;6(4):1338-47.

[50] Sakamoto T, Horiguchi H, Oguma E, Kayama F. Effects of diverse dietary phytoestrogens on cell growth, cell cycle and apoptosis in estrogen-receptor-positive breast cancer cells. J Nutr Biochem. 2010;21(9):856-64.

[51] Lee DH, Lee TH, Jung CH, Kim YH. Wogonin induces apoptosis by activating the AMPK and p53 signaling pathways in human glioblastoma cells. Cell Signal. 2012;24(11):2216-25.

[52] Basile V, Ferrari E, Lazzari S, Belluti S, Pignedoli F, Imbriano C. Curcumin derivatives: molecular basis of their anti-cancer activity. Biochemical pharmacology. 2009;78(10):1305-15.

[53] Hastak K, Agarwal MK, Mukhtar H, Agarwal ML. Ablation of either p21 or Bax prevents p53-dependent apoptosis induced by green tea polyphenol epigallocatechin-3-gallate. FASEB journal : official publication of the Federation of American Societies for Experimental Biology. 2005;19(7):789-91.

[54] Choudhuri T, Pal S, Das T, Sa G. Curcumin selectively induces apoptosis in deregulated cyclin D1-expressed cells at G2 phase of cell cycle in a p53-dependent manner. The Journal of biological chemistry. 2005;280(20):20059-68.

[55] Thakur VS, Ruhul Amin AR, Paul RK, Gupta K, Hastak K, Agarwal MK, et al. p53-Dependent p21-mediated growth arrest pre-empts and protects HCT116 cells from PUMA-mediated apoptosis induced by EGCG. Cancer letters. 2010;296(2):225-32.

[56] Lee DH, Rhee JG, Lee YJ. Reactive oxygen species up-regulate p53 and Puma; a possible mechanism for apoptosis during combined treatment with TRAIL and wogonin. British journal of pharmacology. 2009;157(7):1189-202.

[57] Lee HP, Li TM, Tsao JY, Fong YC, Tang CH. Curcumin induces cell apoptosis in human chondrosarcoma through extrinsic death receptor pathway. International immunopharmacology. 2012;13(2):163-9.

[58] Amin AR, Wang D, Zhang H, Peng S, Shin HJ, Brandes JC, et al. Enhanced anti-tumor activity by the combination of the natural compounds (-)-epigallocatechin-3-gallate and luteolin: potential role of p53. The Journal of biological chemistry. 2010;285(45):34557-65.

[59] Hossain MM, Banik NL, Ray SK. Synergistic anti-cancer mechanisms of curcumin and paclitaxel for growth inhibition of human brain tumor stem cells and LN18 and U138MG cells. Neurochem Int. 2012;61(7):1102-13.

[60] Zanotto-Filho A, Braganhol E, Edelweiss MI, Behr GA, Zanin R, Schroder R, et al. The curry spice curcumin selectively inhibits cancer cells growth in vitro and in preclinical model of glioblastoma. J Nutr Biochem. 2012;23(6):591-601.

[61] Nautiyal J, Kanwar SS, Yu Y, Majumdar AP. Combination of dasatinib and curcumin eliminates chemo-resistant colon cancer cells. J Mol Signal. 2011;6:7.

[62] Chan JY, Phoo MS, Clement MV, Pervaiz S, Lee SC. Resveratrol displays converse dose-related effects on 5-fluorouracil-evoked colon cancer cell apoptosis: the roles of caspase-6 and p53. Cancer Biol Ther. 2008;7(8):1305-12.

[63] Majumdar AP, Banerjee S, Nautiyal J, Patel BB, Patel V, Du J, et al. Curcumin synergizes with resveratrol to inhibit colon cancer. Nutr Cancer. 2009;61(4):544-53.

[64] Watson JL, Hill R, Yaffe PB, Greenshields A, Walsh M, Lee PW, et al. Curcumin causes superoxide anion production and p53-independent apoptosis in human colon cancer cells. Cancer letters. 2010;297(1):1-8.

[65] Sukumari-Ramesh S, Bentley JN, Laird MD, Singh N, Vender JR, Dhandapani KM. Dietary phytochemicals induce p53- and caspase-independent cell death in human neuroblastoma cells. Int J Dev Neurosci. 2011;29(7):701-10.

[66] Gupta K, Thakur VS, Bhaskaran N, Nawab A, Babcook MA, Jackson MW, et al. Green Tea Polyphenols Induce p53-Dependent and p53-Independent Apoptosis in Prostate Cancer Cells through Two Distinct Mechanisms. PloS one. 2012;7(12):e52572.

[67] Onoda C, Kuribayashi K, Nirasawa S, Tsuji N, Tanaka M, Kobayashi D, et al. (-)-Epigallocatechin-3-gallate induces apoptosis in gastric cancer cell lines by down-regulating survivin expression. International journal of oncology. 2011;38(5):1403-8.

[68] Chiu TL, Su CC. Curcumin inhibits proliferation and migration by increasing the Bax to Bcl-2 ratio and decreasing NF-kappaBp65 expression in breast cancer MDA-MB-231 cells. International journal of molecular medicine. 2009;23(4):469-75.

[69] Ferraz da Costa DC, Casanova FA, Quarti J, Malheiros MS, Sanches D, Dos Santos PS, et al. Transient Transfection of a Wild-Type p53 Gene Triggers Resveratrol-Induced Apoptosis in Cancer Cells. PloS one. 2012;7(11):e48746.

[70] Thiel G, Cibelli G. Regulation of life and death by the zinc finger transcription factor Egr-1. Journal of cellular physiology. 2002;193(3):287-92.

[71] Choi BH, Kim CG, Bae YS, Lim Y, Lee YH, Shin SY. p21 Waf1/Cip1 expression by curcumin in U-87MG human glioma cells: role of early growth response-1 expression. Cancer research. 2008;68(5):1369-77.

[72] Whitlock NC, Bahn JH, Lee SH, Eling TE, Baek SJ. Resveratrol-induced apoptosis is mediated by early growth response-1, Kruppel-like factor 4, and activating transcription factor 3. Cancer Prev Res (Phila). 2011;4(1):116-27.

[73] Moon Y, Lee M, Yang H. Involvement of early growth response gene 1 in the modulation of microsomal prostaglandin E synthase 1 by epigallocatechin gallate in A549 human pulmonary epithelial cells. Biochemical pharmacology. 2007;73(1):125-35.

[74] Zhang Q, Zhao XH, Wang ZJ. Cytotoxicity of flavones and flavonols to a human esophageal squamous cell carcinoma cell line (KYSE-510) by induction of G2/M arrest and apoptosis. Toxicol In Vitro. 2009;23(5):797-807.

[75] Shammas MA, Neri P, Koley H, Batchu RB, Bertheau RC, Munshi V, et al. Specific killing of multiple myeloma cells by (-)-epigallocatechin-3-gallate extracted from green tea: biologic activity and therapeutic implications. Blood. 2006;108(8):2804-10.

[76] Achour M, Mousli M, Alhosin M, Ibrahim A, Peluso J, Muller CD, et al. Epigallocatechin-3-gallate up-regulates tumor suppressor gene expression via a reactive oxygen

species-dependent down-regulation of UHRF1. Biochem Biophys Res Commun. 2013;430(1):208-12.

[77] Thangasamy T, Sittadjody S, Mitchell GC, Mendoza EE, Radhakrishnan VM, Limesand KH, et al. Quercetin abrogates chemoresistance in melanoma cells by modulating deltaNp73. BMC cancer. 2010;10:282.

[78] Walter A, Etienne-Selloum N, Brasse D, Khallouf H, Bronner C, Rio MC, et al. Intake of grape-derived polyphenols reduces C26 tumor growth by inhibiting angiogenesis and inducing apoptosis. FASEB journal : official publication of the Federation of American Societies for Experimental Biology. 2010;24(9):3360-9.

[79] Sharif T, Alhosin M, Auger C, Minker C, Kim JH, Etienne-Selloum N, et al. Aronia melanocarpa juice induces a redox-sensitive p73-related caspase 3-dependent apoptosis in human leukemia cells. PloS one. 2012;7(3):e32526.

[80] Sharif T, Auger C, Alhosin M, Ebel C, Achour M, Etienne-Selloum N, et al. Red wine polyphenols cause growth inhibition and apoptosis in acute lymphoblastic leukaemia cells by inducing a redox-sensitive up-regulation of p73 and down-regulation of UHRF1. Eur J Cancer. 2010;46(5):983-94.

[81] Somasagara RR, Hegde M, Chiruvella KK, Musini A, Choudhary B, Raghavan SC. Extracts of strawberry fruits induce intrinsic pathway of apoptosis in breast cancer cells and inhibits tumor progression in mice. PloS one. 2012;7(10):e47021.

[82] Azmi AS, Ali S, Banerjee S, Bao B, Maitah MN, Padhye S, et al. Network modeling of CDF treated pancreatic cancer cells reveals a novel c-myc-p73 dependent apoptotic mechanism. Am J Transl Res. 2011;3(4):374-82.

[83] Singh M, Singh N. Curcumin counteracts the proliferative effect of estradiol and induces apoptosis in cervical cancer cells. Mol Cell Biochem. 2011;347(1-2):1-11.

[84] Amin AR, Thakur VS, Paul RK, Feng GS, Qu CK, Mukhtar H, et al. SHP-2 tyrosine phosphatase inhibits p73-dependent apoptosis and expression of a subset of p53 target genes induced by EGCG. Proceedings of the National Academy of Sciences of the United States of America. 2007;104(13):5419-24.

[85] Zhang Z, Wang H, Li M, Agrawal S, Chen X, Zhang R. MDM2 is a negative regulator of p21WAF1/CIP1, independent of p53. The Journal of biological chemistry. 2004;279(16):16000-6.

[86] Shangary S, Wang S. Targeting the MDM2-p53 interaction for cancer therapy. Clin Cancer Res. 2008;14(17):5318-24.

[87] Li M, Zhang Z, Hill DL, Wang H, Zhang R. Curcumin, a dietary component, has anticancer, chemosensitization, and radiosensitization effects by down-regulating the MDM2 oncogene through the PI3K/mTOR/ETS2 pathway. Cancer research. 2007;67(5):1988-96.

[88] Liu H, Liang Y, Wang L, Tian L, Song R, Han T, et al. In Vivo and In Vitro Suppression of Hepatocellular Carcinoma by EF24, a Curcumin Analog. PloS one. 2012;7(10):e48075.

[89] Hastak K, Gupta S, Ahmad N, Agarwal MK, Agarwal ML, Mukhtar H. Role of p53 and NF-kappaB in epigallocatechin-3-gallate-induced apoptosis of LNCaP cells. Oncogene. 2003;22(31):4851-9.

[90] Manna S, Banerjee S, Mukherjee S, Das S, Panda CK. Epigallocatechin gallate induced apoptosis in Sarcoma180 cells in vivo: mediated by p53 pathway and inhibition in U1B, U4-U6 UsnRNAs expression. Apoptosis. 2006;11(12):2267-76.

[91] Dai JM, Wang ZY, Sun DC, Lin RX, Wang SQ. SIRT1 interacts with p73 and suppresses p73-dependent transcriptional activity. Journal of cellular physiology. 2007;210(1):161-6.

[92] Young LF, Hantz HL, Martin KR. Resveratrol modulates gene expression associated with apoptosis, proliferation and cell cycle in cells with mutated human c-Ha-Ras, but does not alter c-Ha-Ras mRNA or protein expression. J Nutr Biochem. 2005;16(11):663-74.

[93] Moll UM, Zaika A. Nuclear and mitochondrial apoptotic pathways of p53. FEBS Lett. 2001;493(2-3):65-9.

[94] Liontas A, Yeger H. Curcumin and resveratrol induce apoptosis and nuclear translocation and activation of p53 in human neuroblastoma. Anticancer research. 2004;24(2B):987-98.

[95] Shankar S, Srivastava RK. Involvement of Bcl-2 family members, phosphatidylinositol 3'-kinase/AKT and mitochondrial p53 in curcumin (diferulolylmethane)-induced apoptosis in prostate cancer. International journal of oncology. 2007;30(4):905-18.

[96] Dai C, Gu W. p53 post-translational modification: deregulated in tumorigenesis. Trends Mol Med. 2010;16(11):528-36.

[97] Olsson A, Manzl C, Strasser A, Villunger A. How important are post-translational modifications in p53 for selectivity in target-gene transcription and tumour suppression? Cell death and differentiation. 2007;14(9):1561-75.

[98] Bartek J, Bartkova J, Lukas J. DNA damage signalling guards against activated oncogenes and tumour progression. Oncogene. 2007;26(56):7773-9.

[99] Aziz K, Nowsheen S, Pantelias G, Iliakis G, Gorgoulis VG, Georgakilas AG. Targeting DNA damage and repair: embracing the pharmacological era for successful cancer therapy. Pharmacology & therapeutics. 2012;133(3):334-50.

[100] Ide H, Yu J, Lu Y, China T, Kumamoto T, Koseki T, et al. Testosterone augments polyphenol-induced DNA damage response in prostate cancer cell line, LNCaP. Cancer science. 2011;102(2):468-71.

[101] Kang SU, Lee BS, Lee SH, Baek SJ, Shin YS, Kim CH. Expression of NSAID-activated gene-1 by EGCG in head and neck cancer: involvement of ATM-dependent p53 expression. J Nutr Biochem. 2012.

[102] Ye R, Goodarzi AA, Kurz EU, Saito S, Higashimoto Y, Lavin MF, et al. The isoflavonoids genistein and quercetin activate different stress signaling pathways as shown by analysis of site-specific phosphorylation of ATM, p53 and histone H2AX. DNA repair. 2004;3(3):235-44.

[103] Heiss EH, Schilder YD, Dirsch VM. Chronic treatment with resveratrol induces redox stress- and ataxia telangiectasia-mutated (ATM)-dependent senescence in p53-positive cancer cells. The Journal of biological chemistry. 2007;282(37):26759-66.

[104] Heo JI, Oh SJ, Kho YJ, Kim JH, Kang HJ, Park SH, et al. ATM mediates interdependent activation of p53 and ERK through formation of a ternary complex with p-p53 and p-ERK in response to DNA damage. Mol Biol Rep. 2012;39(8):8007-14.

[105] She QB, Chen N, Dong Z. ERKs and p38 kinase phosphorylate p53 protein at serine 15 in response to UV radiation. The Journal of biological chemistry. 2000;275(27): 20444-9.

[106] Watson JL, Greenshields A, Hill R, Hilchie A, Lee PW, Giacomantonio CA, et al. Curcumin-induced apoptosis in ovarian carcinoma cells is p53-independent and involves p38 mitogen-activated protein kinase activation and downregulation of Bcl-2 and survivin expression and Akt signaling. Mol Carcinog. 2010;49(1):13-24.

[107] Weir NM, Selvendiran K, Kutala VK, Tong L, Vishwanath S, Rajaram M, et al. Curcumin induces G2/M arrest and apoptosis in cisplatin-resistant human ovarian cancer cells by modulating Akt and p38 MAPK. Cancer Biol Ther. 2007;6(2):178-84.

[108] She QB, Bode AM, Ma WY, Chen NY, Dong Z. Resveratrol-induced activation of p53 and apoptosis is mediated by extracellular-signal-regulated protein kinases and p38 kinase. Cancer research. 2001;61(4):1604-10.

[109] Lin HY, Sun M, Tang HY, Simone TM, Wu YH, Grandis JR, et al. Resveratrol causes COX-2- and p53-dependent apoptosis in head and neck squamous cell cancer cells. J Cell Biochem. 2008;104(6):2131-42.

[110] Lin C, Crawford DR, Lin S, Hwang J, Sebuyira A, Meng R, et al. Inducible COX-2-dependent apoptosis in human ovarian cancer cells. Carcinogenesis. 2011;32(1):19-26.

[111] Conforti F, Sayan AE, Sreekumar R, Sayan BS. Regulation of p73 activity by post-translational modifications. Cell Death Dis. 2012;3:e285.

[112] Luo J, Su F, Chen D, Shiloh A, Gu W. Deacetylation of p53 modulates its effect on cell growth and apoptosis. Nature. 2000;408(6810):377-81.

[113] Goodman RH, Smolik S. CBP/p300 in cell growth, transformation, and development. Genes Dev. 2000;14(13):1553-77.

[114] Iyer NG, Ozdag H, Caldas C. p300/CBP and cancer. Oncogene. 2004;23(24):4225-31.

[115] Barlev NA, Liu L, Chehab NH, Mansfield K, Harris KG, Halazonetis TD, et al. Acetylation of p53 activates transcription through recruitment of coactivators/histone acetyltransferases. Molecular cell. 2001;8(6):1243-54.

[116] Iyer NG, Chin SF, Ozdag H, Daigo Y, Hu DE, Cariati M, et al. p300 regulates p53-dependent apoptosis after DNA damage in colorectal cancer cells by modulation of PUMA/p21 levels. Proceedings of the National Academy of Sciences of the United States of America. 2004;101(19):7386-91.

[117] Costanzo A, Merlo P, Pediconi N, Fulco M, Sartorelli V, Cole PA, et al. DNA damage-dependent acetylation of p73 dictates the selective activation of apoptotic target genes. Molecular cell. 2002;9(1):175-86.

[118] Narayanan BA, Narayanan NK, Re GG, Nixon DW. Differential expression of genes induced by resveratrol in LNCaP cells: P53-mediated molecular targets. International journal of cancer Journal international du cancer. 2003;104(2):204-12.

[119] Tang HY, Shih A, Cao HJ, Davis FB, Davis PJ, Lin HY. Resveratrol-induced cyclooxygenase-2 facilitates p53-dependent apoptosis in human breast cancer cells. Molecular cancer therapeutics. 2006;5(8):2034-42.

[120] Sen GS, Mohanty S, Hossain DM, Bhattacharyya S, Banerjee S, Chakraborty J, et al. Curcumin enhances the efficacy of chemotherapy by tailoring p65NFkappaB-p300 cross-talk in favor of p53-p300 in breast cancer. The Journal of biological chemistry. 2011;286(49):42232-47.

[121] Brooks CL, Gu W. How does SIRT1 affect metabolism, senescence and cancer? Nature reviews Cancer. 2009;9(2):123-8.

[122] Stunkel W, Campbell RM. Sirtuin 1 (SIRT1): the misunderstood HDAC. Journal of biomolecular screening. 2011;16(10):1153-69.

[123] Herranz D, Serrano M. SIRT1: recent lessons from mouse models. Nature reviews Cancer. 2010;10(12):819-23.

[124] Saunders LR, Verdin E. Sirtuins: critical regulators at the crossroads between cancer and aging. Oncogene. 2007;26(37):5489-504.

[125] Vaziri H, Dessain SK, Ng Eaton E, Imai SI, Frye RA, Pandita TK, et al. hSIR2(SIRT1) functions as an NAD-dependent p53 deacetylase. Cell. 2001;107(2):149-59.

[126] Luo J, Nikolaev AY, Imai S, Chen D, Su F, Shiloh A, et al. Negative control of p53 by Sir2alpha promotes cell survival under stress. Cell. 2001;107(2):137-48.

[127] Ikenoue T, Inoki K, Zhao B, Guan KL. PTEN acetylation modulates its interaction with PDZ domain. Cancer research. 2008;68(17):6908-12.

[128] Marshall GM, Liu PY, Gherardi S, Scarlett CJ, Bedalov A, Xu N, et al. SIRT1 promotes N-Myc oncogenesis through a positive feedback loop involving the effects of MKP3 and ERK on N-Myc protein stability. PLoS genetics. 2011;7(6):e1002135.

[129] Frazzi R, Valli R, Tamagnini I, Casali B, Latruffe N, Merli F. Resveratrol-mediated apoptosis of hodgkin lymphoma cells involves SIRT1 inhibition and FOXO3a hyper-acetylation. International journal of cancer Journal international du cancer. 2012.

[130] Howitz KT, Bitterman KJ, Cohen HY, Lamming DW, Lavu S, Wood JG, et al. Small molecule activators of sirtuins extend Saccharomyces cerevisiae lifespan. Nature. 2003;425(6954):191-6.

[131] Baur JA, Ungvari Z, Minor RK, Le Couteur DG, de Cabo R. Are sirtuins viable targets for improving healthspan and lifespan? Nature reviews Drug discovery. 2012;11(6): 443-61.

[132] Pizarro JG, Verdaguer E, Ancrenaz V, Junyent F, Sureda F, Pallas M, et al. Resveratrol inhibits proliferation and promotes apoptosis of neuroblastoma cells: role of sirtuin 1. Neurochemical research. 2011;36(2):187-94.

[133] Singh NP, Singh UP, Hegde VL, Guan H, Hofseth L, Nagarkatti M, et al. Resveratrol (trans-3,5,4'-trihydroxystilbene) suppresses EL4 tumor growth by induction of apoptosis involving reciprocal regulation of SIRT1 and NF-kappaB. Molecular nutrition & food research. 2011;55(8):1207-18.

[134] Han DW, Lee MH, Kim B, Lee JJ, Hyon SH, Park JC. Preventive Effects of Epigallocatechin-3-O-Gallate against Replicative Senescence Associated with p53 Acetylation in Human Dermal Fibroblasts. Oxid Med Cell Longev. 2012;2012:850684.

[135] Kai L, Samuel SK, Levenson AS. Resveratrol enhances p53 acetylation and apoptosis in prostate cancer by inhibiting MTA1/NuRD complex. International journal of cancer Journal international du cancer. 2010;126(7):1538-48.

[136] Toh Y, Pencil SD, Nicolson GL. A novel candidate metastasis-associated gene, mta1, differentially expressed in highly metastatic mammary adenocarcinoma cell lines. cDNA cloning, expression, and protein analyses. The Journal of biological chemistry. 1994;269(37):22958-63.

[137] Xue Y, Wong J, Moreno GT, Young MK, Cote J, Wang W. NURD, a novel complex with both ATP-dependent chromatin-remodeling and histone deacetylase activities. Molecular cell. 1998;2(6):851-61.

[138] Moon HE, Cheon H, Lee MS. Metastasis-associated protein 1 inhibits p53-induced apoptosis. Oncology reports. 2007;18(5):1311-4.

[139] Thakur VS, Gupta K, Gupta S. Green tea polyphenols increase p53 transcriptional activity and acetylation by suppressing class I histone deacetylases. International journal of oncology. 2012;41(1):353-61.

[140] Thakur VS, Gupta K, Gupta S. Green tea polyphenols causes cell cycle arrest and apoptosis in prostate cancer cells by suppressing class I histone deacetylases. Carcinogenesis. 2012;33(2):377-84.

[141] Mellert HS, Stanek TJ, Sykes SM, Rauscher FJ, 3rd, Schultz DC, McMahon SB. Deacetylation of the DNA-binding domain regulates p53-mediated apoptosis. The Journal of biological chemistry. 2011;286(6):4264-70.

[142] Uramoto H, Wetterskog D, Hackzell A, Matsumoto Y, Funa K. p73 competes with co-activators and recruits histone deacetylase to NF-Y in the repression of PDGF beta-receptor. J Cell Sci. 2004;117(Pt 22):5323-31.

[143] Finzer P, Krueger A, Stohr M, Brenner D, Soto U, Kuntzen C, et al. HDAC inhibitors trigger apoptosis in HPV-positive cells by inducing the E2F-p73 pathway. Oncogene. 2004;23(28):4807-17.

[144] Spiegel S, Milstien S, Grant S. Endogenous modulators and pharmacological inhibitors of histone deacetylases in cancer therapy. Oncogene. 2012;31(5):537-51.

[145] Shu L, Khor TO, Lee JH, Boyanapalli SS, Huang Y, Wu TY, et al. Epigenetic CpG demethylation of the promoter and reactivation of the expression of Neurog1 by curcumin in prostate LNCaP cells. Aaps J. 2011;13(4):606-14.

[146] Perron NR, Brumaghim JL. A review of the antioxidant mechanisms of polyphenol compounds related to iron binding. Cell biochemistry and biophysics. 2009;53(2): 75-100.

[147] Kang NJ, Shin SH, Lee HJ, Lee KW. Polyphenols as small molecular inhibitors of signaling cascades in carcinogenesis. Pharmacology & therapeutics. 2011;130(3):310-24.

[148] Cerutti PA. Prooxidant states and tumor promotion. Science. 1985;227(4685):375-81.

[149] Yoo MH, Xu XM, Carlson BA, Patterson AD, Gladyshev VN, Hatfield DL. Targeting thioredoxin reductase 1 reduction in cancer cells inhibits self-sufficient growth and DNA replication. PloS one. 2007;2(10):e1112.

[150] Sun G, Kemble DJ. To C or not to C: direct and indirect redox regulation of Src protein tyrosine kinase. Cell cycle. 2009;8(15):2353-5.

[151] Vaughn AE, Deshmukh M. Glucose metabolism inhibits apoptosis in neurons and cancer cells by redox inactivation of cytochrome c. Nature cell biology. 2008;10(12): 1477-83.

[152] Salvioli S, Storci G, Pinti M, Quaglino D, Moretti L, Merlo-Pich M, et al. Apoptosis-resistant phenotype in HL-60-derived cells HCW-2 is related to changes in expression of stress-induced proteins that impact on redox status and mitochondrial metabolism. Cell death and differentiation. 2003;10(2):163-74.

[153] Khromova NV, Kopnin PB, Stepanova EV, Agapova LS, Kopnin BP. p53 hot-spot mutants increase tumor vascularization via ROS-mediated activation of the HIF1/VEGF-A pathway. Cancer letters. 2009;276(2):143-51.

[154] Kim TH, Hur EG, Kang SJ, Kim JA, Thapa D, Lee YM, et al. NRF2 blockade suppresses colon tumor angiogenesis by inhibiting hypoxia-induced activation of HIF-1alpha. Cancer research. 2011;71(6):2260-75.

[155] Park IJ, Lee YK, Hwang JT, Kwon DY, Ha J, Park OJ. Green tea catechin controls apoptosis in colon cancer cells by attenuation of H2O2-stimulated COX-2 expression via the AMPK signaling pathway at low-dose H2O2. Annals of the New York Academy of Sciences. 2009;1171:538-44.

[156] Su CC, Lin JG, Li TM, Chung JG, Yang JS, Ip SW, et al. Curcumin-induced apoptosis of human colon cancer colo 205 cells through the production of ROS, Ca2+ and the activation of caspase-3. Anticancer research. 2006;26(6B):4379-89.

[157] Min NY, Kim JH, Choi JH, Liang W, Ko YJ, Rhee S, et al. Selective death of cancer cells by preferential induction of reactive oxygen species in response to (-)-epigallocatechin-3-gallate. Biochem Biophys Res Commun. 2012;421(1):91-7.

[158] Shankar S, Chen Q, Siddiqui I, Sarva K, Srivastava RK. Sensitization of TRAIL-resistant LNCaP cells by resveratrol (3, 4', 5 tri-hydroxystilbene): molecular mechanisms and therapeutic potential. J Mol Signal. 2007;2:7.

[159] Maillet A, Pervaiz S. Redox regulation of p53, redox effectors regulated by p53: a subtle balance. Antioxidants & redox signaling. 2012;16(11):1285-94.

[160] Tierney BJ, McCann GA, Cohn DE, Eisenhauer E, Sudhakar M, Kuppusamy P, et al. HO-3867, a STAT3 inhibitor induces apoptosis by inactivation of STAT3 activity in BRCA1-mutated ovarian cancer cells. Cancer Biol Ther. 2012;13(9):766-75.

[161] Kryston TB, Georgiev AB, Pissis P, Georgakilas AG. Role of oxidative stress and DNA damage in human carcinogenesis. Mutat Res. 2011;711(1-2):193-201.

[162] Meplan C, Richard MJ, Hainaut P. Redox signalling and transition metals in the control of the p53 pathway. Biochemical pharmacology. 2000;59(1):25-33.

[163] Dhandapani KM, Mahesh VB, Brann DW. Curcumin suppresses growth and chemoresistance of human glioblastoma cells via AP-1 and NFkappaB transcription factors. J Neurochem. 2007;102(2):522-38.

[164] Fang Y, DeMarco VG, Nicholl MB. Resveratrol enhances radiation sensitivity in prostate cancer by inhibiting cell proliferation and promoting cell senescence and apoptosis. Cancer science. 2012;103(6):1090-8.

[165] Hsieh TC, Huang YC, Wu JM. Control of prostate cell growth, DNA damage and repair and gene expression by resveratrol analogues, in vitro. Carcinogenesis. 2011;32(1):93-101.

[166] Li GX, Chen YK, Hou Z, Xiao H, Jin H, Lu G, et al. Pro-oxidative activities and dose-response relationship of (-)-epigallocatechin-3-gallate in the inhibition of lung cancer cell growth: a comparative study in vivo and in vitro. Carcinogenesis. 2010;31(5): 902-10.

[167] Lee SC, Chan JY, Pervaiz S. Spontaneous and 5-fluorouracil-induced centrosome amplification lowers the threshold to resveratrol-evoked apoptosis in colon cancer cells. Cancer letters. 2010;288(1):36-41.

[168] Selvendiran K, Ahmed S, Dayton A, Kuppusamy ML, Rivera BK, Kalai T, et al. HO-3867, a curcumin analog, sensitizes cisplatin-resistant ovarian carcinoma, leading to therapeutic synergy through STAT3 inhibition. Cancer Biol Ther. 2011;12(9): 837-45.

[169] Dayton A, Selvendiran K, Meduru S, Khan M, Kuppusamy ML, Naidu S, et al. Amelioration of doxorubicin-induced cardiotoxicity by an anticancer-antioxidant dual-function compound, HO-3867. J Pharmacol Exp Ther. 2011;339(2):350-7.

[170] Stearns ME, Wang M. Synergistic Effects of the Green Tea Extract Epigallocatechin-3-gallate and Taxane in Eradication of Malignant Human Prostate Tumors. Transl Oncol. 2011;4(3):147-56.

[171] Chen TC, Wang W, Golden EB, Thomas S, Sivakumar W, Hofman FM, et al. Green tea epigallocatechin gallate enhances therapeutic efficacy of temozolomide in orthotopic mouse glioblastoma models. Cancer letters. 2011;302(2):100-8.

[172] Amin AR, Khuri FR, Chen ZG, Shin DM. Synergistic growth inhibition of squamous cell carcinoma of the head and neck by erlotinib and epigallocatechin-3-gallate: the role of p53-dependent inhibition of nuclear factor-kappaB. Cancer Prev Res (Phila). 2009;2(6):538-45.

[173] Howells LM, Sale S, Sriramareddy SN, Irving GR, Jones DJ, Ottley CJ, et al. Curcumin ameliorates oxaliplatin-induced chemoresistance in HCT116 colorectal cancer cells in vitro and in vivo. International journal of cancer Journal international du cancer. 2011;129(2):476-86.

[174] Sharif T, Auger C, Bronner C, Alhosin M, Klein T, Etienne-Selloum N, et al. Selective proapoptotic activity of polyphenols from red wine on teratocarcinoma cell, a model of cancer stem-like cell. Invest New Drugs. 2011;29(2):239-47.

[175] Rashid A, Liu C, Sanli T, Tsiani E, Singh G, Bristow RG, et al. Resveratrol enhances prostate cancer cell response to ionizing radiation. Modulation of the AMPK, Akt and mTOR pathways. Radiation oncology. 2011;6:144.

[176] Lin HY, Shih A, Davis FB, Tang HY, Martino LJ, Bennett JA, et al. Resveratrol induced serine phosphorylation of p53 causes apoptosis in a mutant p53 prostate cancer cell line. The Journal of urology. 2002;168(2):748-55.

[177] Hsieh TC, Wong C, John Bennett D, Wu JM. Regulation of p53 and cell proliferation by resveratrol and its derivatives in breast cancer cells: an in silico and biochemical

approach targeting integrin alphavbeta3. International journal of cancer Journal international du cancer. 2011;129(11):2732-43.

[178] Singh N, Zaidi D, Shyam H, Sharma R, Balapure AK. Polyphenols sensitization potentiates susceptibility of MCF-7 and MDA MB-231 cells to Centchroman. PloS one. 2012;7(6):e37736.

[179] Lin HY, Lansing L, Merillon JM, Davis FB, Tang HY, Shih A, et al. Integrin alphaVbeta3 contains a receptor site for resveratrol. FASEB journal : official publication of the Federation of American Societies for Experimental Biology. 2006;20(10):1742-4.

[180] Zhang S, Cao HJ, Davis FB, Tang HY, Davis PJ, Lin HY. Oestrogen inhibits resveratrol-induced post-translational modification of p53 and apoptosis in breast cancer cells. British journal of cancer. 2004;91(1):178-85.

[181] Gogada R, Prabhu V, Amadori M, Scott R, Hashmi S, Chandra D. Resveratrol induces p53-independent, X-linked inhibitor of apoptosis protein (XIAP)-mediated Bax protein oligomerization on mitochondria to initiate cytochrome c release and caspase activation. The Journal of biological chemistry. 2011;286(33):28749-60.

[182] De Amicis F, Giordano F, Vivacqua A, Pellegrino M, Panno ML, Tramontano D, et al. Resveratrol, through NF-Y/p53/Sin3/HDAC1 complex phosphorylation, inhibits estrogen receptor alpha gene expression via p38MAPK/CK2 signaling in human breast cancer cells. FASEB journal : official publication of the Federation of American Societies for Experimental Biology. 2011;25(10):3695-707.

[183] Casanova F, Quarti J, da Costa DC, Ramos CA, da Silva JL, Fialho E. Resveratrol chemosensitizes breast cancer cells to melphalan by cell cycle arrest. J Cell Biochem. 2012;113(8):2586-96.

[184] Maccario C, Savio M, Ferraro D, Bianchi L, Pizzala R, Pretali L, et al. The resveratrol analog 4,4'-dihydroxy-trans-stilbene suppresses transformation in normal mouse fibroblasts and inhibits proliferation and invasion of human breast cancer cells. Carcinogenesis. 2012;33(11):2172-80.

[185] Al-Abd AM, Mahmoud AM, El-Sherbiny GA, El-Moselhy MA, Nofal SM, El-Latif HA, et al. Resveratrol enhances the cytotoxic profile of docetaxel and doxorubicin in solid tumour cell lines in vitro. Cell proliferation. 2011;44(6):591-601.

[186] Zhou JH, Cheng HY, Yu ZQ, He DW, Pan Z, Yang DT. Resveratrol induces apoptosis in pancreatic cancer cells. Chinese medical journal. 2011;124(11):1695-9.

[187] Lin H, Xiong W, Zhang X, Liu B, Zhang W, Zhang Y, et al. Notch-1 activation-dependent p53 restoration contributes to resveratrol-induced apoptosis in glioblastoma cells. Oncology reports. 2011;26(4):925-30.

[188] Rusin M, Zajkowicz A, Butkiewicz D. Resveratrol induces senescence-like growth inhibition of U-2 OS cells associated with the instability of telomeric DNA and upregulation of BRCA1. Mechanisms of ageing and development. 2009;130(8):528-37.

[189] Cecchinato V, Chiaramonte R, Nizzardo M, Cristofaro B, Basile A, Sherbet GV, et al. Resveratrol-induced apoptosis in human T-cell acute lymphoblastic leukaemia MOLT-4 cells. Biochemical pharmacology. 2007;74(11):1568-74.

[190] Faber AC, Chiles TC. Resveratrol induces apoptosis in transformed follicular lymphoma OCI-LY8 cells: evidence for a novel mechanism involving inhibition of BCL6 signaling. International journal of oncology. 2006;29(6):1561-6.

[191] George J, Singh M, Srivastava AK, Bhui K, Roy P, Chaturvedi PK, et al. Resveratrol and black tea polyphenol combination synergistically suppress mouse skin tumors growth by inhibition of activated MAPKs and p53. PloS one. 2011;6(8):e23395.

[192] Rajasekaran D, Elavarasan J, Sivalingam M, Ganapathy E, Kumar A, Kalpana K, et al. Resveratrol interferes with N-nitrosodiethylamine-induced hepatocellular carcinoma at early and advanced stages in male Wistar rats. Molecular medicine reports. 2011;4(6):1211-7.

[193] Kalra N, Roy P, Prasad S, Shukla Y. Resveratrol induces apoptosis involving mitochondrial pathways in mouse skin tumorigenesis. Life sciences. 2008;82(7-8):348-58.

[194] Yan G, Graham K, Lanza-Jacoby S. Curcumin enhances the anticancer effects of trichostatin a in breast cancer cells. Mol Carcinog. 2012.

[195] Altenburg JD, Bieberich AA, Terry C, Harvey KA, Vanhorn JF, Xu Z, et al. A synergistic antiproliferation effect of curcumin and docosahexaenoic acid in SK-BR-3 breast cancer cells: unique signaling not explained by the effects of either compound alone. BMC cancer. 2011;11:149.

[196] Banerjee M, Singh P, Panda D. Curcumin suppresses the dynamic instability of microtubules, activates the mitotic checkpoint and induces apoptosis in MCF-7 cells. The FEBS journal. 2010;277(16):3437-48.

[197] Kizhakkayil J, Thayyullathil F, Chathoth S, Hago A, Patel M, Galadari S. Modulation of curcumin-induced Akt phosphorylation and apoptosis by PI3K inhibitor in MCF-7 cells. Biochem Biophys Res Commun. 2010;394(3):476-81.

[198] Maher DM, Bell MC, O'Donnell EA, Gupta BK, Jaggi M, Chauhan SC. Curcumin suppresses human papillomavirus oncoproteins, restores p53, Rb, and PTPN13 proteins and inhibits benzo[a]pyrene-induced upregulation of HPV E7. Mol Carcinog. 2011;50(1):47-57.

[199] Pan W, Yang H, Cao C, Song X, Wallin B, Kivlin R, et al. AMPK mediates curcumin-induced cell death in CaOV3 ovarian cancer cells. Oncology reports. 2008;20(6): 1553-9.

[200] Shi M, Cai Q, Yao L, Mao Y, Ming Y, Ouyang G. Antiproliferation and apoptosis induced by curcumin in human ovarian cancer cells. Cell Biol Int. 2006;30(3):221-6.

[201] Chendil D, Ranga RS, Meigooni D, Sathishkumar S, Ahmed MM. Curcumin confers radiosensitizing effect in prostate cancer cell line PC-3. Oncogene. 2004;23(8): 1599-607.

[202] Pichu S, Krishnamoorthy S, Shishkov A, Zhang B, McCue P, Ponnappa BC. Knock-down of ki-67 by dicer-substrate small interfering RNA sensitizes bladder cancer cells to curcumin-induced tumor inhibition. PloS one. 2012;7(11):e48567.

[203] Tian B, Wang Z, Zhao Y, Wang D, Li Y, Ma L, et al. Effects of curcumin on bladder cancer cells and development of urothelial tumors in a rat bladder carcinogenesis model. Cancer letters. 2008;264(2):299-308.

[204] Guo LD, Chen XJ, Hu YH, Yu ZJ, Wang D, Liu JZ. Curcumin Inhibits Proliferation and Induces Apoptosis of Human Colorectal Cancer Cells by Activating the Mitochondria Apoptotic Pathway. Phytotherapy research : PTR. 2012. DOI: 10.1002/ptr. 4731

[205] Song G, Mao YB, Cai QF, Yao LM, Ouyang GL, Bao SD. Curcumin induces human HT-29 colon adenocarcinoma cell apoptosis by activating p53 and regulating apoptosis-related protein expression. Braz J Med Biol Res. 2005;38(12):1791-8.

[206] Mosieniak G, Adamowicz M, Alster O, Jaskowiak H, Szczepankiewicz AA, Wilczynski GM, et al. Curcumin induces permanent growth arrest of human colon cancer cells: link between senescence and autophagy. Mechanisms of ageing and development. 2012;133(6):444-55.

[207] Gogada R, Amadori M, Zhang H, Jones A, Verone A, Pitarresi J, et al. Curcumin induces Apaf-1-dependent, p21-mediated caspase activation and apoptosis. Cell cycle. 2011;10(23):4128-37.

[208] William BM, Goodrich A, Peng C, Li S. Curcumin inhibits proliferation and induces apoptosis of leukemic cells expressing wild-type or T315I-BCR-ABL and prolongs survival of mice with acute lymphoblastic leukemia. Hematology. 2008;13(6):333-43.

[209] Bush JA, Cheung KJ, Jr., Li G. Curcumin induces apoptosis in human melanoma cells through a Fas receptor/caspase-8 pathway independent of p53. Exp Cell Res. 2001;271(2):305-14.

[210] Wang L, Shen Y, Song R, Sun Y, Xu J, Xu Q. An anticancer effect of curcumin mediated by down-regulating phosphatase of regenerating liver-3 expression on highly metastatic melanoma cells. Molecular pharmacology. 2009;76(6):1238-45.

[211] Su CC, Wang MJ, Chiu TL. The anti-cancer efficacy of curcumin scrutinized through core signaling pathways in glioblastoma. International journal of molecular medicine. 2010;26(2):217-24.

[212] Liu E, Wu J, Cao W, Zhang J, Liu W, Jiang X, et al. Curcumin induces G2/M cell cycle arrest in a p53-dependent manner and upregulates ING4 expression in human glioma. Journal of neuro-oncology. 2007;85(3):263-70.

[213] Roy AM, Baliga MS, Katiyar SK. Epigallocatechin-3-gallate induces apoptosis in estrogen receptor-negative human breast carcinoma cells via modulation in protein expression of p53 and Bax and caspase-3 activation. Molecular cancer therapeutics. 2005;4(1):81-90.

[214] Meeran SM, Patel SN, Chan TH, Tollefsbol TO. A novel prodrug of epigallocatechin-3-gallate: differential epigenetic hTERT repression in human breast cancer cells. Cancer Prev Res (Phila). 2011;4(8):1243-54.

[215] Gupta S, Ahmad N, Nieminen AL, Mukhtar H. Growth inhibition, cell-cycle dysregulation, and induction of apoptosis by green tea constituent (-)-epigallocatechin-3-gallate in androgen-sensitive and androgen-insensitive human prostate carcinoma cells. Toxicology and applied pharmacology. 2000;164(1):82-90.

[216] Singh M, Singh R, Bhui K, Tyagi S, Mahmood Z, Shukla Y. Tea polyphenols induce apoptosis through mitochondrial pathway and by inhibiting nuclear factor-kappaB and Akt activation in human cervical cancer cells. Oncol Res. 2011;19(6):245-57.

[217] Kim YW, Bae SM, Lee JM, Namkoong SE, Han SJ, Lee BR, et al. Activity of green tea polyphenol epigallocatechin-3-gallate against ovarian carcinoma cell lines. Cancer Res Treat. 2004;36(5):315-23.

[218] Kuo PL, Lin CC. Green tea constituent (-)-epigallocatechin-3-gallate inhibits Hep G2 cell proliferation and induces apoptosis through p53-dependent and Fas-mediated pathways. J Biomed Sci. 2003;10(2):219-27.

[219] Huang CH, Tsai SJ, Wang YJ, Pan MH, Kao JY, Way TD. EGCG inhibits protein synthesis, lipogenesis, and cell cycle progression through activation of AMPK in p53 positive and negative human hepatoma cells. Molecular nutrition & food research. 2009;53(9):1156-65.

[220] Lee JH, Jeong YJ, Lee SW, Kim D, Oh SJ, Lim HS, et al. EGCG induces apoptosis in human laryngeal epidermoid carcinoma Hep2 cells via mitochondria with the release of apoptosis-inducing factor and endonuclease G. Cancer letters. 2010;290(1):68-75.

[221] Yamauchi R, Sasaki K, Yoshida K. Identification of epigallocatechin-3-gallate in green tea polyphenols as a potent inducer of p53-dependent apoptosis in the human lung cancer cell line A549. Toxicol In Vitro. 2009;23(5):834-9.

[222] Lee MH, Han DW, Hyon SH, Park JC. Apoptosis of human fibrosarcoma HT-1080 cells by epigallocatechin-3-O-gallate via induction of p53 and caspases as well as suppression of Bcl-2 and phosphorylated nuclear factor-kappaB. Apoptosis. 2011;16(1):75-85.

[223] Harakeh S, Abu-El-Ardat K, Diab-Assaf M, Niedzwiecki A, El-Sabban M, Rath M. Epigallocatechin-3-gallate induces apoptosis and cell cycle arrest in HTLV-1-positive and -negative leukemia cells. Med Oncol. 2008;25(1):30-9.

[224] Kaur M, Mandair R, Agarwal R, Agarwal C. Grape seed extract induces cell cycle arrest and apoptosis in human colon carcinoma cells. Nutr Cancer. 2008;60 Suppl 1:2-11.

[225] Lin YS, Chen SF, Liu CL, Nieh S. The chemoadjuvant potential of grape seed procyanidins on p53-related cell death in oral cancer cells. J Oral Pathol Med. 2012;41(4): 322-31.

[226] Kaur M, Agarwal R, Agarwal C. Grape seed extract induces anoikis and caspase-mediated apoptosis in human prostate carcinoma LNCaP cells: possible role of ataxia telangiectasia mutated-p53 activation. Molecular cancer therapeutics. 2006;5(5): 1265-74.

[227] Roy P, George J, Srivastava S, Tyagi S, Shukla Y. Inhibitory effects of tea polyphenols by targeting cyclooxygenase-2 through regulation of nuclear factor kappa B, Akt and p53 in rat mammary tumors. Invest New Drugs. 2011;29(2):225-31.

[228] Gu Q, Hu C, Chen Q, Xia Y, Feng J, Yang H. Development of a rat model by 3,4-benzopyrene intra-pulmonary injection and evaluation of the effect of green tea drinking on p53 and bcl-2 expression in lung carcinoma. Cancer Detect Prev. 2009;32(5-6): 444-51.

[229] Bhattacharyya A, Choudhuri T, Pal S, Chattopadhyay S, G KD, Sa G, et al. Apoptogenic effects of black tea on Ehrlich's ascites carcinoma cell. Carcinogenesis. 2003;24(1):75-80.

[230] Schwartz JL, Baker V, Larios E, Chung FL. Molecular and cellular effects of green tea on oral cells of smokers: a pilot study. Molecular nutrition & food research. 2005;49(1):43-51.

[231] Tsao AS, Liu D, Martin J, Tang XM, Lee JJ, El-Naggar AK, et al. Phase II randomized, placebo-controlled trial of green tea extract in patients with high-risk oral premalignant lesions. Cancer Prev Res (Phila). 2009;2(11):931-41.

[232] Wang X, Wang R, Hao MW, Dong K, Wei SH, Lin F, et al. The BH3-only protein PUMA is involved in green tea polyphenol-induced apoptosis in colorectal cancer cell lines. Cancer Biol Ther. 2008;7(6):902-8.

Chapter 7

START-GAP/DLC Family Proteins: Molecular Mechanisms for Anti-Tumor Activities

Hitoshi Yagisawa

Additional information is available at the end of the chapter

http://dx.doi.org/10.5772/55268

1. Introduction

The Rho family of GTPases belongs to the superfamily named "Ras-like" proteins, which consists of over 150 varieties in mammals [1]. Significant progress has recently been made in understanding the biological functions mediated by this family of small (~21 kDa) G proteins (guanine nucleotide-binding proteins). Rho GTPases affect crucial biological processes such as transcriptional regulation, cell cycle progression, apoptosis and membrane trafficking [2, 3]. Thus far, a total of 23 Rho proteins have been identified [4], among which RhoA, Rac1 and Cdc42 are characterized in detail. Rho GTPases are also involved in the cytoskeleton formation of the cell via the regulation of actin dynamics [5, 6]. RhoA induces stress fiber formation and focal adhesion assembly, thereby regulating cell shape, attachment and motility, whereas Rac1 promotes extension of lamellipodia and membrane ruffling [7]. Cdc42 has been shown to play a role in the formation of filopodia [8].

Like other small G proteins of the Ras-like protein family, Rho GTPases act by switching between an inactive GDP-bound and an active GTP-bound form, with the latter form capable of interacting with a myriad of downstream effectors (so far, more than 70 proteins have been identified [4]) to be activated by them. The activation of Rho GTPases is stimulated by guanine nucleotide exchange factors (GEFs) that exchange GDP for GTP and is inhibited by GTPase-activating proteins (GAPs) that hydrolyze the GTP to GDP [9]. Rho GTPases are also negatively regulated by guanine nucleotide dissociation inhibitors (GDIs), which bind to the GDP-bound form and not only prevent nucleotide exchange, but also remove Rho GTPases from the plasma membrane to the cytoplasm [4, 10, 11]. Taken together, cell morphology requires spatiotemporally restricted regulation of Rho GTPases through these regulatory proteins. Nonetheless, detailed insight into regulation of Rho GTPases has not been provided.

The effect on the wide spectrum of biological functions suggests the involvement of Rho GTPases and their regulators in cancer progression. Findings from extensive in vitro and in vivo studies shows that deregulated signaling of Rho GTPases may lead to tumorigenesis [12] and thus Rho GTPases are taken as potential targeting candidates for cancer therapy [13]. Since no constitutively active Rho mutants have been reported in human cancers, it is likely that aberrant Rho GTPase signaling in malignancy is caused by the alterations of their regulators [4, 12]. Extensive studies so far have revealed that regulators of Rho proteins are over- or underexpressed in various types of human cancer cells [14-17]. The most common alteration reported for Rho regulators in cancer is inactivation of RhoGAPs, especially of the START-GAP/DLC family RhoGAPs [18]. "START" stands for "steroidogenic acute regulatory protein (STAR)-related lipid transfer" and "DLC" stands for "Deleted in Liver Cancer (DLC)," a gene (or its product) found to be commonly deleted in liver tumors. The START-GAP/DLC family proteins have become the focus of attention on their roles in tumorigenesis. This type of genetic loss was also found in a number of other cancers [17, 19-21].

START-GAP1/DLC1 was originally cloned from rat cDNA library as a binding partner of phospholipase C-δ1 (PLCδ1) [22]. It has been shown that the C-terminal region of START-GAP1/DLC1 and the PH domain of PLCδ1 are responsible for the interaction [23, 24]. START-GAP1/DLC1 enhances the activity of PLCδ1, which generates two second-messengers, inositol 1,4,5-trisphospate (Ins(1,4,5)P_3) and diacylglycerol via hydrolysis of phosphatidylinositol 4,5-bisphosphate (PtdIns(4,5)P_2). Ins(1,4,5)P_3 is accepted by receptors on ER, resulting the elevation of intracellular calcium concentration, whereas diacylglycerol acts as the activator for PKC [25, 26]. Indeed, microinjection of START-GAP1/DLC1 into the cytosol elevated intracellular calcium concentration [23].

As mentioned in detail in the following sections, each member of the START-GAP/DLC multi-domain protein family contains one GAP domain for Rho GTPases. Overexpression of START-GAP1/DLC1 in cultured cells was first to demonstrate induction of drastic morphological changes accompanied by the disruption of actin stress fibers and elevation of intracellular Ca^{2+} concentrations [23]. START-GAP1/DLC1 was therefore originally designated ARP (adaptor for both Rho and PLC) [22], but later the name p122RhoGAP (or just p122) was used to avoid confusion with another ARP (actin related proteins) [23, 24, 27, 28]. We have then introduced the new name, START-GAP1, based on the characteristic domain structure of this Rho GTPase family [29-32]. Meanwhile the antioncogenic properties of the protein family have been revealed and the DLC nomenclature was introduced [21]. In this chapter, the combination of both the structure- and function-based nomenclatures is used throughout, since the use of either of them would not be appropriate to reflect the fundamental properties of the protein family accurately.

In the following sections we will focus on the structure, localization and expression-function relationship of the START-GAP/DLC family proteins in physiological conditions and in human diseases.

2. The START-GAP/DLC gene family

In mammalian genome, there are three genes encoding structurally-related RhoGAPs containing the START domain (Figure 1). There are three groups of START domain-containing RhoGAP proteins in vertebrates, while a worm (*Caenorhabditis elegans)* or a fly (*Drosophila melanogaster)* possesses only one START-containing RhoGAP.

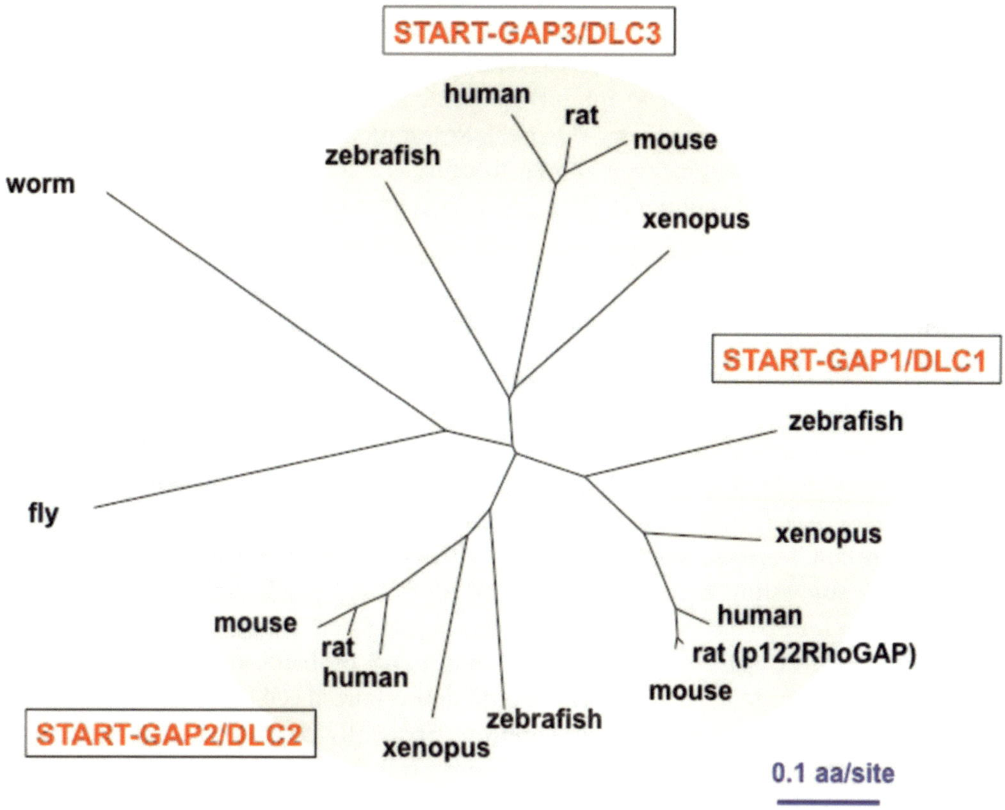

Figure 1. A phylogenetic analysis of the START domain-containing RhoGAPs generated with a "Treeview" after CLUSTAL W analysis (http://clustalw.ddbj.nig.ac.jp/). Modified from Kawai et al. [31].

2.1. Human START-GAP/DLC genes and their expression in various tumor cells

There are about 70 human genes encoding RhoGAPs that share a conserved GAP domain and are capable of switching off the Rho signal [33]. The *START-GAP1/DLC1* (also named as STARD12 or ARHGAP7) gene is localized on chromosome 8p21-22 and encodes a 1,091-amino acid protein with a molecular mass of 122 kDa. Using the quantitative RT-PCR method, Ko et al. have reported that START-GAP1/DLC1 is widely expressed in normal tissues, with high abundance in the lung and ovary, and moderately in the thyroid, spleen, intestine and kidney [34]. START-GAP1/DLC1 has also been known as a tumor suppressor gene product. It is frequently underexpressed or not expressed in several tumor cells and inhibits cell growth, invasion and metastasis [35, 36]. Studies have indicated that downexpression of START-GAP1/ DLC1 either by genomic deletion or DNA methylation [37] is associated with a variety of cancer types including lung [38], breast, prostate, kidney, colon, uterus, ovary, and stomach [38] [39] [40]. These phenotypes require the GAP activity of START-GAP1/DLC1 [35].

Negative regulation of the Rho/Rho-kinase (ROCK)/myosin light chain (MLC) pathway in hepatocellular carcinoma (HCC) cell lines by START-GAP1/DLC1 was shown to be RhoGAP-dependent [23, 41, 42]. The RhoGAP defective mutant failed to inhibit stress fiber formation in HCC lines [41], whereas the overexpression of START-GAP1/DLC1 resulted in morphological change with disruption of actin stress fibers [23, 42]. Using various cancer cell models, START-GAP1/DLC1 was shown to inhibit cell proliferation, suppress cell migration and invasion, and induce apoptosis [35, 36, 41, 43-45]. Restoration of START-GAP1/DLC1 expression in metastatic cell lines has been shown to cause the inhibition of cell migration and invasion as well as a significant reduction in metastases in nude mice [45].

Underexpression of START-GAP1/DLC1 was associated with either heterozygous deletions of the *START-GAP1/DLC1* gene or hypermethylation of the gene promoter region [17, 37, 46-49]. This protein is therefore thought to be under the epigenetic regulation for expression.

The *START-GAP1/DLC1* gene is transcribed from two different promoters, resulting in transcripts encoding three isoforms [47]. To date, most of studies on START-GAP1/DLC1 have focused on the so-called isoform 2. Low et al. have recently identified a new isoform of START-GAP1/DLC1, isoform 4 (DLC1-i4), using 5'-RACE method [50]. This novel isoform encodes a 1,125-amino acid protein with distinct N-terminus as compared with other three isoforms. Similar to them, DLC1-i4 is expressed ubiquitously in normal tissues and immortalized normal epithelial cells, suggesting a role as a major START-GAP1/DLC1 transcript. Differential expression of the four START-GAP1/DLC1 isoforms, however, is found in tumor cell lines: Isoform 1 (the longest isoform) and isoform 3 (short and probably nonfunctional) share a promoter and are silenced in almost all cancer and immortalized cell lines, whereas isoform 2 and isoform 4 utilize different promoters and are frequently downregulated. Isoform 4 is significantly downregulated in multiple carcinoma cell lines, including nasopharyngeal, esophageal, gastric, breast, colorectal, cervical and lung carcinomas. Ectopic expression of DLC1-i4 suppresses tumor cell colony formation. Differential expression of the isoforms suggests interplay in modulating the complex activities of START-GAP1/DLC1 during carcinogenesis.

There are two additional members of the START-GAP/DLC family. *START-GAP2/DLC2* (or *STARD13*) gene is located on chromosome 13q12 [51] and *START-GAP3/DLC3* (or *STARD8/ KIAA0189*) gene is located on the X chromosome at q13 band [31, 52]. The *START-GAP2/ DLC2* encodes a 1,113-amino acid protein with a molecular mass of 125 kDa, whereas the protein product of the START-GAP3/DLC3 transcript has 1,103 amino acids with a molecular mass of 121 kDa. START-GAP2/DLC2 has a broad tissue distribution, with the highest levels in the brain, heart and liver [15, 53]. Human START-GAP2/DLC2 protein shares 51% identity and 64% similarity to human START-GAP1/DLC1 at the level of the amino acid sequence [51]. Introduction of human START-GAP2/DLC2 into mouse fibroblasts suppress Ras signaling and cell transformation in a GAP dependent manner, suggesting the role of START-GAP2/DLC2 for growth suppression and carcinogenesis [51].

START-GAP3/DLC3 is also detected in a variety of human tissues with high abundance in the lung, kidney and placenta [53].

Following the findings of dysregulation of the START-GAP1/DLC1 gene function in a variety of solid tumors, downregulation of START-GAP2/DLC2 and START-GAP3/DLC3 genes was also shown to be involved in human cancer development.

START-GAP2/DLC2 was found to be downregulated in breast, lung, ovarian, renal, uterine, gastric, colon, rectal, and liver tumors [15, 52, 54]. The comparison of *START-GAP1/DLC1* and *START-GAP2/DLC2* gene expression in the same cell lines revealed that START-GAP2/DLC2 is more frequently downregulated than START-GAP1/DLC1 in HCC cell lines [15]. Moreover, the overexpression of START-GAP2/DLC2 suppresses cell proliferation, motility and anchorage-independent growth in the human hepatoma cell line, HepG2 [55]. START-GAP2/DLC2 was also reported to have an inhibitory effect on the growth of breast cancer cells in vitro [56].

Decreased START-GAP3/DLC3 expression in primary tumors from kidney, lung, uterine, ovary and breast has been reported [52]. Kawai et al. have demonstrated that START-GAP3/DLC3 serves as a GAP for both RhoA and Cdc42 in in vitro assays. Furthermore, the overexpression of START-GAP3/DLC3 in HeLa cells disrupts actin stress fibers and changes cell morphology in a GAP-dependent manner [31]. Ectopic expression of START-GAP3/DLC3 in human breast and prostate cancer cell lines inhibits cell proliferation, colony formation and growth in soft agar [52].

The structures of the three human START-GAP/DLC transcripts are depicted in Figure 2.

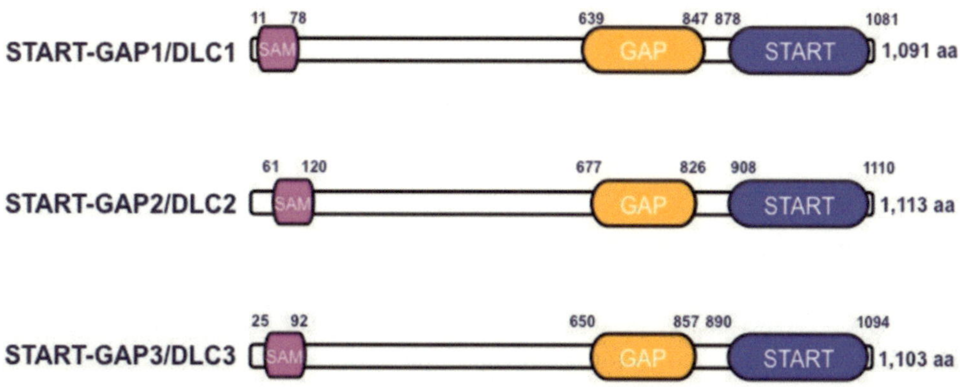

Figure 2. Schematic representation of the START-GAP/DLC family proteins. Each member of the family comprises of three distinct domains, namely the sterile α motif (SAM), RhoGAP (GAP) domain and START (steroidogenic acute regulatory protein (STAR)-related lipid transfer) domain.

2.2. Gene knockout studies of START-GAP/DLC proteins

Using a mouse model a gene knockout study of START-GAP1/DLC1 has been carried out [57]. The mouse *START-GAP1/DLC1* gene was inactivated by homologous recombination. Mice heterozygous for the targeted allele were phenotypically normal, but homozygous mutant embryos did not survive beyond 10.5 days postcoitum. Cultured fibroblasts from START-GAP1/DLC1-deficient embryos displayed alterations in the organization of actin filaments and focal adhesions [57]. In addition, a gene knockdown of START-GAP1/DLC1 in the background of c-*myc* overexpression promotes the formation of liver tumors [58].

Although the START-GAP1/DLC1 gene deficient mice were embryonic lethal, deletion of the *START-GAP2/DLC2* gene from mice resulted in survival to adulthood, indicating that the gene, unlike the *START-GAP1/DLC1* gene, was dispensable for embryonic development [59]. Neither did the authors observe a higher incidence of liver tumor formation in the *START-GAP2/DLC2* gene knockout mice. Nevertheless, they reported smaller phenotype with less formation of adipose tissue [59].

To the best of our knowledge, no reports describing the gene knockout study of the START-GAP3/DLC3 gene are currently available.

2.3. Homologs of the START-GAP/DLC family

As mentioned earlier, there are homologs of the mammalian START-GAP/DLC family proteins in invertebrates in the BLAST database. A *Drosophila* ortholog of START-GAP1/DLC1, *Crossveinless-c* (*Cv-c or RhoGAP88C*), was identified in search for genes that regulate *Drosophila* morphogenesis [60]. The function of *Cv-c* has been revealed to be a key regulator for unidirectional growth of dendritic branches of the fly via downregulating the activity of Rho1, the *Drosophila* Rho GTPase [61]. In the *Cv-c* mutant, two subclass of multidendritic sensory neurons formed dorsally directed branches; however, dendritic branches had a difficulty in growing along the anterior–posterior (A–P) body axis, suggesting that *Cv-c* contributes to sprouting and subsequent growth of the A–P-oriented branches through negative regulation of Rho1 [61]. Thus *Cv-c* plays a key role in directional dendritic growth presumably via its RhoGAP activity, localizing the GAP activity to sites undergoing cytoskeleton rearrangements during morphogenesis.

3. Structure and function of the domains

The START-GAP/DLC family proteins are composed of multi-domain structures. START-GAP/DLC proteins have three distinct domains: The sterile α motif (SAM) localized at its N-terminus, a conserved RhoGAP (GAP) domain in the middle and the steroidogenic acute regulatory (StAR)-related lipid transfer (START) domain at the C-terminus [51-53]. (Figure 2)

Although the START-GAP/DLC family proteins contain a potential lipid-binding START domain, the proteins are produced in soluble forms. The intracellular localization of these proteins, therefore, is determined by their specific interactions with target proteins resided at various cellular structures. Each domain of the START-GAP/DLC family proteins may contribute to different subcellular localization patterns.

3.1. The GAP domain

The biological activity of START-GAP1/DLC1 is mainly executed by the GAP domain (~150-200 amino acid), which promotes the hydrolysis of GTP bound to the Rho GTPases. This catalytic activity is mediated by the 'arginine-finger' present in the GAP domain [62]. START-GAP1/DLC1 and START-GAP3/DLC3 contain a conserved 'arginine-finger' at position 677 and 688, respectively [52, 53]. In in vitro assays both the full-length START-GAP1/DLC1 and the isolated GAP domain reveal activity on RhoA, RhoB and RhoC, to a lesser extent on Cdc42, and no activity on Rac1 [17, 32, 63]. By inactivating these small GTPases, START-GAP1/DLC1 affects cell morphology and control actin cytoskeletal remodeling [22, 23]. Similar to START-GAP1/DLC1, both START-GAP2/DLC2 and START-GAP3/DLC3 contain a RhoGAP domain and exhibit the GAP activity for RhoA and Cdc42 but not Rac1 in vitro [31, 32, 51].

3.2. The sterile α motif (SAM)

The SAM domain (~70 amino acids) has been found in signaling proteins (e.g., p53 related proteins p73 and p63, Eph-related tyrosine kinases and Ets transcription factors) [64, 65]. The SAM domain is thought to act as a protein interaction module via homo- or hetero-oligomerization with other SAM domains [64]. As an example, the EphA2 receptor that plays key roles in many physiological and pathological events including cancer has the SAM domain. Recently, a structural study of the EphA2 receptor SAM domain has validated structural elements relevant for the heterotypic SAM-SAM interactions: two SAM domains interact with a head-to-tail topology characteristic of several SAM-SAM complexes [66]. The SAM domain even interacts with RNA [67].

Structural studies of the SAM domain of START-GAP2/DLC2 have suggested that it binds to lipids such as phosphatidylglycerol [68]. Nevertheless, we know little about exact roles of the SAM domain in START-GAP/DLC function. Kim et al. have shown that the expression of the amino-terminal domain of START-GAP1/DLC1 acts as a dominant negative and profoundly inhibits cell migration by displacing endogenous START-GAP1/DLC1 from focal adhesions [69]. The SAM domain of START-GAP1/DLC1 may serve as an autoinhibitory domain for intrinsic RhoGAP catalytic activity. Eukaryotic elongation factor-1A1 (EF1A1) was found to be a target of the SAM domain of START-GAP1/DLC1 [70]. EF1A1 is involved in protein synthesis [71] and also in transporting β-actin mRNA [72]. EF1A1 is a regulator of cell growth and the cytoskeletal network controlling the actin network through its G-actin-binding activity [73], F-actin-bundling activity [74] and by stabilizing microtubules [75]. EF1A1 is overexpressed in various human cancers, including pancreas, lung, prostate, breast and colon cancers [71]. The SAM domain of START-GAP1/DLC1 adopts a four-helix fold similarly to the SAM domain of START-GAP2/DLC2, but it utilizes a unique motif on a hydrophobic surface to bind directly to EF1A1 [70]. Importantly, the SAM domain is necessary for START-GAP1/DLC1 to translocate EF1A1 to the membrane periphery and ruffles upon fibroblast growth factor stimulation, acting as an auxiliary oncogenic switch to the GAP domain [70].

3.3. The START domain

The START domain (~210 amino acids) is a well-conserved lipid binding domain, which is primarily found in proteins that transfer lipids between organelles and are involved in lipid metabolism as well as in modulation of signaling events involved in lipid processing [76, 77]. The mammalian START domain protein family is well characterized and is composed of 15 members that are classified into 6 subfamilies based on the sequence and ligand specificity: STARD1/3 and STARD4/5/6 subfamilies bind cholesterol and oxysterols, STARD2 (PCTP: phosphatidylcholine transfer protein)/7/10/11(CERT: ceramide transfer protein) subfamily binds phospholipids and ceramides/sphingolipids, STARD14 binds possibly fatty acids. They all may have roles in non-vesicular lipid transport. STARD14/15 subfamily consists of proteins with the thioesterase activity such as the Acyl-CoA thioesterase (ACOT) family proteins. The START-GAP/DLC family proteins fall into the STARD8/12/13 subfamily [76, 77]. Lipid binding properties of START-GAP1/DLC1 (STARD12), START-GAP2/DLC2 (STARD13) and START-GAP3/DLC3 (STARD8) have not well characterized yet.

Among these proteins, STARD1 (or steroidogenic acute regulatory protein, StAR) and STARD3 (or metastatic lymph node 64 kDa protein, MLN64) appear to be the most well characterized [78]. Both STARD1 and STARD3 bind cholesterol [79], and are also known to play a role in lipid transport into mitochondria [80, 81]. In particular, STARD1 localizes to the mitochondria and stimulates the translocation of cholesterol from the outer to the inner mitochondrial membranes [79]. It has been suggested that STARD1 is an essential component in steroid hormone production in steroidogenic cell [82, 83]. It is noteworthy, therefore, that START-GAP2/DLC2 has been found to localize in mitochondria and was found in proximity to the lipid droplets through the START domain [84]. Future research is required in order to establish the lipid ligand of the START domain of START-GAP2/DLC2 and clarify whether the START domain plays a role in mitochondrial lipid transport. START-GAP2/DLC2 also mediates ceramide activation of phosphatidylglycerolphosphate (PGP) synthase and drug response in Chinese hamster ovary cells [85].

The crystal structures for the START domains of human STARD3/MLN64 (PDB entry: IEM2) [86] and murine STARD4 (PDB entry: 1JSS) [87] were the first to be solved and showed an α-β helix-grip fold with a nine-stranded anti-parallel β-sheet forming a U-shaped hydrophobic cleft that binds the ligand and is flanked by N- and C-terminal α helices. The C-terminal α helix is proposed to serve as a 'cap' to the ligand-binding site, with lipid access to the binding pocket requiring a conformational change in the START domain and movement of the C-terminal helix. To date, the crystal structures for a limited number of the START domain-containing proteins were solved: human STARD1/StAR (PDB entry: 3P0L) [88], human STARD5 (PDB entry: 2R55) [88], human STARD2/PCTP (PDB entry: 1LN1) [87], STARD11/CERT (PDB entry: 2E3R) [89], human STARD13/START-GAP2/DLC2 (PDB entry: 2PSO) [88] and human STARD14 (PDB entry: 3FO5) [88]. The data confirm the basic helix-grip fold structure across the five mammalian subfamilies that defines this family of proteins [88].

http://dx.doi.org/10.5772/55268

3.4. The FAT region

Between the SAM and GAP domains there is a long unstructured region (~190 amino acids) termed the FAT (focal adhesion targeting) region, due to the fact that its presence determines of focal adhesion localization of the START-GAP/DLC proteins [27]. START-GAP1/DLC1, START-GAP2/DLC2 and START-GAP3/DLC3 are recruited to focal adhesion sites via their FAT sequence, which binds to the *Src* homology 2 (SH2) domains of focal adhesion proteins and interacts with tensins [32, 43, 90]. The same region of START-GAP1/DLC1, however, has been found to possess the ability to interact with the PTB domain of tensin2 [30, 91, 92] and tensin1 [43]. The role of the interaction with the PTB domain in the localization of START-GAP1/DLC1 to focal adhesions remains controversial and requires further investigation. Using pull-down assay, Kawai et al. reported the interaction between the other members of the START-GAP/DLC family, START-GAP2/DLC2 and START-GAP3/DLC3, and tensin2 PTB [30].

Since the START-GAP/DLC family proteins are rich in serine residues, especially in their FAT regions, it is natural to propose that these proteins could be phosphorylated by the AGC (protein kinases A, G, and C) family protein kinases. Indeed, a numerous potential phosphorylation motifs by these kinases can be found in the START-GAP/DLC family proteins. Protein kinase B (PKB or Akt) is among the member of the kinase family. Since PKB/Akt plays an essential role in the actions of growth factors as well as the regulation of many other cellular processes, such as apoptosis and anoikis, neuronal development and degeneration, and the cell cycle [93], its involvement in function of the START-GAP/DLC family proteins is argued [94].

Hers et al. have demonstrated that Ser^{322} in the FAT region of rat START-GAP1/DLC1 is phosphorylated upon insulin stimulation of intact cells and that this site is directly phosphorylated in vitro by PKB/Akt and ribosomal S6 kinase (RSK1), another member of the AGC family of protein kinases [95], suggesting the phosphorylation via both the PKB/Akt and MAPK kinase (MEK)/extracellular signal-regulated kinase (ERK)/RSK pathways by growth factors. In other words, this site has the potential to integrate the activities of two different signal transduction pathways in a manner dependent on the cellular context. As Ser^{322} falls within the FAT region, its phosphorylation may be involved in regulating the targeting of STASRT-GAP1/DLC1 to focal adhesions. However, despite the profound morphological changes, both $S^{322}A$ and $S^{322}D$ mutants showed similar localizations to focal adhesions as the wild-type STASRT-GAP1/DLC1. The function of the phosphorylation in signaling events downstream of PKB/Akt, such as GLUT4 translocation, the activation of RhoA effectors and cellular transformation, therefore awaits further studies.

A recent report by Ko et al. [96] has also postulated a central role of PKB/Akt phosphorylation of human START-GAP1/DLC1 in the regulation of its tumor suppressive activity, but it argues against the results obtained from the previously-mentioned study on rodent STASRT-GAP1/DLC1. Although human START-GAP1/DLC1 has three characteristic phospho-PKB/Akt substrate motifs, the authors showed that only Ser^{567} was phosphorylated by PKB/Akt. Only active PKB/Akt was able to interact with STASRT-GAP1/DLC1. Since phosphorylated START-GAP1/DLC1 forms a more stable interaction with PKB/Akt, there seems to be a cooperative increase in binding. Furthermore, Ko et al. showed that unphosphorylated START-GAP1/DLC1 is sufficient to suppress proliferation and anchorage-independent cell growth. Using a *ras* transformed, *p53*-deficient murine hepatoma line, the authors demonstrated that only wild-type START-GAP1/DLC1 or a phosphorylation-dead mutant ($S^{567}A$) was able to inhibit tumor formation in nude mice, whereas the $S^{567}D$ mutant, simulating constitutive phosphorylation, did not inhibit tumor growth. PKB/Akt, suggesting that all START-GAP/DLC family members may share common mechanisms of post-translational regulation, also phosphorylated human START-GAP2/DLC2 in the corresponding motif.

The central region of START-GAP1/DLC1 containing the FAT region was reported to target caveolae by interacting with caveolin-1 [28, 92].

4. Regulation of intracellular localization

Generally, members of the START-GAP family, START-GAP1/DLC1, START-GAP2/DLC2 and START-GAP3/DLC3, START-GAP/DLCs do not localize evenly in the cytoplasm. Rather, they are localized in the specialized place in intracellular spaces. All three members are localized to focal adhesions of attached cells. A conserved region among them (the FAT region), responsible for targeting to focal adhesions, has now been identified. It is now established that the tensin family, which is the major component of the focal adhesion complex, is responsible for recruiting the START-GAP/DLC family proteins to focal adhesions. Nevertheless, many proteins are now revealed to interact with START-GAP/DLCs. Vinculin, another member in the focal adhesion complex, can also interacts with the START-GAP/DLC family proteins. In addition, PLCδ1 interacts with all three isoforms of START-GAP/DLCs and make a molecular complex in lipid rafts upon stimulation with extracellular stimuli. Moreover, 14-3-3 binds to START-GAP1/DLC1. Binding of 14-3-3 proteins often sequesters the target protein in a particular subcellular compartment and the release of 14-3-3 proteins then allows the target to be relocated. Since START-GAP1/DLC1 has a nuclear localization signal (NLS) sequence and found in the nucleus, it is suggested that it also interacts with importins.

Thus expected function of the START-GAP/DLC family proteins varies from one cell type to next, depending on the spatiotemporal regulation according to the binding proteins. Nevertheless, as a whole, they regulate cell shapes and motility via remodeling of actin cytoskeleton.

4.1. Focal adhesion targeting via interaction with tensin and vinculin

START-GAP1/DLC1 is localized in focal adhesions via the FAT region located in its N-terminal half and interacts with tensin family proteins, that constitutes focal adhesion components. Evidences that the interaction between START-GAP1/DLC1 and tensin2 occurs in a PTB domain-dependent manner have been provided. It was revealed that FAT3, the third subregion of the FAT region divided into five (39 amino acids), binds directly to the PTB domain of tensin2 [30]. This interaction does not require protein phosphorylation, since the interaction was detected with proteins expressed in bacterial expression system.

http://dx.doi.org/10.5772/55268

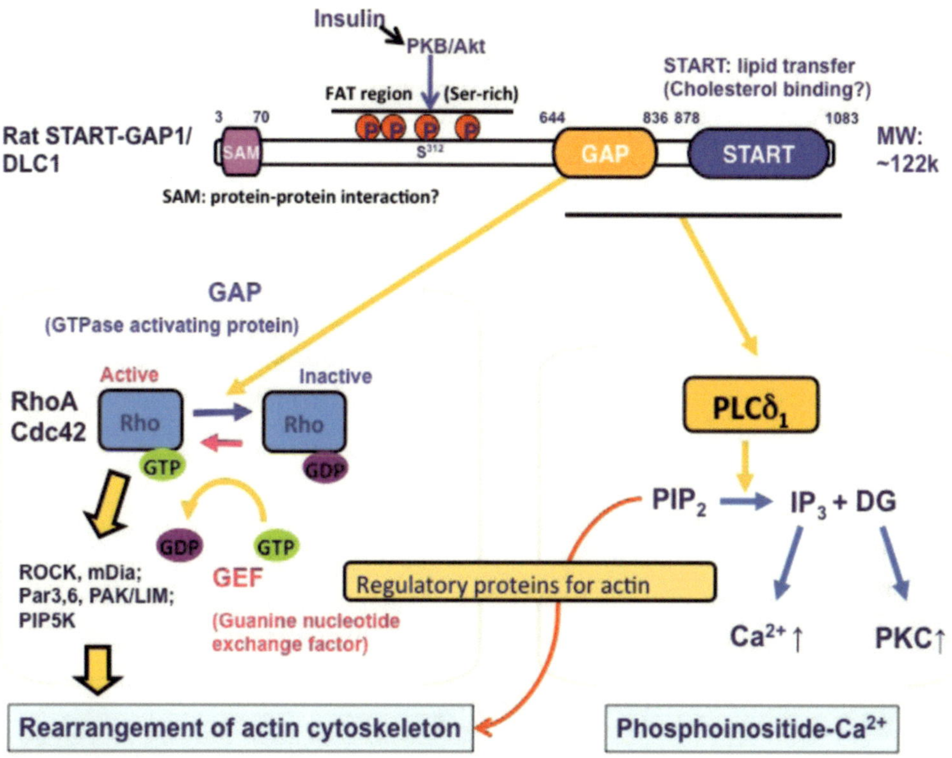

Figure 3. START-GAP1/DLC1 contributes to at least two signaling pathways. START-GAP1/DLC1 consists of about 11 hundred amino acids (for rat: 1,083 and for human: 1, 093) and contain the sterile α motif at the N-terminus, which is known to function in protein-protein interaction. In the C-terminal half there are "GAP domain" followed by the START domain, which is generally thought as a lipid binding or transfer domain. The GAP domain shows a GTPase-activating function, specific for Rho family GTPases, RhoA and Cdc42 but not Rac1 [31], converting "the active, GTP-bound form" to "the inactive, GDP-bound form." The GAP activity of START-GAP1/DLC1 therefore may be implicated in control of the cytoskeleton, cell polarity and cell migration. As for phosphoinositide signaling, one of the downstream effectors of RhoA is PIP5K that generates PtdIns(4,5)P_2. So by inhibition of activated Rho, PtdIns(4,5)P_2 generation is expected to be reduced. The C-terminal half of START-GAP1 is responsible for stimulation of PLCδ1 and breaks down PtdIns(4,5)P_2. This not only causes the generation of two second messengers, Ins(1,4,5)P_3 and diacylgrycerol, but also alters the activity of several PtdIns(4,5)P_2-dependent actin-regulating proteins. As a whole, in microenvironment surrounding active START-GAP1/DLC1, RhoA and Cdc42 are inacitvated and a loss of PtdIns(4,5)P_2 is expected. The region between the SAM and GAP domains is rich in serine residues and does not fall on any known protein domains and has been thought to form disordered conformation.

START-GAP2/DLC2 and START-GAP3/DLC3, as well as STRT-GAP1/DLC1, bind to the PTB domain of tensin2, presumably due to the presence of highly conserved residues in the center of FAT3. Deletion of this sub-region abrogates the interaction with the tensin PTB domain. The tensin2 PTB domain seems to determine the subcellular localization of FAT3. Nevertheless, our study with deletion mutants revealed that FAT3 is essential but not sufficient for the focal adhesion localization of START-GAP1/DLC1. These results suggest that the interaction between the tensin PTB domain and FAT3 contributes to START-GAP1/DLC1 localization but only partially. Other factors could affect the START-GAP1/DLC1 localization.

Amino acid sequence of START-GAP3/DLC3 contains a segment similar to the START-GAP1/DLC1 tensin binding site (^{353}STYDNL^{358}) and full-length START-GAP3/DLC3 was shown to bind the SH2 and PTB domains of tensin1 [43].

We noticed that tensin2 does not always colocalize with START-GAP1/DLC1 and that this interaction was insufficient for targeting START-GAP1/DLC1 to focal adhesions. We thereby explored if there is another molecule that interacts with START-GAP1/DLC1 by the GST pull-down assay using GST-START-GAP1/DLC1 transfected HeLa cells. As a result, we found that START-GAP1/DLC2 can bind with the C-terminus (730-1066) of vinculin. Moreover, START-GAP1/DLC1 and vinculin were found colocalized in foal adhesions. The domain in START-GAP1/DLC1 required for interaction with vinculin was narrowed down to residues 460-470 including an LD motif, a motif in paxillin required for interaction with vinculin. A deletion mutant and point mutants of this motif were fully localized to focal adhesions, although they lost binding ability to vinculin. It is likely that when START-GAP1, vinculin and tensin2 form a stable complex, they are localized to focal adhesions.

PtdIns(4,5)P_2 has an important role in regulating cytoskeleton assembly by inducing conformational changes in actin binding proteins such as vinculin and talin [97]. Thus, START-GAP1/DLC1 could influence cytoskeletal dynamics by altering local PtdIns(4,5)P_2 levels by activating PLCδ1 in the vicinity of focal adhesions as well as by regulating Rho GTPase activity there.

4.2. Raft localization via interaction with PLCδ1

The C-terminal region of START-GAP1/DLC1 covering the GAP and START domain is known to interact with PLCδ1, enhancing its activity [22]. Although START-GAP1/DLC1 binds with PLCδ1, it is not usually localized at the plasma membrane where PLCδ1 targets itself through its PH domain under unstimulated conditions.

Caveolae are plasma membrane domains that appear as flask-shaped invaginations at the cell surface and are enriched in cholesterol and sphingolipids. Yamaga et al. have shown that endogenous START-GAP1/DLC1 was found to sediment with caveolin-1 in low density cholesterol-rich membrane fractions [28], and both endogenous and exogenous START-GAP1/DLC1 were co-immunoprecipitated with caveolin-1 [28, 92]. Since multiple functions have been proposed for caveolae [98] and caveolin-1 has been found in membrane subdomains other than caveolae, including focal adhesions [98] the physiological significance of the interaction between START-GAP1/DLC1 and caveolin-1 requires further investigation.

In their previous study [28], Yamaga et al. provided supportive evidences for START-GAP1/DLC1 localization in caveola, which is cholesterol-enriched membrane microdomain, using an expression system of GFP-tagged proteins, an immunoprecipitation assay and a sucrose density gradient centrifugation analysis of cell lysates. GFP-tagged START-GAP1/DLC1 was observed as patch-like structures at the cell surface and in the cytoplasm of attached cells. The patches were dependent on the levels of the membrane cholesterol and co-localized with caveolin-1. START-GAP1/DLC1 interacts with caveolin-1 in vivo and is fractionated into low-density caveolin-enriched membrane fractions. These results support the idea that START-GAP1/DLC1 is targeted to caveolae via binding to caveolin-1. Yamaga et al. also found the C-terminal half of START-GAP1/DLC1 was responsible for its patch-like distribution. The authors found amino acid sequences that resemble putative caveolin-binding motifs (ΦXΦXXXXΦ, ΦXXXXΦXXΦ or ΦXΦXXXXΦXXΦ), where Φ is an aromatic residue and X is any amino acid [99] in the N-terminal region (^{14}WLRVTGFPQY^{23} and ^{93}WTFQRDSKRWSRLEEFDVF^{111}) and in the GAP domain (^{690}YVNYEGQSAY^{699} and ^{725}FLQIYQY^{731}) of START-GAP1/DLC1. Since GFP-tagged START-GAP1/DLC1–534ΔC does not seem to reside in caveolin-1-containing membranes, the sequences in the N-terminal region may not function as caveolin-binding motifs. Either (or both) the sequence(s) in the GAP domain, however, could be important for caveolin-1 binding and therefore responsible for the distribution of START-GAP1/DLC1. It is therefore possible that the RhoGAP domain of START-GAP1/DLC1 contributes not only to the catalytic activity but also to the intracellular localization of START-GAP1/DLC1. The exact roles of the N-terminal region have to be clarified. GFP-tagged START domain was also observed as patches, but unlike GFP-tagged full-length protein or GFP-tagged GAP domain, it was just partially co-localized with caveolin-1. The patch-like localization of GFP-tagged START domain was similar to the distribution of cholesterol visualized by filipin, a fluorescent molecule that specifically binds to free cholesterol. The result suggests that the START domain can associate with cholesterol but it is not sufficient to recruit START-GAP1/DLC1 to caveolin-1-enriched membrane microdomains.

Recruitment and activation of PLCδ1 in lipid rafts by a muscarinic agonist was examined using PC12 cells [24]. Sucrose density gradient centrifugation analyses of cell lysates revealed that only a small amount of PLCδ1 was recovered in low-density membrane fractions. The amount, however, significantly increased when cells were treated with a muscarinic agonist carbachol. Immunoprecipitation studies demonstrated that carbachol also enhanced the interaction between PLCδ1 and START-GAP1/DLC1, which is constitutively localized in lipid rafts. The PH domain of PLCδ1 is likely responsible for the interaction. Since carbachol elevates intracellular Ca^{2+}levels in PC12 cells, Yamaga et al. next examined whether a rise of intracellular Ca^{2+} levels participates in the carbachol-induced raft recruitment of PLCδ1. After treatment with a Ca^{2+} ionophore ionomycin PLCδ1 was also translocated to lipid rafts in PC12 cells. Chelating extracellular Ca^{2+} by EGTA did not inhibit the carbachol-induced translocation of PLCδ1 to lipid rafts, whereas treatment of cells with thapsigargin to block the intracellular Ca^{2+} mobilization inhibited its translocation. These results suggest that PLCδ1 is recruited to lipid rafts and binds to preexisting START-GAP1/DLC1 in Ca^{2+} dependent manner, and this process can be triggered by external stimuli activating GPCRs (Figure 4).

Although the physiological function and significance of START-GAP1/DLC1 in caveolar localization remain unclear, it appears that caveolae are one of the compartments whereby START-GAP1/DLC1 may exhibit its tumor-suppressive role and possibly vasoregulatory role which will be mentioned in Section 7, and affect the cytoskeletal reorganization by its RhoGAP activity.

4.3. START-GAP2/DLC2 localization in mitochondria

Apart from focal adhesion localization, START-GAP/DLCs could be localized in mitochondria. Ng et al. has disclosed that START-GAP2/DLC2 is targeted to mitochondria through the

START domain using Huh-7 hepatoma cells [84]. They could observe the expression of ectopic START-GAP2/DLC2 in the cytoplasm especially in a punctate structure, suggesting that START-GAP2/DLC2 was concentrated in cytoplasmic speckles. It is noteworthy that the localization patterns of full-length START-GAP2/DLC2, START-GAP2/DLC2-ΔSAM lacking the N-terminal SAM domain and START-GAP2/DLC2-START, that consists of only from the START domain were very similar, indicating that the START domain is responsible for the intracellular localization. Analysis of the amino acid sequence has revealed that no noticeable localization signal in the START domain. START-GAP2/DLC2-START was in dot-like structure throughout the cytosol, although some cells showed aggregates in the perinuclear region, while START-GAP2/DLC2-ΔSTART, a mutant lacking the START domain, was found to distribute homogenously in the cytoplasm. Notably, almost all START-GAP2/DLC2 signals overlapped those of mitochondria, but not all mitochondria were targeted by START-GAP2/DLC2. They also biochemically fractionated mitochondrial fraction from myc-tagged START-GAP2/DLC2 expressed Huh-7 cells and found that it was found in both the cytoplasmic and mitochondrial fraction. Taken together, the START domain plays a pivotal role in intracellular localization and function of START-GAP2/DLC2. Whether these results could also apply for endogenous START-GAP1/DLC1 or START-GAP3/DLC3 in other cell types awaits future experiments.

4.4. START-GAP2/DLC2 localization around lipid droplets

In addition to demonstrate the mitochondrial localization of START-GAP2/DLC2, Ng et al. also examined whether START-GAP2/DLC2 and/or the START domain of START-GAP2/DLC2 can interact with intracellular lipids [84]. Since localization pattern of START-GAP2/DLC2 and its START domain are very similar, they examined whether lipophilic dyes such as Nile red and Sudan III overlap with signal of the START domain of START-GAP2/DLC2. The speckles containing the START domain were localized around the lipid droplets stained by Nile red. Nevertheless, the two signals did not overlap substantially with each other, despite that a major portion of the START domain was found to be in proximity to the lipid droplets. The authors suggested that the START domain of START-GAP2/DLC2 likely serves a lipid related function in the cell and it targets START-GAP2/DLC2 to areas proximal to the lipid droplets [84].

4.5. START-GAP1/DLC1 localization in the nucleus

Recently, using lung carcinoma cells Yuan et al. showed that START-GAP1/DLC1 harbors a functional bipartite nuclear localization signal (NLS) in serine-rich domain in the FAT region, which works together with the GAP domain to mediate START-GAP1/DLC1 nuclear import and subsequent apoptosis [38]. The potential NLS sequence was found as amino acids ^{415}RRENSSDSPKELKRRNS^{431} including a pat7 NLS spanning residues ^{423}PKELKRR^{429} [100]. A deletion mutant START-GAP1/DLC1-Δ372 lacking the N-terminus 372 amino acids accumulates in the nucleus, suggesting the presence of nuclear export signal (NES) in this region, whereas mutants with disruption of this NLS sequence could only localize in the cytosol.

The function of START-GAP1/DLC1 in the nucleus remains to be defined. Some Rho GTPases have C-terminal polybasic region that could be function as an NLS [101]. Therefore these GTPases may serve as substrates for START-GAP1/DLC1 in the nucleus. In the nucleus START-GAP1/DLC1 may also interact with non-GTPase substrates such as PLCδ1 [22], which is under nuclear import-export equilibrium and accumulate in the nucleus under specific phases of the cell cycle [102, 103] or extracellular stimuli that causes aberrant increase in intracellular Ca^{2+} [103-106]. Nuclear translocation of START-GAP1/DLC1 was proposed to be associated with apoptosis by a yet unknown mechanism [38].

START-GAP1/DLC1 interacts with 14-3-3 proteins, resulting inhibition of the GAP activity and block nucleo-cytoplasmic shuttling [100]. In GST pull-down assays, START-GAP1/DLC1 interacted with all 14-3-3 isoforms except 14-3-3σ. The other six 14-3-3 isoforms are ubiquitously expressed, readily form homo- and heterodimers, and could therefore potentially participate in the regulation of START-GAP1/DLC1 function. 14-3-3 proteins often regulate cellular processes by modulating target protein localization. The pat7 NLS spanning residues of START-GAP1/DLC1, ^{423}PKELKRR^{429}, was demonstrated to be masked by phorbolester-induced phosphorylation and the 14-3-3 interaction. Inactivation of this NLS by exchange of critical arginine residues impaired but did not prevent nuclear import of START-GAP1/DLC1. This suggests the presence of another NLS that contributes to START-GAP1/DLC1 nuclear shuttling.

Taken together, it was suggested that START-GAP1/DLC1 functions both as a cytoplasmic and nuclear tumor suppressor.

5. Negative regulation of carcinogenesis not dependent on the RhoGAP activity

5.1. RhoGAP activity-dependent pathway

As mentioned earlier, the START-GAP/DLC family proteins may exert their suppressive function by decreasing the levels of active, GTP-bound Rho proteins or inhibiting the GDP-GTP cycling process, affecting cytoskeletal remodeling, cell shape, motility, proliferation and apoptosis. Nevertheless, molecular mechanisms through which START-GAP/DLC family proteins are capable of suppressing cell motility and cell growth are still unclear.

Holeiter et al. have shown that enhanced migration of cells lacking START-GAP1/DLC1 is dependent on the Rho effector, Dia1, and does not require the activity of ROCK [107]. Leung et al. provide evidence for a key mechanism through which the START-GAP/DLC family proteins act as tumor suppressors [108]. In this study, the authors demonstrated that the expression of START-GAP2/DLC2 is involved in the inactivation of the Raf/MEK/ERK/RSK pathway, which is crucial for cell proliferation. This inhibitory activity of START-GAP2/DLC2 is attributed to RhoGAP function [108].

By binding to PLCδ1, START-GAP1/DLC1 enhances the hydrolysis of PtdIns(4,5)P_2, which interacts with a variety of actin regulatory proteins that affect the actin cytoskeleton [22, 23,

28]. In addition to the PLCδ1 activation and the modulation of RhoGAP activities, proper focal adhesion localization and interaction with tensins have been demonstrated to be essential for the growth-suppression function of START-GAP1/DLC1 [43, 90, 92, 109].

START-GAP1/DLC1 overexpression induces a significant reduction of stress fibers and filopodia, as well as membrane blabbing and cellular decomposition, nuclear condensation or fragmentation in NCI-H358 cells known as highly sensitive cells to tumor suppression functions of START-GAP1/DLC1 [110]. These effects were due to the collapse of actin cytoskeleton during apoptosis [111].

To overcome the embryonic lethality of homozygous START-GAP1/DLC1 knockdown [57], Zender et al. combined in vivo RNAi and a 'mosaic' mouse model of HCC to address the impact of the loss of START-GAP1/DLC1 on liver carcinogenesis [112]. Genetically modified liver progenitors ($p53^{-/-}$ cells infected with a retrovirus expressing *c-myc* and another expressing a START-GAP1/DLC1 shRNA) were transplanted into the liver of syngenic mice to assess their ability to generate tumors in situ. In contrast to control shRNA, START-GAP1/DLC1 shRNA accelerates the formation of liver tumors, which mimics aggressive HCC. Conversely, reintroduction of START-GAP1/DLC1 in hepatoma cells co-expressing oncogenic Ras results in a dramatic reduction of tumor growth in situ. This study demonstrates that START-GAP1/DLC1 loss, when combined with other oncogenic lesions, efficiently promotes the development of HCC.

Activation of RhoA, and thereby of its downstream effector ROCK, is both necessary and sufficient to promote HCC in vivo. Therefore, START-GAP1/DLC1, due to its RhoGAP activity, is capable to antagonize the activities of RhoA and its downstream effectors in HCC [58, 113, 114]. Moreover, RhoA is required for tumorigenesis induced by the loss of START-GAP1/DLC1 and constitutively active RhoA mimics loss of START-GAP1/DLC1 in promoting HCC in the 'mosaic' mouse model [58].

START-GAP1/DLC1 was also implicated in tumor metastasis. The expression levels of START-GAP1/DLC1 mRNA are significantly lower in highly invasive tumors than in less invasive ones [115]. Restoration of START-GAP1/DLC1 expression in vitro reduces the migration and the invasiveness of HCC cells [35, 41, 113]. This mechanism also seems to be dependent on RhoGAP activity and its downstream effectors ROCK and MLC [42].

Finally, reexpression of START-GAP1/DLC1 in HCC cells downregulated the expression of osteopontin and matrix metalloproteinase-9, which are found overexpressed in most primary metastatic liver tumors [114]. START-GAP1/DLC1 restoration also suppresses the distant dissemination of cells from subcutaneous tumors developing after inoculation of HCC cell lines in mice. This process is also dependent on the RhoA activity and reorganization of actin cytoskeleton in tumor cells [114].

5.2. RhoGAP activity-independent pathway

Angiogenesis is another form of cancer development. In addition to its direct effect via activation of Rho pathway, START-GAP1/DLC1 is responsible for regulation of angiogenesis in an indirect manner [116]. START-GAP1/DLC1 negatively regulates angiogenesis in a paracrine fashion. START-GAP1/DLC1 silencing promoted pro-angiogenic responses through vascular endothelial growth factor upregulation, accompanied by the accumulation of hypoxia-inducible factor 1 and its nuclear localization.

6. Stability of START-GAP/DLCs

Stability of proteins may also contribute to the expression levels of the START-GAP/DLC family proteins. Generally, degradation of proteins is controlled by ubiquitination-dependent and independent proteolysis. START-GAP1/DLC1 seems to be ubiqutinylated and processed at least partly by ubiquitinylation-dependent proteasomal degradation [117]. Nevertheless, the sites and the mode of ubiquitinylation (mono- or poly-) on the molecule are unclear. In addition, a proinflammatory protein S100A10, a key cell surface receptor for plasminogen and a regulator of proteolysis, was found to be a novel binding partner of START-GAP1/DLC1 [118]. S100A10 colocalizes with START-GAP1/DLC1 in the cytoplasm via interaction between the C-terminus of S100A10 and the central domain of START-GAP1/DLC1, regulating tumor cell invasion [118]. These results strongly suggest that proteolysis dependent on post-transcriptional modification of START-GAP1/DLC1 plays an important role in the regulation of its tumor suppressive activity.

7. Possible roles of START-GAP1/DLC1 in the development of vascular diseases

Recently, another potential role of START-GAP1/DLC1 in the pathogenesis of a human disease has emerged. It has been known that the PLC activity with its effect on the regulation of intracellular Ca^{2+} levels has potential roles in cardiac regulation. PLCδ1 activity has shown to be enhanced in patients with coronary artery spasm, CSA (coronary spastic angina) [119, 120]. In this context, a new finding that START-GAP1/DLC1 protein levels and mRNA levels in fibroblasts from Japanese patients with CSA were enhanced compared with levels in control subjects is noteworthy [121]. The authors also found in the START-GAP1/DLC1 promoter analysis, the -228G/A and -1466C/T variants found associated with CSA patients using 5'-RACE method revealed the increase in luciferase activity [121]. The incidence of -228G-A was more frequent in male patients with CSA than in male control subjects, suggesting that this variant is a possible candidate responsible for upregulation of START-GAP1/DLC1 protein in CSA. Thus, it was postulated that START-GAP1/DLC1 is up-regulated in patients with coronary spasm, possibly by mutations in the promoter region, causing increased $[Ca^{2+}]_i$ to acetylcholine, and thereby seems to be related to enhanced coronary vasomotility.

As stated in the earlier section, START-GAP1/DLC1 plays two important roles in signaling pathways: One of the dual functions is the ability to enhance the PtdIns(4,5)P_2-hydrolyzing activity of PLCδ1 and the other is the GAP activity specific for RhoA and Cdc42. The former leads to both the protein kinase C-mediated pathway and Ca^{2+}-dependent pathway, resulting in enhanced myosin light chain phosphorylation that plays an important role in the constrictor response of the coronary artery smooth muscle to transmitters such as serotonin and histamine. Homma and Emori showed that recombinant PLCδ1 catalyzes the hydrolysis of PtdIns(4,5)P_2 in a Ca^{2+}-dependent manner; in the presence of START-GAP1/DLC1, its activity is 5- to 10-fold increased in the range of physiological Ca^{2+} concentration [22]. Thus, it is conceivable that upregulation of START-GAP1/DLC1 protein observed in patients with CSA is responsible for the high activity of PLCδ1. These results are consistent with the previous finding of the $H^{257}R$ variant of PLCδ1, in which the conformational change is associated with upregulation of PLCδ1 activity [120]. These characteristics seem to explain the pathogenesis of CSA in which both the basal vascular tone and the vasoconstrictor response to the diverse stimuli were enhanced. Previously, PLC activity was shown to positively correlate not only with basal coronary artery tone but also with the maximal and averaged constrictor responses of the coronary artery to acetylcholine [119]. This finding also suggests a critical role of START-GAP1/DLC1 protein in the genesis of coronary spasm.

The START-GAP1/DLC1 protein has another function, a GAP activity for Rho GTPases. Alterations in RhoA/ROCK pathway have been implicated in the development of a variety of cardiovascular diseases, including hypertension, atherosclerosis, and cerebral and coronary vasospasm [122, 123].

It was reported that the ROCK inhibitor fasudil attenuated the constrictor response of the coronary artery to ACh and prevented the occurrence of chest pain in patients with CSA [124]. By its GAP activity, START-GAP1/DLC1 may antagonize the development of coronary spasm like the ROCK inhibitor. On the other hand, Ca^{2+} mobilization induced by PLCδ1 activation may upregulate Rho and ROCK in the vascular smooth muscle [125]. Thus, the role of START-GAP1/DLC1 in the regulation of Rho is complicated, and the relation of START-GAP1/DLC1 to the genesis of coronary spasm via Rho activity remains to be determined.

The mechanisms of enhancement of START-GAP1/DLC1 promoter activity by the -228G/A variant may be related to transcription factor SP because this variance causes loss of binding to SP1 in its region [121]. Upregulation of START-GAP1/DLC1 protein in the coronary arteries of CSA patients should be confirmed. It remains unclear whether upregulation of START-GAP2/DLC2 or START-GAP3/DLC3 is associated with CSA.

Regarding the caveola localization of START-GAP1/DLC1, the recent report by Nuno et al. that RhoA colocalization with caveolin-1 in caveolae regulates vascular contractions to serotonin is of interest [126]. Caveolins are not only structural components of caveola micro-domains, but also regulate assembly and activation of a variety of receptors and signaling molecules such as PtdIns(4,5)P_2, glucose transporter 4 (GLUT4), epidermal growth factor receptors (EGFR) and endothelial nitric oxide synthase (eNOS). Localization of RhoA to caveolae versus noncaveolar lipid rafts differentially regulates its activation and contractions to RhoA-dependent agonists with greater activation associated with its localization to noncaveolar rafts. Presumably, translocalization of START-GAP1/DLC1 to caveolae by agonists reported by Yamaga et al. [24, 28] may also contribute to modulate the activity of RhoA/ROCK pathway in vascular cells.

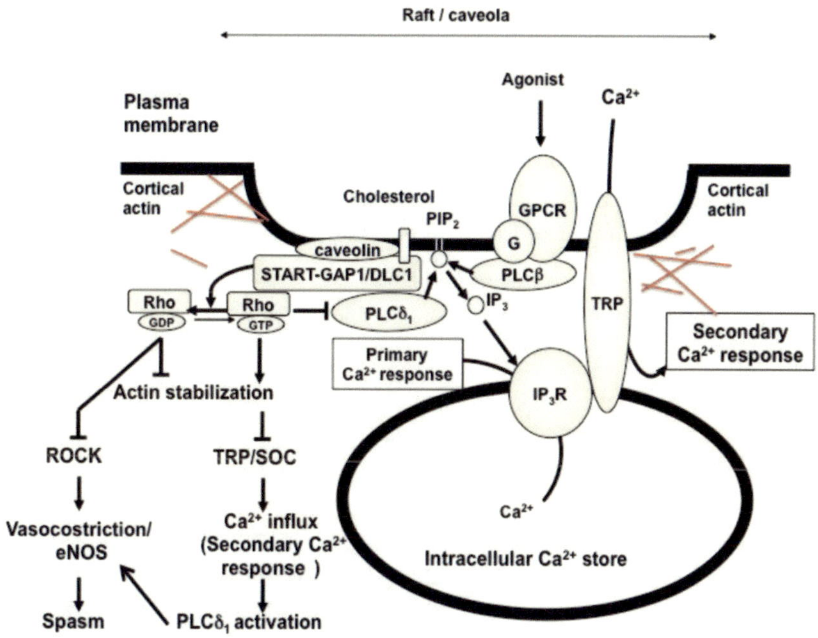

Figure 4. Roles of START-GAP1/DLC1 in a lipid raft/caveola and its possible involvement in the pathogenesis of coronary vasospasm. When G protein coupled receptors (GPCRs) such as muscarinic receptors are activated by agonists they initially stimulate β-type PLC resulting in the production of Ins(1,4,5)P_3 from PtdIns(4,5)P_2. Ins(1,4,5)P_3 binds to the receptors (IP3R or ER Ca^{2+} pump) at the intracellular Ca^{2+} stores such as ER and mobilizes Ca^{2+} from the stores. PLCδ1 is localized at the plasma membrane and in the cytosol in unstimulated cells, whereas START-GAP1/DLC1 is localized in lipid rafts via binding with cholesterol (or with caveolin if the lipid rafts are caveolae). The agonist-induced primary increase in Ca^{2+} recruits PLCδ1 into lipid rafts from other parts of the plasma membrane or from the cytosol by unknown mechanisms. PLCδ1 in the lipid rafts then binds START-GAP1/DLC1 to be activated in the presence of Ca^{2+}, resulting in a robust hydrolysis of PtdIns(4,5)P_2 that forms clusters in lipid rafts. Released Ins(1,4,5)P_3 then empties the internal Ca^{2+} stores via activation of IP3R followed by stimulation of TRP channels, a component of store-operated channels (SOCs) to increase Ca^{2+} influx (secondary Ca^{2+}). Thus the recruitment of PLCδ1 to rafts confers a general positive feedback mechanism for phosphoinositide/Ca^{2+} signaling in the cells. START-GAP1/DLC1 also modulates the RhoA/ROCK signaling pathway. Modified from [24].

8. Conclusion

The START-GAP/DLC family proteins are a group of Rho GTPases whose aberrant function suggests dysregulation of numerous cell processes that can develop diseases such as cancers and vascular spasm. Both in vitro and in vivo studies have provided strong evidence that START-GAP/DLCs have roles in regulating actin cytoskeleton organization. Binding to tensins and other molecules may target the RhoGAP activity of the START-GAP/DLCs to particular subcellular domains. Accumulating reports indicate that the START-GAP/DLC proteins are regulated in various ways at a genetic level by gene deletion, at an epigenetic level by the aberrant promoter methylation, or at a cellular level by the regulation of localization and stability. More information is required on the mechanisms responsible for these regulations, including post-translational modifications such as phosphorylation and ubiquitination. In addition, dissection of the interacting partners of START-GAP/DLCs by proteomic analyses will reveal detailed signaling pathways that regulate cell shape, motility and proliferation. Future studies on the difference among START-GAP/DLCs may reveal which member of this family plays a key role in individual types of cancer or in vascular abnormality and whether restoration of the proper function of the START-GAP/DLCs is capable of restraining cell transformation or abnormal signaling. These efforts would result in identification of the member that becomes the subject of future studies as a potential drug target for therapy.

Nomenclature

CSA: cardiac spastic angina, DLC: deleted in liver cancer, GAP: GTPase activating protein, HCC: hepatocellular carcinoma, Ins(1,4,5)P_3: inositol 1,4,5-trisphospate (IP_3), NLS: nuclear localization signal, PLC: phosphoinositide-specific phospholipase C, PKB: protein kinase B, PtdIns(4,5)P_2: phosphatidylinositol 4,5-bisphosphate (PIP_2), ROCK: Rho-kinase, START: steroidogenic acute regulatory protein (STAR)-related lipid transfer

Acknowledgements

I am grateful to my collaborators who co-authored several papers introduced in this chapter with me. Work in our laboratory was supported by grants from the Japan Society of Promotion of Science (Grant-in-Aid for Scientific Research: #19570184 and #22501016), University of Hyogo Special Grant for Research and Education (FY2012) and Narishige Zoological Science Award (FY2012).

Author details

Hitoshi Yagisawa*

Address all correspondence to: yagisawa@sci.u-hyogo.ac.jp

Graduate School of Life Science, University of Hyogo, Hyogo-ken, Japan

References

[1] Boureux A, Vignal E, Faure S, Fort P. Evolution of the Rho family of ras-like GTPases in eukaryotes. Molecular biology and evolution. 2007;24(1):203-16.

[2] Ridley AJ, Hall A. The small GTP-binding protein rho regulates the assembly of focal adhesions and actin stress fibers in response to growth factors. Cell. 1992;70(3): 389-99.

[3] Ridley AJ, Paterson HF, Johnston CL, Diekmann D, Hall A. The small GTP-binding protein rac regulates growth factor-induced membrane ruffling. Cell. 1992;70(3): 401-10.

[4] Grise F, Bidaud A, Moreau V. Rho GTPases in hepatocellular carcinoma. Biochim Biophys Acta. 2009;1795(2):137-51.

[5] Ridley AJ. Rho GTPases and actin dynamics in membrane protrusions and vesicle trafficking. Trends Cell Biol. 2006;16(10):522-9.

[6] Hall A. Rho GTPases and the actin cytoskeleton. Science. 1998;279(5350):509-14.

[7] Ridley AJ, Schwartz MA, Burridge K, Firtel RA, Ginsberg MH, Borisy G, et al. Cell migration: integrating signals from front to back. Science. 2003;302(5651):1704-9.

[8] Wennerberg K, Der CJ. Rho-family GTPases: it's not only Rac and Rho (and I like it). J Cell Sci. 2004;117(Pt 8):1301-12.

[9] Moon SY, Zheng Y. Rho GTPase-activating proteins in cell regulation. Trends Cell Biol. 2003;13(1):13-22.

[10] Zheng Y. Dbl family guanine nucleotide exchange factors. Trends Biochem Sci. 2001;26(12):724-32.

[11] Olofsson B. Rho guanine dissociation inhibitors: pivotal molecules in cellular signalling. Cell Signal. 1999;11(8):545-54.

[12] Ellenbroek SI, Collard JG. Rho GTPases: functions and association with cancer. Clinical & experimental metastasis. 2007;24(8):657-72.

[13] Mardilovich K, Olson MF, Baugh M. Targeting Rho GTPase signaling for cancer therapy. Future oncology. 2012;8(2):165-77.

[14] Xu XR, Huang J, Xu ZG, Qian BZ, Zhu ZD, Yan Q, et al. Insight into hepatocellular carcinogenesis at transcriptome level by comparing gene expression profiles of hepatocellular carcinoma with those of corresponding noncancerous liver. Proc Natl Acad Sci U S A. 2001;98(26):15089-94.

[15] Ullmannova V, Popescu NC. Expression profile of the tumor suppressor genes DLC-1 and DLC-2 in solid tumors. Int J Oncol. 2006;29(5):1127-32.

[16] Plaumann M, Seitz S, Frege R, Estevez-Schwarz L, Scherneck S. Analysis of DLC-1 expression in human breast cancer. J Cancer Res Clin Oncol. 2003;129(6):349-54.

[17] Wong CM, Lee JM, Ching YP, Jin DY, Ng IO. Genetic and epigenetic alterations of DLC-1 gene in hepatocellular carcinoma. Cancer Res. 2003;63(22):7646-51.

[18] Lahoz A, Hall A. DLC1: a significant GAP in the cancer genome. Genes Dev. 2008;22(13):1724-30.

[19] Wistuba, II, Behrens C, Virmani AK, Milchgrub S, Syed S, Lam S, et al. Allelic losses at chromosome 8p21-23 are early and frequent events in the pathogenesis of lung cancer. Cancer Res. 1999;59(8):1973-9.

[20] Chinen K, Isomura M, Izawa K, Fujiwara Y, Ohata H, Iwamasa T, et al. Isolation of 45 exon-like fragments from 8p22-->p21.3, a region that is commonly deleted in hepatocellular, colorectal, and non-small cell lung carcinomas. Cytogenet Cell Genet. 1996;75(2-3):190-6.

[21] Yuan BZ, Miller MJ, Keck CL, Zimonjic DB, Thorgeirsson SS, Popescu NC. Cloning, characterization, and chromosomal localization of a gene frequently deleted in human liver cancer (DLC-1) homologous to rat RhoGAP. Cancer Res. 1998;58(10): 2196-9.

[22] Homma Y, Emori Y. A dual functional signal mediator showing RhoGAP and phospholipase C-delta stimulating activities. EMBO J. 1995;14(2):286-91.

[23] Sekimata M, Kabuyama Y, Emori Y, Homma Y. Morphological changes and detachment of adherent cells induced by p122, a GTPase-activating protein for Rho. J Biol Chem. 1999;274(25):17757-62.

[24] Yamaga M, Kawai K, Kiyota M, Homma Y, Yagisawa H. Recruitment and activation of phospholipase C (PLC)-delta1 in lipid rafts by muscarinic stimulation of PC12 cells: contribution of p122RhoGAP/DLC1, a tumor-suppressing PLCdelta1 binding protein. Adv Enzyme Regul. 2008;48:41-54.

[25] Rebecchi MJ, Pentyala SN. Structure, function, and control of phosphoinositide-specific phospholipase C. Physiol Rev. 2000;80(4):1291-335.

[26] Nishizuka Y. Protein kinase C and lipid signaling for sustained cellular responses. Faseb J. 1995;9(7):484-96.

[27] Kawai K, Yamaga M, Iwamae Y, Kiyota M, Kamata H, Hirata H, et al. A PLCdelta1-binding protein, p122RhoGAP, is localized in focal adhesions. Biochem Soc Trans. 2004;32(Pt 6):1107-9.

[28] Yamaga M, Sekimata M, Fujii M, Kawai K, Kamata H, Hirata H, et al. A PLCdelta1-binding protein, p122/RhoGAP, is localized in caveolin-enriched membrane domains and regulates caveolin internalization. Genes Cells. 2004;9(1):25-37.

[29] Kawai K, Iwamae Y, Yamaga M, Kiyota M, Ishii H, Hirata H, et al. Focal adhesion-localization of START-GAP1/DLC1 is essential for cell motility and morphology. Genes Cells. 2009;14(2):227-41.

[30] Kawai K, Kitamura SY, Maehira K, Seike J, Yagisawa H. START-GAP1/DLC1 is localized in focal adhesions through interaction with the PTB domain of tensin2. Adv Enzyme Regul. 2010;50(1):202-15.

[31] Kawai K, Kiyota M, Seike J, Deki Y, Yagisawa H. START-GAP3/DLC3 is a GAP for RhoA and Cdc42 and is localized in focal adhesions regulating cell morphology. Biochem Biophys Res Commun. 2007;364(4):783-9.

[32] Kawai K, Seike J, Iino T, Kiyota M, Iwamae Y, Nishitani H, et al. START-GAP2/DLC2 is localized in focal adhesions via its N-terminal region. Biochem Biophys Res Commun. 2009;380(4):736-41.

[33] Tcherkezian J, Lamarche-Vane N. Current knowledge of the large RhoGAP family of proteins. Biol Cell. 2007;99(2):67-86.

[34] Ko FC, Yeung YS, Wong CM, Chan LK, Poon RT, Ng IO, et al. Deleted in liver cancer 1 isoforms are distinctly expressed in human tissues, functionally different and under differential transcriptional regulation in hepatocellular carcinoma. Liver Int. 2010;30(1):139-48.

[35] Zhou X, Thorgeirsson SS, Popescu NC. Restoration of DLC-1 gene expression induces apoptosis and inhibits both cell growth and tumorigenicity in human hepatocellular carcinoma cells. Oncogene. 2004;23(6):1308-13.

[36] Yuan BZ, Zhou X, Durkin ME, Zimonjic DB, Gumundsdottir K, Eyfjord JE, et al. DLC-1 gene inhibits human breast cancer cell growth and in vivo tumorigenicity. Oncogene. 2003;22(3):445-50.

[37] Yuan BZ, Durkin ME, Popescu NC. Promoter hypermethylation of DLC-1, a candidate tumor suppressor gene, in several common human cancers. Cancer Genet Cytogenet. 2003;140(2):113-7.

[38] Yuan BZ, Jefferson AM, Millecchia L, Popescu NC, Reynolds SH. Morphological changes and nuclear translocation of DLC1 tumor suppressor protein precede apoptosis in human non-small cell lung carcinoma cells. Exp Cell Res. 2007;313(18): 3868-80.

[39] Ullmannova V, Popescu NC. Inhibition of cell proliferation, induction of apoptosis, reactivation of DLC1, and modulation of other gene expression by dietary flavone in breast cancer cell lines. Cancer Detect Prev. 2007;31(2):110-8.

[40] Zheng SL, Mychaleckyj JC, Hawkins GA, Isaacs SD, Wiley KE, Turner A, et al. Evaluation of DLC1 as a prostate cancer susceptibility gene: mutation screen and association study. Mutat Res. 2003;528(1-2):45-53.

[41] Wong CM, Yam JW, Ching YP, Yau TO, Leung TH, Jin DY, et al. Rho GTPase-activating protein deleted in liver cancer suppresses cell proliferation and invasion in hepatocellular carcinoma. Cancer Res. 2005;65(19):8861-8.

[42] Wong CC, Wong CM, Ko FC, Chan LK, Ching YP, Yam JW, et al. Deleted in liver cancer 1 (DLC1) negatively regulates Rho/ROCK/MLC pathway in hepatocellular carcinoma. PLoS One. 2008;3(7):e2779.

[43] Qian X, Li G, Asmussen HK, Asnaghi L, Vass WC, Braverman R, et al. Oncogenic inhibition by a deleted in liver cancer gene requires cooperation between tensin binding and Rho-specific GTPase-activating protein activities. Proc Natl Acad Sci U S A. 2007;104(21):9012-7.

[44] Ng IO, Liang ZD, Cao L, Lee TK. DLC-1 is deleted in primary hepatocellular carcinoma and exerts inhibitory effects on the proliferation of hepatoma cell lines with deleted DLC-1. Cancer Res. 2000;60(23):6581-4.

[45] Goodison S, Yuan J, Sloan D, Kim R, Li C, Popescu NC, et al. The RhoGAP protein DLC-1 functions as a metastasis suppressor in breast cancer cells. Cancer Res. 2005;65(14):6042-53.

[46] Kim TY, Jong HS, Song SH, Dimtchev A, Jeong SJ, Lee JW, et al. Transcriptional silencing of the DLC-1 tumor suppressor gene by epigenetic mechanism in gastric cancer cells. Oncogene. 2003;22(25):3943-51.

[47] Seng TJ, Low JS, Li H, Cui Y, Goh HK, Wong ML, et al. The major 8p22 tumor suppressor DLC1 is frequently silenced by methylation in both endemic and sporadic nasopharyngeal, esophageal, and cervical carcinomas, and inhibits tumor cell colony formation. Oncogene. 2007;26(6):934-44.

[48] Guan M, Zhou X, Soulitzis N, Spandidos DA, Popescu NC. Aberrant methylation and deacetylation of deleted in liver cancer-1 gene in prostate cancer: potential clinical applications. Clin Cancer Res. 2006;12(5):1412-9.

[49] Zhang Q, Ying J, Zhang K, Li H, Ng KM, Zhao Y, et al. Aberrant methylation of the 8p22 tumor suppressor gene DLC1 in renal cell carcinoma. Cancer Lett. 2007;249(2): 220-6.

[50] Low JS, Tao Q, Ng KM, Goh HK, Shu XS, Woo WL, et al. A novel isoform of the 8p22 tumor suppressor gene DLC1 suppresses tumor growth and is frequently silenced in multiple common tumors. Oncogene. 2011.

[51] Ching YP, Wong CM, Chan SF, Leung TH, Ng DC, Jin DY, et al. Deleted in liver cancer (DLC) 2 encodes a RhoGAP protein with growth suppressor function and is underexpressed in hepatocellular carcinoma. J Biol Chem. 2003;278(12):10824-30.

[52] Durkin ME, Ullmannova V, Guan M, Popescu NC. Deleted in liver cancer 3 (DLC-3), a novel Rho GTPase-activating protein, is downregulated in cancer and inhibits tumor cell growth. Oncogene. 2007;26(31):4580-9.

[53] Durkin ME, Yuan BZ, Zhou X, Zimonjic DB, Lowy DR, Thorgeirsson SS, et al. DLC-1:a Rho GTPase-activating protein and tumour suppressor. J Cell Mol Med. 2007;11(5):1185-207.

[54] Xiaorong L, Wei W, Liyuan Q, Kaiyan Y. Underexpression of deleted in liver cancer 2 (DLC2) is associated with overexpression of RhoA and poor prognosis in hepatocellular carcinoma. BMC Cancer. 2008;8:205.

[55] Leung TH, Ching YP, Yam JW, Wong CM, Yau TO, Jin DY, et al. Deleted in liver cancer 2 (DLC2) suppresses cell transformation by means of inhibition of RhoA activity. Proc Natl Acad Sci U S A. 2005;102(42):15207-12.

[56] Nagaraja GM, Kandpal RP. Chromosome 13q12 encoded Rho GTPase activating protein suppresses growth of breast carcinoma cells, and yeast two-hybrid screen shows its interaction with several proteins. Biochem Biophys Res Commun. 2004;313(3): 654-65.

[57] Durkin ME, Avner MR, Huh CG, Yuan BZ, Thorgeirsson SS, Popescu NC. DLC-1, a Rho GTPase-activating protein with tumor suppressor function, is essential for embryonic development. FEBS letters. 2005;579(5):1191-6.

[58] Xue W, Krasnitz A, Lucito R, Sordella R, Vanaelst L, Cordon-Cardo C, et al. DLC1 is a chromosome 8p tumor suppressor whose loss promotes hepatocellular carcinoma. Genes Dev. 2008;22(11):1439-44.

[59] Yau TO, Leung TH, Lam S, Cheung OF, Tung EK, Khong PL, et al. Deleted in liver cancer 2 (DLC2) was dispensable for development and its deficiency did not aggravate hepatocarcinogenesis. PLoS One. 2009;4(8):e6566.

[60] Denholm B, Brown S, Ray RP, Ruiz-Gomez M, Skaer H, Hombria JC. crossveinless-c is a RhoGAP required for actin reorganisation during morphogenesis. Development. 2005;132(10):2389-400.

[61] Sato D, Sugimura K, Satoh D, Uemura T. Crossveinless-c, the Drosophila homolog of tumor suppressor DLC1, regulates directional elongation of dendritic branches via down-regulating Rho1 activity. Genes Cells. 2010;15(5):485-500.

[62] Ahmadian MR, Stege P, Scheffzek K, Wittinghofer A. Confirmation of the arginine-finger hypothesis for the GAP-stimulated GTP-hydrolysis reaction of Ras. Nat Struct Biol. 1997;4(9):686-9.

[63] Healy KD, Hodgson L, Kim TY, Shutes A, Maddileti S, Juliano RL, et al. DLC-1 suppresses non-small cell lung cancer growth and invasion by RhoGAP-dependent and independent mechanisms. Mol Carcinog. 2008;47(5):326-37.

[64] Stapleton D, Balan I, Pawson T, Sicheri F. The crystal structure of an Eph receptor SAM domain reveals a mechanism for modular dimerization. Nat Struct Biol. 1999;6(1):44-9.

[65] Chi SW, Ayed A, Arrowsmith CH. Solution structure of a conserved C-terminal domain of p73 with structural homology to the SAM domain. EMBO J. 1999;18(16): 4438-45.

[66] Mercurio FA, Marasco D, Pirone L, Pedone EM, Pellecchia M, Leone M. Solution structure of the first Sam domain of Odin and binding studies with the EphA2 receptor. Biochemistry. 2012;51(10):2136-45.

[67] Qiao F, Bowie JU. The many faces of SAM. Science's STKE : signal transduction knowledge environment. 2005;2005(286):re7.

[68] Li H, Fung KL, Jin DY, Chung SS, Ching YP, Ng IO, et al. Solution structures, dynamics, and lipid-binding of the sterile alpha-motif domain of the deleted in liver cancer 2. Proteins. 2007;67(4):1154-66.

[69] Kim TY, Healy KD, Der CJ, Sciaky N, Bang YJ, Juliano RL. Effects of structure of Rho GTPase-activating protein DLC-1 on cell morphology and migration. J Biol Chem. 2008;283(47):32762-70.

[70] Zhong D, Zhang J, Yang S, Soh UJ, Buschdorf JP, Zhou YT, et al. The SAM domain of the RhoGAP DLC1 binds EF1A1 to regulate cell migration. J Cell Sci. 2009;122(Pt 3): 414-24.

[71] Thornton S, Anand N, Purcell D, Lee J. Not just for housekeeping: protein initiation and elongation factors in cell growth and tumorigenesis. Journal of molecular medicine. 2003;81(9):536-48.

[72] Liu G, Grant WM, Persky D, Latham VM, Jr., Singer RH, Condeelis J. Interactions of elongation factor 1alpha with F-actin and beta-actin mRNA: implications for anchoring mRNA in cell protrusions. Mol Biol Cell. 2002;13(2):579-92.

[73] Murray JW, Edmonds BT, Liu G, Condeelis J. Bundling of actin filaments by elongation factor 1 alpha inhibits polymerization at filament ends. J Cell Biol. 1996;135(5): 1309-21.

[74] Gross SR, Kinzy TG. Translation elongation factor 1A is essential for regulation of the actin cytoskeleton and cell morphology. Nature structural & molecular biology. 2005;12(9):772-8.

[75] Moore RC, Durso NA, Cyr RJ. Elongation factor-1alpha stabilizes microtubules in a calcium/calmodulin-dependent manner. Cell motility and the cytoskeleton. 1998;41(2):168-80.

[76] Alpy F, Tomasetto C. Give lipids a START: the StAR-related lipid transfer (START) domain in mammals. J Cell Sci. 2005;118(Pt 13):2791-801.

[77] Clark BJ. The mammalian START domain protein family in lipid transport in health and disease. The Journal of endocrinology. 2012;212(3):257-75.

[78] Iyer LM, Koonin EV, Aravind L. Adaptations of the helix-grip fold for ligand binding and catalysis in the START domain superfamily. Proteins. 2001;43(2):134-44.

[79] Soccio RE, Adams RM, Romanowski MJ, Sehayek E, Burley SK, Breslow JL. The cholesterol-regulated StarD4 gene encodes a StAR-related lipid transfer protein with two

closely related homologues, StarD5 and StarD6. Proc Natl Acad Sci U S A. 2002;99(10):6943-8.

[80] Strauss JF, 3rd, Kishida T, Christenson LK, Fujimoto T, Hiroi H. START domain proteins and the intracellular trafficking of cholesterol in steroidogenic cells. Mol Cell Endocrinol. 2003;202(1-2):59-65.

[81] Zhang M, Liu P, Dwyer NK, Christenson LK, Fujimoto T, Martinez F, et al. MLN64 mediates mobilization of lysosomal cholesterol to steroidogenic mitochondria. J Biol Chem. 2002;277(36):33300-10.

[82] Stocco DM. StAR protein and the regulation of steroid hormone biosynthesis. Annu Rev Physiol. 2001;63:193-213.

[83] Lavigne P, Najmanivich R, Lehoux JG. Mammalian StAR-related lipid transfer (START) domains with specificity for cholesterol: structural conservation and mechanism of reversible binding. Subcell Biochem. 2010;51:425-37.

[84] Ng DC, Chan SF, Kok KH, Yam JW, Ching YP, Ng IO, et al. Mitochondrial targeting of growth suppressor protein DLC2 through the START domain. FEBS Lett. 2006;580(1):191-8.

[85] Hatch GM, Gu Y, Xu FY, Cizeau J, Neumann S, Park JS, et al. StARD13(Dlc-2) RhoGap mediates ceramide activation of phosphatidylglycerolphosphate synthase and drug response in Chinese hamster ovary cells. Mol Biol Cell. 2008;19(3):1083-92.

[86] Tsujishita Y, Hurley JH. Structure and lipid transport mechanism of a StAR-related domain. Nat Struct Biol. 2000;7(5):408-14.

[87] Romanowski MJ, Soccio RE, Breslow JL, Burley SK. Crystal structure of the Mus musculus cholesterol-regulated START protein 4 (StarD4) containing a StAR-related lipid transfer domain. Proc Natl Acad Sci U S A. 2002;99(10):6949-54.

[88] Thorsell AG, Lee WH, Persson C, Siponen MI, Nilsson M, Busam RD, et al. Comparative structural analysis of lipid binding START domains. PLoS One. 2011;6(6):e19521.

[89] Kudo N, Kumagai K, Matsubara R, Kobayashi S, Hanada K, Wakatsuki S, et al. Crystal structures of the CERT START domain with inhibitors provide insights into the mechanism of ceramide transfer. Journal of molecular biology. 2010;396(2):245-51.

[90] Liao YC, Si L, deVere White RW, Lo SH. The phosphotyrosine-independent interaction of DLC-1 and the SH2 domain of cten regulates focal adhesion localization and growth suppression activity of DLC-1. J Cell Biol. 2007;176(1):43-9.

[91] Chan LK, Ko FC, Sze KM, Ng IO, Yam JW. Nuclear-Targeted Deleted in Liver Cancer 1 (DLC1) Is Less Efficient in Exerting Its Tumor Suppressive Activity Both In Vitro and In Vivo. PLoS One. 2011;6(9):e25547.

[92] Yam JW, Ko FC, Chan CY, Jin DY, Ng IO. Interaction of deleted in liver cancer 1 with tensin2 in caveolae and implications in tumor suppression. Cancer Res. 2006;66(17): 8367-72.

[93] Brazil DP, Yang ZZ, Hemmings BA. Advances in protein kinase B signalling: AKTion on multiple fronts. Trends Biochem Sci. 2004;29(5):233-42.

[94] Wuestefeld T, Zender L. DLC1 and liver cancer: the Akt connection. Gastroenterology. 2010;139(4):1093-6.

[95] Hers I, Wherlock M, Homma Y, Yagisawa H, Tavare JM. Identification of p122RhoGAP (deleted in liver cancer-1) Serine 322 as a substrate for protein kinase B and ribosomal S6 kinase in insulin-stimulated cells. J Biol Chem. 2006;281(8):4762-70.

[96] Ko FC, Chan LK, Tung EK, Lowe SW, Ng IO, Yam JW. Akt phosphorylation of Deleted in Liver Cancer 1 abrogates its suppression of liver cancer tumorigenesis and metastasis. Gastroenterology. 2010.

[97] Niggli V. Regulation of protein activities by phosphoinositide phosphates. Annu Rev Cell Dev Biol. 2005;21:57-79.

[98] Parton RG, Simons K. The multiple faces of caveolae. Nat Rev Mol Cell Biol. 2007;8(3):185-94.

[99] Couet J, Li S, Okamoto T, Ikezu T, Lisanti MP. Identification of peptide and protein ligands for the caveolin-scaffolding domain. Implications for the interaction of caveolin with caveolae-associated proteins. The Journal of biological chemistry. 1997;272(10):6525-33.

[100] Scholz RP, Regner J, Theil A, Erlmann P, Holeiter G, Jahne R, et al. DLC1 interacts with 14-3-3 proteins to inhibit RhoGAP activity and block nucleocytoplasmic shuttling. J Cell Sci. 2009;122(Pt 1):92-102.

[101] Williams CL. The polybasic region of Ras and Rho family small GTPases: a regulator of protein interactions and membrane association and a site of nuclear localization signal sequences. Cell Signal. 2003;15(12):1071-80.

[102] Stallings JD, Tall EG, Pentyala S, Rebecchi MJ. Nuclear translocation of phospholipase C-delta1 is linked to the cell cycle and nuclear phosphatidylinositol 4,5-bisphosphate. J Biol Chem. 2005;280(23):22060-9.

[103] Yagisawa H, Okada M, Naito Y, Sasaki K, Yamaga M, Fujii M. Coordinated intracellular translocation of phosphoinositide-specific phospholipase C-delta with the cell cycle. Biochimica et biophysica acta. 2006;1761(5-6):522-34.

[104] Okada M, Taguchi K, Maekawa S, Fukami K, Yagisawa H. Calcium fluxes cause nuclear shrinkage and the translocation of phospholipase C-delta1 into the nucleus. Neurosci Lett. 2010;472(3):188-93.

[105] Yamaga M, Fujii M, Kamata H, Hirata H, Yagisawa H. Phospholipase C-delta1 contains a functional nuclear export signal sequence. The Journal of biological chemistry. 1999;274(40):28537-41.

[106] Yagisawa H, Yamaga M, Okada M, Sasaki K, Fujii M. Regulation of the intracellular localization of phosphoinositide-specific phospholipase Cdelta(1). Advances in enzyme regulation. 2002;42:261-84.

[107] Holeiter G, Heering J, Erlmann P, Schmid S, Jahne R, Olayioye MA. Deleted in liver cancer 1 controls cell migration through a Dia1-dependent signaling pathway. Cancer Res. 2008;68(21):8743-51.

[108] Leung TH, Yam JW, Chan LK, Ching YP, Ng IO. Deleted in liver cancer 2 suppresses cell growth via the regulation of the Raf-1-ERK1/2-p70S6K signalling pathway. Liver Int. 2010.

[109] Chan LK, Ko FC, Ng IO, Yam JW. Deleted in liver cancer 1 (DLC1) utilizes a novel binding site for Tensin2 PTB domain interaction and is required for tumor-suppressive function. PLoS One. 2009;4(5):e5572.

[110] Yuan BZ, Jefferson AM, Baldwin KT, Thorgeirsson SS, Popescu NC, Reynolds SH. DLC-1 operates as a tumor suppressor gene in human non-small cell lung carcinomas. Oncogene. 2004;23(7):1405-11.

[111] Hacker G. The morphology of apoptosis. Cell Tissue Res. 2000;301(1):5-17.

[112] Zender L, Spector MS, Xue W, Flemming P, Cordon-Cardo C, Silke J, et al. Identification and validation of oncogenes in liver cancer using an integrative oncogenomic approach. Cell. 2006;125(7):1253-67.

[113] Kim TY, Lee JW, Kim HP, Jong HS, Jung M, Bang YJ. DLC-1, a GTPase-activating protein for Rho, is associated with cell proliferation, morphology, and migration in human hepatocellular carcinoma. Biochem Biophys Res Commun. 2007;355(1):72-7.

[114] Zhou X, Zimonjic DB, Park SW, Yang XY, Durkin ME, Popescu NC. DLC1 suppresses distant dissemination of human hepatocellular carcinoma cells in nude mice through reduction of RhoA GTPase activity, actin cytoskeletal disruption and down-regulation of genes involved in metastasis. Int J Oncol. 2008;32(6):1285-91.

[115] Song LJ, Ye SL, Wang KF, Weng YQ, Liang CM, Sun RX, et al. [Relationship between DLC-1 expressions and metastasis in hepatocellular carcinoma]. Zhonghua Gan Zang Bing Za Zhi. 2005;13(6):428-31.

[116] Shih YP, Liao YC, Lin Y, Lo SH. DLC1 negatively regulates angiogenesis in a paracrine fashion. Cancer Res. 2010;70(21):8270-5.

[117] Chung GE, Yoon JH, Lee JH, Kim HY, Myung SJ, Yu SJ, et al. Ursodeoxycholic acid-induced inhibition of DLC1 protein degradation leads to suppression of hepatocellular carcinoma cell growth. Oncol Rep. 2011;25(6):1739-46.

[118] Yang X, Popescu NC, Zimonjic DB. DLC1 interaction with S100A10 mediates inhibition of in vitro cell invasion and tumorigenicity of lung cancer cells through a Rho-GAP-independent mechanism. Cancer Res. 2011;71(8):2916-25.

[119] Okumura K, Osanai T, Kosugi T, Hanada H, Ishizaka H, Fukushi T, et al. Enhanced phospholipase C activity in the cultured skin fibroblast obtained from patients with coronary spastic angina: possible role for enhanced vasoconstrictor response. Journal of the American College of Cardiology. 2000;36(6):1847-52.

[120] Nakano T, Osanai T, Tomita H, Sekimata M, Homma Y, Okumura K. Enhanced activity of variant phospholipase C-delta1 protein (R257H) detected in patients with coronary artery spasm. Circulation. 2002;105(17):2024-9.

[121] Murakami R, Osanai T, Tomita H, Sasaki S, Maruyama A, Itoh K, et al. p122 protein enhances intracellular calcium increase to acetylcholine: its possible role in the pathogenesis of coronary spastic angina. Arterioscler Thromb Vasc Biol. 2010;30(10): 1968-75.

[122] Kandabashi T, Shimokawa H, Miyata K, Kunihiro I, Kawano Y, Fukata Y, et al. Inhibition of myosin phosphatase by upregulated rho-kinase plays a key role for coronary artery spasm in a porcine model with interleukin-1beta. Circulation. 2000;101(11): 1319-23.

[123] Sato M, Tani E, Fujikawa H, Kaibuchi K. Involvement of Rho-kinase-mediated phosphorylation of myosin light chain in enhancement of cerebral vasospasm. Circulation research. 2000;87(3):195-200.

[124] Masumoto A, Mohri M, Shimokawa H, Urakami L, Usui M, Takeshita A. Suppression of coronary artery spasm by the Rho-kinase inhibitor fasudil in patients with vasospastic angina. Circulation. 2002;105(13):1545-7.

[125] Sakurada S, Takuwa N, Sugimoto N, Wang Y, Seto M, Sasaki Y, et al. Ca2+-dependent activation of Rho and Rho kinase in membrane depolarization-induced and receptor stimulation-induced vascular smooth muscle contraction. Circulation research. 2003;93(6):548-56.

[126] Nuno DW, England SK, Lamping KG. RhoA localization with caveolin-1 regulates vascular contractions to serotonin. American journal of physiology Regulatory, integrative and comparative physiology. 2012;303(9):R959-67.

Chapter 8

MIG-6 and SPRY2 in the Regulation of Receptor Tyrosine Kinase Signaling: Balancing Act via Negative Feedback Loops

Yu-Wen Zhang and George F. Vande Woude

Additional information is available at the end of the chapter

http://dx.doi.org/10.5772/54393

1. Introduction

Tumor suppressor genes (TSGs) function in concert with diverse parts of the cellular machinery and integrate the signaling networks in a cell [1]. TSGs act to safeguard the networks, to fine-tune signaling outputs, and to maintain tissue homeostasis. The loss of tumor suppressor activity or inactivation of a TSG is often due to genetic alterations such as a loss-of-function mutation or a deletion in the gene; alternatively, epigenetic silencing can result from methylation or histone modification in the TSG's promoter regulatory elements [1-3]. Cells with a loss or a significant reduction of a particular tumor suppressor's activity are prone to develop neoplasia in the tissues/organs where the TSG is expressed [1, 2].

The properties and modes of action of TSGs can be very distinct from one class to another. Most TSGs encode proteins that participate in controlling cell cycle progression, inducing apoptosis, repairing damaged DNAs, or performing other important functions [1]. TSGs can also be a source of microRNAs, a class of small hairpin RNAs that are transcribed in many cells and may act as tumor suppressors by regulating the expression of their targeted genes [4]. In this chapter, we will focus on one class of the TSGs, represented by the mitogen-inducible gene 6 (MIG-6) and Sprouty 2 (SPRY2), whose activities are crucial in regulating receptor tyrosine kinases signaling through negative feedback loops.

Receptor tyrosine kinases (RTKs) are important cellular components, and there are nearly 60 members encoded in the genome [5, 6]. They all possess a single transmembrane domain linking their extracellular ligand-binding region to the cytoplasmic region in which the catalytic kinase domain and the domain for docking of downstream signaling molecules reside. Upon binding of the ligand to its physiologic RTK partner, the receptors dimerize, resulting in autophosphorylation of key tyrosine residues in the kinase domain. This leads to a conformational change in the receptor and the recruitment of downstream signaling molecules to its docking domain or in close proximity for phosphorylation and activation. In this cellular process, the RTK plays a central role by relaying external stimuli (ligands) to internal signaling cascades such as the RAS-MAPK or PI3K-AKT pathways, translating the signal input into biological actions ranging from mitogenesis, to motility, morphogenesis, metabolism, and many others [5, 6].

RTK signaling is essential in many developmental processes and in normal physiology, and the actions of RTKs must be controlled temporally and spatially [5, 6]. Their actions are tightly regulated at several molecular levels to ensure appropriate cellular responses. Among the mechanisms that keep RTK signaling in "check and balance" are receptor endocytosis/degradation, dephosphorylation by protein tyrosine phosphatases (PTPs), and negative feedback regulation. Signal overactivity caused by inappropriate RTK activation can lead to serious pathological outcomes, particularly cancer. Thus, many RTKs such as epidermal growth factor receptor (EGFR) and the N-methyl-N'-nitroso-guanidine human osteosarcoma (MNNG HOS) transforming gene (MET) have been classified as oncogenes, because activating mutations, amplifications or other anomalies in these receptors have been identified in various human cancers, and their roles in the development and progression of tumor malignancy have been well documented [7, 8]. For example, aberrant activation of MET can result in deregulated cell proliferation, transformation, and promotion of tumor cell invasion and metastasis [8].

Unlike the rapid attenuation resulting from receptor endocytosis/degradation or PTP-mediated dephosphorylation, negative feedback regulation of RTK signaling by MIG-6 and SPRY2 is a delayed event because it requires *de novo* mRNA and protein syntheses. The expression of both MIG-6 and SPRY2 can be induced by ligands of many RTKs including epidermal growth factor (EGF), hepatocyte growth factor (HGF), and fibroblast growth factor (FGF) [9, 10]. In turn, MIG-6 and SPRY2 exerts their inhibitory activities on RTK signaling by either directly affecting the receptor itself or by modulating the signaling molecules downstream of the RTK. In this review, we will highlight the current understanding of how MIG-6 and SPRY2 regulates RTK signaling via negative feedback loops, and shed lights on why the loss of their tumor suppressor activities may affect RTK signaling in cancer cells, as well as their impact on cancer therapy.

2. The features and functions of MIG-6 and SPRY2

2.1. MIG-6

MIG-6 (also known as *gene 33*, *ERRFI1* or *RALT*) is a unique and immediate early response gene that is not present in relatively simple organisms like *Drosophila* and *C. elegans*. It emerges in the more complex, higher-order species [9, 11], underlying its importance in evolution. It encodes a 58 kDa nonkinase protein that resides in the cytoplasm and functions as a scaffolding adaptor for modulating signal transduction. Structurally, MIG-6 has several functional motifs/domains that are crucial for interaction with other signaling molecules [9, 12]. The Cdc42/Rac-interaction and binding (CRIB) domain of MIG-6 (Figure 1A) shares consensus sequences with many other proteins that associate with Cdc42 or Rac small GTPases, which are important regulators of actin cytoskeleton remodeling and signal transduction [11, 13]. CRIB domain mediates the binding of MIG-6 to active (GTP-bound) Cdc42 and negatively regulates HGF-induced Cdc42 activation and cell migration [12, 14]. This domain has also been shown to play a role in regulating transactivation of nuclear factor κB (NFκB) by sequestering the inhibitor of κBα (IκBα) [15, 16]. Within its middle region, MIG-6 has several proline-rich motifs that likely mediate its binding to various Src homology-3 (SH3] domain-containing proteins such as GRB2, Src, PI3K p85 and PLC-γ [14, 17]. MIG-6 interacts with 14-3-3ζ via the 14-3-3 protein binding motif [12]. MIG-6 also possesses two PEST sequences, and is targeted by ubiquitination and proteasome degradation [18]. The ErbB-binding region (EBR), a large portion of the carboxyl terminus in MIG-6, is required for physical interaction with EGFR family receptors, which resulted in attenuation of EGFR/ErbBs signaling [17, 19, 20]. The EBR domain shares a high homology with the activated Cdc42-associated kinase 1 (ACK1), a non-receptor tyrosine kinase that also interacts with and regulates EGFR [21].

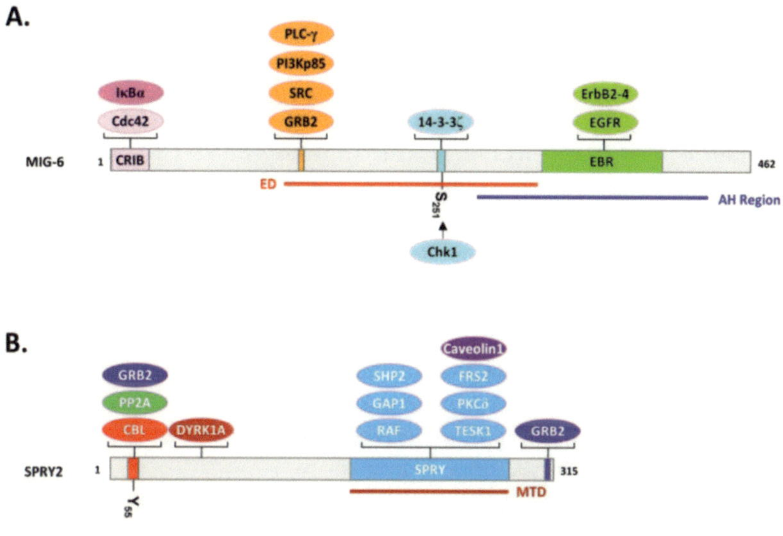

Figure 1. MIG-6 and SPRY2 structures. A. MIG-6 protein structural features and its interacting partners. The CRIB domain (amino acids 1-38) interacts with Cdc42 and IκBα. The orange box indicates the SH3-domain binding motif that likely mediates interactions with SH3-domain-containing proteins such as GRB2 and PI3Kp85. The cyan box (amino acids 246-253) marks the 14-3-3 binding motif in which serine residue 251 can be phosphorylated by Chk1 kinase. The EBR domain (amino acids 337-412) binds to EGFR and other ErbB members. The red bar (ED) indicates the MIG-6 endocytic domain (amino acids 143-323); the blue bar indicates the Ack homology (AH) region (amino acids 264-424). B. Structural features of SPRY2 and its partner molecules. The red box shows the SH2-domain binding motif (amino acids 50-60) that binds to CBL and GRB2; the key tyrosine residue Y_{55} is also indicated. The conserved cysteine-rich SPRY domain (amino acids 178-293) is crucial for its ability to interact with signaling molecules such as FRS2 and SHP2. The SPRY domain is also responsible for membrane translocation (MTD). There is an SH3-domain binding motif (amino acids 303-309) shown in blue at the C-terminal end of SPRY2 that also binds to GRB2.

MIG-6 is expressed in many tissues/organs, with high expression in liver and kidney and low to moderate expression in brain, lung, heart, and other tissues [22, 23]. Its expression can be induced by diverse factors ranging from hormones and growth factors, to chemical agents, to stress stimuli [9]. The induction of MIG-6 expression by growth factors is mainly mediated by the RAS-MEK-MAPK/ERK pathway, while other inducers may involve other pathways such as PI3K [9]. MIG-6 has also been reported to be a G-actin-regulated target gene, because the actin-MAL-serum response factor (SRF) cascade mediates MIG-6 induction by serum or lipid agonists such as lysophosphatidic acid (LPA) or sphingosine 1-phosphate (S1P) [24]. MIG-6 may play a crucial role in patho-physiological conditions such as myocardial ischemic injury and infarction [25], liver regeneration [26-28], joint mechanical injury [29], and diabetic nephropathy and hypertension [12]. Its activity is required for skin morphogenesis [30, 31] and lung development in mice [32], and it plays an important role in the maintenance of tissue homeostasis in joints, the lungs and the uterus [32-35].

2.2. SPRY2

In term of evolution, the *Spry* gene emerged far earlier than *Mig-6*. *SPRY2* is the mammalian homolog of *Drosophila melanogaster Spry* (*dSpry*), and is one of the four *SPRY* genes in the human genome [10]. The dSpry protein is 63 kDa, while its mammalian counterparts are 32–34 kDa, but they all contain a functional cysteine-rich region in their C-terminus (designated the SPRY domain) and an SH2-binding motif carrying a conserved tyrosine at the N-terminus (Tyr55 residue in human SPRY2) (Figure 1B). These conserved regions are essential for SPRY2 to fully execute its inhibitory function in the regulation of RTK signaling [10, 36, 37]. The SPRY domain is also found in the SPRED (Sprouty-related EVH1 domain-containing protein) family, which like SPRY proteins inhibits RTK signaling upon stimulation by growth factors [10, 38]. The SPRY domain mediates the binding of SPRY2 to many signaling molecules including GAP1, FRS2, SHP2, RAF, PKCδ, TESK1 and caveolin1 [10, 39]. The SPRY domain is also required for translocation of SPRY2 to the membrane during its activation. The Tyr55 residue is essential for the SPRY2 protein's interaction with CBL, PP2A and GRB2 [10, 39]. A cryptic SH3-binding motif (PxxPxR) in the C-terminal end of SPRY2 (but not present in other SPRY members) has been shown to mediate GRB2 interaction as well [39-41]. The dual-specificity tyrosine phosphorylation–regulated kinase (DYRK1A) interacts with and phosphorylates Thr75 on SPRY2 [42]. SPRY2 is targeted for ubiquitination and proteasome degradation by at least two ubiquitin E3 ligases: CBL-mediated Tyr55 phosphorylation-dependent ubiquitination and SIAH2-mediated Tyr55 phosphorylation-independent ubiquitination [43, 44]. On the other hand, SPRY2 protein can be stabilized by phosphorylation of its serines 112 and 121 residues by the mitogen-activated protein kinase-interacting kinase 1 (Mnk1], thereby decreasing growth factor–induced degradation [45].

In *Drosophila, dSpry* expression is detected at the tips of branching lung buds and is induced by branchless, the *Drosophila* Fgf. Losing *dSpry* leads to excessive tracheal branching as a result of increased Fgf signaling activity [46]. The *Xenopus* homolog, *xSpry2,* is expressed in a pattern resembling that of *Xenopus* Fgf8 and inhibits Fgf-mediated gastrulation [47]. During mammalian embryogenesis, Spry2 expression tends to localize closest to the sites where Fgf

http://dx.doi.org/10.5772/54393

activity is needed for organ/tissue development, underlying the importance of this molecule as an intrinsic Fgf signaling regulator in organogenesis [48-50]. Mouse *Spry2* is highly expressed in the terminal buds of peripheral mesenchyme in the embryonic lung, adjacent to that of Fgf10, a key mouse lung-branching morphogen [50]. Ectopic expression of Spry2 in the mouse pulmonary epithelium results in decreased branching; exogenous Fgf10 produces greater lung branching and higher Spry2 expression [50]. Spry2 also plays a role in mouse kidney development; its ectopic expression in the ureteric bud leads to postnatal kidney failure due to deficiency in ureteric branching [51]. Moreover, Spry2 deficiency in mice results in defects of the auditory sensory epithelium development in the inner ear, and leads to enteric neuronal hyperplasia and esophageal achalasia [52, 53]. The lack of phenotypes in other Spry2-deficient tissues is likely due to compensatory roles of other Spry family members, because there are overlapping expressions of Spry1, 2 and 4 in many tissues during the development [54]. In adult mice, Spry2 expression is abundant in the brain, lung, heart, kidney, skeletal muscle and mammary glands [48, 55].

Beyond being an intrinsic inhibitor for Fgf signaling, Spry also regulates signaling driven by other RTKs like Egfr. The *dSpry* gene is required for eye and wing development in *Drosophila*, antagonizing Egfr signaling for neuronal induction in the retina and for vein formation in the wings [56, 57]. Loss of *dSpry* results in excess photoreceptors, cone cells and pigment cells in the retina, while its overexpression leads to phenotypes that mimic loss of Egfr signaling [56, 57].

3. Negative feedback regulation of RTK signaling by MIG-6 and SPRY2

3.1. Regulation of RTK pathways by MIG-6

Many growth factors can induce MIG-6 expression, including EGF, FGF, and HGF [9]. Upon induction, MIG-6 proteins rapidly and transiently accumulate in the cytoplasm, where they feed back to inhibit the activated RTK signaling (Figure 2]. The inhibition of EGFR/ErbB signaling by MIG-6 can occur at two molecular levels: one on the receptor itself, and another on the signaling molecules downstream of the receptor (Figure 2). Through its C-terminal EBR domain, MIG-6 directly binds to EGFR and other ErbB members [17, 19, 20, 58]. The interaction involves the kinase domain of EGFR or ErbB2 and requires their catalytic activities, but does not involve their C-terminal regions in which there are tyrosine residues essential for activating downstream signaling [20, 58]. Crystal structures reveal that the MIG-6 EBR domain binds to the distal surface of the carboxy-terminal lobe (C-lobe) in the kinase domain of EGFR [59]. The C-lobe is crucial in asymmetric EGFR dimer formation [60]; binding of MIG-6 to the C-lobe blocks the dimer interface thereby preventing EGFR activation [59]. MIG-6 coupling also promotes clathrin-mediated endocytosis of EGFR [61], an important mechanism for timely attenuation of ligand-induced EGFR activation [62, 63]. The region responsible for promoting EGFR endocytosis has been mapped to the endocytic domain (ED) of MIG-6 (see Figure 1A) [61], which mediates the binding of MIG-6 to the AP2 adaptor complex, a key component in forming clathrin-coated pits during endocytosis [63]. Interestingly, the non-receptor tyrosine kinase ACK1 also binds to EGFR upon EGF stimulation through a region sharing high homology with the MIG-6 EBR domain [21]. This interaction also regulates clathrin-mediated EGFR endocytosis and degradation [21, 64]. However, it is not clear whether MIG-6 and ACK1 cooperate in regulating EGFR turnover or whether they bind to EGFR in a mutually exclusive way to accomplish individual inhibitory roles under different circumstances. The internalized EGFR is guided to late endosome through the binding of MIG-6 to the endosomal SNARE complex component STX8, en route to degradation in the lysosome [61, 65].

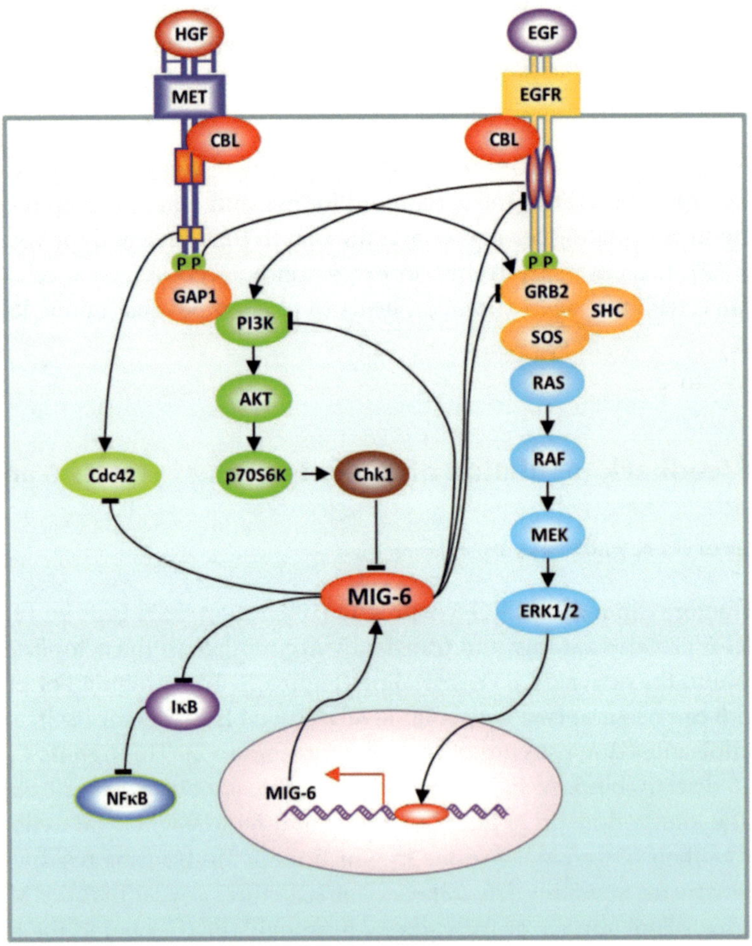

Figure 2. MIG-6 regulates EGFR and MET signaling via a negative feedback loop. Upon ligand stimulation, EGFR and MET activate the RAS-RAF-MEK-ERK pathway and induce *MIG-6* expression. In turn, MIG-6 exerts its inhibitory activity by interacting with signaling molecules to fine-tune RTK signaling and its timely attenuation. The direct MIG-6–EGFR interaction facilitates receptor endocytosis and degradation. This inhibitory activity is unique to the EGFR family and does not extend to other RTKs like MET. The interaction of MIG-6 with signaling molecules such as GRB2 and PI3Kp85 indistinguishably influences the RTK signaling in general, resulting in the inhibition of downstream pathways like RAS-MAPK/ERK and PI3K-AKT.

http://dx.doi.org/10.5772/54393

The magnitude of MIG-6-mediated inhibition of EGFR signaling is likely maximized and integrated by the second level of molecular regulation, that is, direct inhibition of downstream signaling molecules (Figure 2). MIG-6 binds to the SH3 domain-containing protein GRB2 [14, 17], the key molecule in linking activated RTK to the intracellular signaling cascade; GRB2 brings RAS together with SOS to the phosphorylated receptor for activation [66]. Binding of MIG-6 may block GRB2's interaction with activated EGFR and other RTKs or disrupt GRB2-SOS-RAS complex formation, thereby preventing activation of the RAS-MEK-MAPK pathway. MIG-6 also interacts with Src, the PI3K p85 subunit, PLCγ, and Fyn [17], although those interactions were demonstrated in an artificial system and their biological meaning remains to be determined. More complexity is likely in the dynamics of MIG-6-mediated RTK signaling regulation due to the interaction of MIG-6 with 14-3-3 proteins [12, 67], an adaptor family that may interact with diverse signaling molecules and regulate many biological activities [68]. Nonetheless, most of these direct downstream signaling regulations by MIG-6 remain largely speculation and require further investigation.

Assessing the effect of MIG-6 on RTK pathways other than those of the EGFR family, however, may provide insightful answers to such speculation, because direct regulation of the RTK itself by MIG-6 appears to be unique to the EGFR family. For instance, MIG-6 can be induced by HGF and function as negative feedback regulator of the MET pathway by inhibiting HGF-induced cell migration and proliferation [14], yet no physical interaction between MIG-6 and MET has been observed. Through its CRIB domain, MIG-6 can bind to and inhibit the activity of the Cdc42 small GTPase [12, 14], and this inhibitory activity is required for blocking of HGF-induced cell migration [14]. The CRIB domain also interacts with IκBα, thereby activating NFκB for transcriptional regulation of its target gene expression [15, 16]. Whether the inhibition of HGF-induced cell proliferation by MIG-6 is mediated by its ability to bind GRB2 or to bind other downstream molecules is still unknown. Negative feedback inhibition of other RTKs (such as FGFR and IGFR) by MIG-6 is also likely to be mediated by its inhibitory activities on the downstream signaling molecules rather than on the receptor itself.

The inhibitory activity of MIG-6 on RTK signaling seems to be modulated by phosphorylation; Liu et al. recently reported that MIG-6 can be phosphorylated by Chk1 upon EGF stimulation [67]. EGF activates Chk1 via the PI3K-AKT-S6K pathway, which in turn phosphorylates Ser251 of MIG-6 and results in inhibition of MIG-6 [67]. Thus, Chk1 counterbalances the EGFR inhibition of MIG-6, positively regulating EGFR signaling. Interestingly, Ser251 resides in the 14-3-3 binding motif of MIG-6 and its phosphorylation is likely involved in the MIG-6 and 14-3-3ζ interaction, because that interaction is abolished by Chk1 depletion [67]. In addition, two tyrosine residues (Tyr394 and Tyr458) in MIG-6 are phosphorylated by EGFR activation [69-71], but the underlying mechanism and the biological significance remain unclear.

3.2. Regulation of RTK pathways by SPRY2

SPRY2 renders another layer of modulation on RTK signaling via a negative feedback loop [10, 36, 37, 72]; its expression is induced by many activated RTKs including EGFR, FGFR,

MET, and VEGFR [10, 72]. As with MIG-6, SPRY2-mediated regulation can occur on two levels: on the RTK itself and on the downstream signaling molecules (Figure 3). However unlike MIG-6, the most prominent inhibitory activity of SPRY2 is derived from its abilities to interact with downstream signaling molecules centering on the RAS-RAF-MAPK pathway, while its effect on the RTK itself seems to be indirect and may provide some signaling specificity for different RTKs [10, 36, 37].

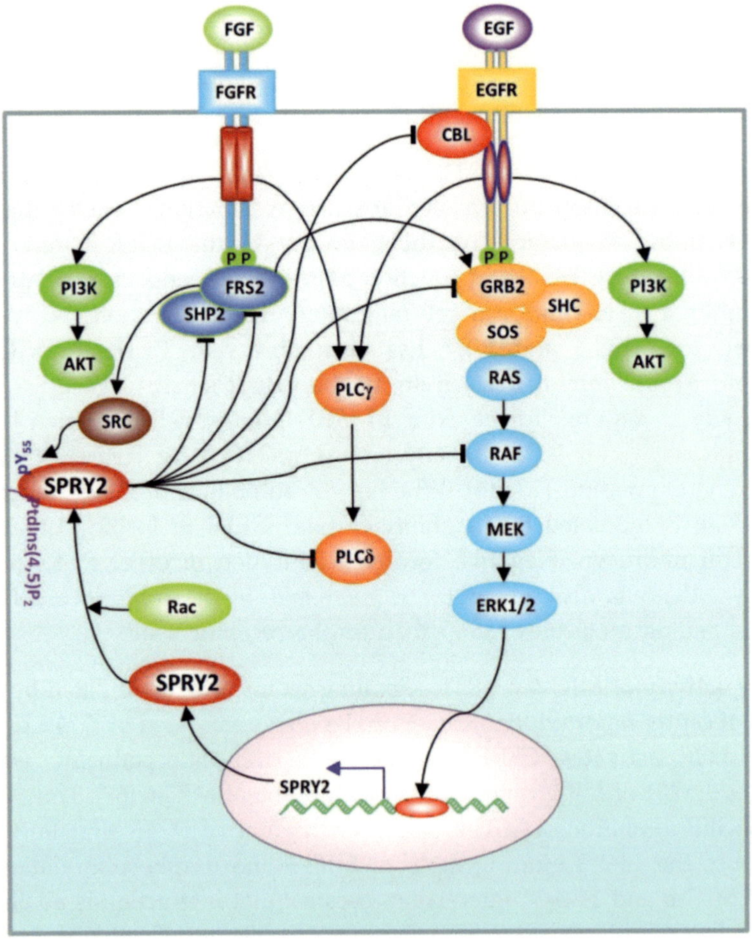

Figure 3. Feedback regulation of EGFR and FGFR signaling by SPRY2. SPRY2, upon induction, translocates to the membrane by binding to PtdIns(4,5)P_2, and is phosphorylated on its Y_{55} residue. This modification is essential for SPRY2's inhibitory activity, which is mediated by interaction with many signaling proteins including FRS2, SHP2, GRB2, RAF and PLCδ. On the other hand, its interaction with CBL E3 ubiquitin ligase may positively regulate EGFR signaling, because such interaction sequesters CBL and prevents CBL-mediated EGFR degradation.

Upon growth factor stimulation, SPRY2 translocates from the cytosol to the plasma membrane where it binds to phosphatidylinositol-4,5-bisphosphate [PtdIns(4,5)P2] via the conserved SPRY domain [73-75]. Membrane translocation appears to be triggered by the activated Rac small GTPase and is essential for SPRY2 function [76]. The phosphorylation of the Tyr55 residue in SPRY2 is likely regulated by a SRC family kinase, which upon FGF stimulation is recruited to activated FGFR by FRS2 [77]. While Tyr55 phosphorylation enables the binding of SPRY2 to GRB2 via the SH2-binding motif in the N-terminus [78], the SH3-binding motif in its C-terminus may also play a role in the SPRY2-GRB2 interaction [39-41] (Figure 1B). The latter is likely regulated by the protein phosphatase 2A (PP2A), a serine/threonine phosphatase that interacts with Tyr55 and dephosphorylates certain serine residues in SPRY2 for permitting access of GRB2 to the SH3-binding motif on SPRY2 [41, 45]. Consequently, the binding of SPRY2 to GRB2 prevents the recruitment of the GRB2-SOS complex to the FRS2 adaptor or SHP2 phosphatase proximal to the activated RTK, thereby inhibiting the activation of RAS and its downstream molecules [10, 36, 39, 77, 78]. The RTK-RAS-RAF-MAPK/ERK pathway can also be inhibited by direct binding of SPRY2 to RAF [75, 79, 80]. On the other hand, TESK1 negatively regulates SPRY2 inhibitory activity by interfering with the SPRY2 interaction with GRB2 and PP2A [81], while DYRK1A binding results in Thr75 phosphorylation on SPRY2 and suppression of SPRY2 inhibitory activity, thereby enhancing FGF-induced ERK activation [42].

The regulation of EGFR signaling by SPRY2 appears to be more complicated than that of FGFR signaling, because SPRY2 can indirectly influence the turnover of EGFR through its interaction with CBL (Figure 3) [10, 36, 37]. This regulation is mediated by the SH2-binding motif, which includes Tyr55 in the N-terminus of SPRY2. When Tyr55 is phosphorylated, SPRY2 interacts with the SH2-domain on CBL, and prevents CBL from binding to the activated EGFR, thereby interfering with CBL-mediated EGFR endocytosis and degradation [43, 82-84]. This action can prolong the activation of EGFR and the downstream RAS-RAF-MAPK pathway, contrary to the direct inhibitory activity of SPRY2 on the downstream signaling molecules. These two opposite activities render SPRY2 a delicate role in fine-tuning EGFR signaling, negative in some situations and positive in others, depending on the threshold and balance of these two activities.

It is conceivable that RTKs such as MET and PDGFR that are also substrates of CBL might also be affected by SPRY2 like that of EGFR [8, 85], while non-CBL-substrate RTKs like FGFR appears unaffected by SPRY2-CBL interaction [36]. However, it is known that MET protein level is not affected by SPRY2 overexpression that inhibits HGF-induced ERK and AKT activation, indicating that the effect of SPRY2-CBL interaction on EGFR and on MET might not be the same [72, 86]. The sequestration of CBL by SPRY2 can also affect downstream signaling molecules, as it may free proteins like GRB2 from CBL-mediated ubiquitination and degradation [85]. Furthermore, SPRY2 itself can be ubiquitinylated and degraded by the CBL binding, thereby influencing the RTK signaling [83].

4. Tumor suppressor role of MIG-6 and SPRY2 in cancer

4.1. MIG-6 as a tumor suppressor gene

Human chromosome 1p36, a locus frequently associated with many human cancers [87-92], harbors the *MIG-6* gene. In fact, loss or reduction of *MIG-6* expression has been observed in non-small cell lung cancer (NSCLC) [35, 93-95], breast carcinoma [30, 96], melanoma and skin cancer [30, 94], ovarian carcinoma [30], pancreatic cancer [30], endometrial cancer [34], thyroid cancer [97, 98], hepatocellular carcinoma [28], and glioblastoma [91]. Prognostically, low *MIG-6* expression is often associated with poor prognosis or poor patient survival [93, 96, 97].

Unlike many other tumor suppressor genes whose expression is directly regulated by epigenetic modification of their promoter regulatory elements [3], the silencing of *MIG-6* expression seems otherwise [94]. Although high in CpG contents, the *MIG-6* promoter appears hypomethylated and is not directly affected by either the DNA methyltransferase inhibitor 5-aza-2'-deoxycytidine [5-aza-dC) or the histone deacetylase inhibitor trichostatin A (TSA), indicating indirect regulation [94]. Interestingly, *MIG-6* expression seems to be differently induced by 5-aza-dC and TSA in different cancer types: it is induced by 5-aza-dC in melanoma but not in NSCLC and neuroblastoma, and by TSA in NSCLC and neuroblastoma but not in melanoma, suggesting a possible tissue-specific transcriptional regulation for this gene [90, 94, 96]. However in some papillary thyroid cancer, it is reported that the *MIG-6* promoter is hypermethylated as determined by methylation-specific PCR [98]. It is unclear at this point what cause such differences.

Besides loss or reduction of expression, *MIG-6* can also be inactivated by genetic mutation, even though this occurs rarely [35, 90, 96]. To date, two homozygous mutations (Asp109 to Asn, and Glu83 to a stop codon) in *MIG-6* were identified in human lung cancer cell lines, while heterozygous germline mutations were found in a primary lung cancer (Ala373 to Val) and a neuroblastoma patient (Asn343 to Ser) [35, 90]. Evidence that *MIG-6* is a bona fide tumor suppressor gene also arises from mouse studies. Mice with targeted disruption of *Mig-6* are prone to neoplastic development ranging from epithelial hyperplasia to carcinoma at multiple sites including the lung, gallbladder, bile duct, uterus, gastrointestinal tract and skin [30, 34, 35]. The carcinogen-induced skin cancer seen in Mig-6-deficient mice is likely mediated by EGFR-ERK/MAPK pathway, because inhibiting EGFR kinase activity with gefitinib or replacing it with a kinase-defective EGFR rescues the phenotype [30].

4.2. SPRY2 as a tumor suppressor gene

There is compelling evidence supporting SPRY2 as a tumor suppressor [99]. The *SPRY2* gene is located on human chromosome 13q31, where loss of heterozygosity (LOH) is observed in prostate cancer and hepatocellular carcinoma [86, 100]. Down-regulation of *SPRY2* expression has been reported in breast cancer [55, 101], hepatocellular carcinoma [86, 102, 103], NSCLC [104], prostate cancer [100, 105, 106], endometrial carcinoma [107], gliomas [108], and B-cell lymphomas [109, 110]. However in colon cancer, both downregulation and upregulation of *SPRY2* have been reported [111, 112]. Low (or no) *SPRY2* expression is associated with advanced tumor stages and poor survival, and it may be a significant prognostic factor in breast cancer [101], hepatocellular carcinoma [86, 103], prostate cancer [100, 105], gliomas [108], and colon cancer [111]. Further, *SPRY2* expression is inversely correlated with the level of miR-21 microRNA expression in gliomas and colon cancer, indicating that *SPRY2* is a target of miR-21[108, 111].

Downregulation of SPRY2 expression may be attributed to DNA methylation; hypermethylation of its promoter has been observed in prostate cancer [100], hepatocellular carcinoma [86], endometrial carcinoma [107] and B-cell lymphomas [109, 110]. However, controversial results have also been reported, since no methylation in *SPRY2* promoter was found in other studies involving different cancer types [55, 102, 106, 111], indicating that other epigenetic mechanisms might as well be responsible for *SPRY2* down-regulation. Thus far, no mutation has been identified in the *SPRY2* gene in any human cancers, and no neoplastic phenotypes have been observed in any Spry2-deficient mice. This may be due to compensatory roles played by other family members such as SPRY1 or SPRY4.

5. The impacts of MIG-6 or SPRY2 activity on RTK signaling in cancer

The loss of MIG-6 and SPRY2 feedback regulation leads to prolonged RTK signaling activation and may contribute to hallmark activities of cancer [113]. Ectopic expression of MIG-6 results in decreased phosphorylation of EGFR/ErbBs and downstream ERK/MAPK and AKT, and it inhibits cell proliferation in several cancer types [19, 20, 65, 96, 114]. In contrast, down-regulation of *MIG-6* expression by small interference RNA (siRNA)-mediated knockdown leads to prolonged activation of the EGFR or ErbB2 pathway and increases ligand-induced proliferation, cell cycle progression, and cell migration [28, 30, 65, 96, 114]. Likewise, *MIG-6* overexpression inhibits HGF-induced cell migration and proliferation, whereas MIG-6 knockdown by siRNA enhances those activities [14]. Intriguingly, it has been shown that in thyroid cancer, *MIG-6* overexpression suppresses MET phosphorylation along with the inhibition of EGFR, ErbB2, and SRC, while its knockdown does the opposite and enhances cell proliferation and invasion [98]. However, it is unclear how MIG-6 affects MET tyrosine phosphorylation given that no physical interaction between the two is observed [14]. The activity of MIG-6 on apoptosis is unsettled: one group showed that MIG-6 inhibits apoptosis of breast cancer cells [115], while another showed that it promotes the death of cardiomyocytes [25]. This discrepancy might be due to the differences in cellular states (cancer cells versus normal cells) or in tissue types (breast versus cardiac).

Nonetheless, a study of NCI-60 cell lines, which cover a broad spectrum of cancer types, revealed that *MIG-6* expression correlated with EGFR expression, indicating an intrinsic activity for MIG-6 in regulating EGFR signaling [116]. *MIG-6* expression has been shown to have a significant effect on ErbB-targeted cancer therapy [96, 116, 117]. MIG-6 synergizes with the EGFR inhibitor gefitinib to suppress the growth of NSCLC cells carrying gefitinib-sensitive EGFR mutations [116, 117]. Further, the loss of *MIG-6* expression renders ErbB2-amplified breast cancer cells more resistant to Herceptin (trastuzumab), a neutralizing antibody against ErbB2/HER2 [96]. It will be interesting to see the influence of *MIG-6* expression on cancer therapies targeting other RTKs such as MET as well.

SPRY2 overexpression inhibits MET-mediated ERK and AKT activation in leiomyosarcoma and hepatocellular carcinoma cells, and it suppresses cell proliferation, migration and invasion induced by HGF [72, 86, 102]. In NSCLC, SPRY2 suppresses ERK but not AKT activation, and it inhibits the migration of the cells expressing wild-type but not constitutively activated K-RAS; however, proliferation and tumorigenesis of cells with either wild-type or mutant K-RAS can be inhibited by *SPRY2* overexpression [104]. SPRY2 has different effects on wild-type and V599E mutant B-RAF in melanoma cells: its downregulation enhances ERK phosphorylation only in melanoma cells carrying wild-type B-RAF, likely because of its ability to interact with wild-type, but not mutant, B-RAF [118]. Overexpression of *SPRY2* suppresses ERK activation in osteosarcoma and B-cell lymphoma, and it inhibits tumor growth and metastasis *in vivo* [109, 119], while suppression of SPRY2 activity by its dominant negative mutant $SPRY2^{Y55F}$ enhances the proliferation and tumorigenesis of breast cancer cells [55]. Surprisingly, SPRY2 has also been reported to enhance cell proliferation and HGF-induced ERK and AKT activation, migration and invasion of colon cancer cells [112], quite opposite to another report showing that in the same cell line, SPRY2 negatively regulates ERK and AKT phosphorylation and inhibits proliferation, migration and tumorigenesis [111]. The discrepancy is quite puzzling, and it is unclear whether it is due to differences in the experimental approaches or to other factors such as clonal effects originating from tumor cell heterogeneity. Using an inducible system in the same cell line might be able to solve the puzzle of whether the effects in those two reports were truly the results of *SPRY2* overexpression.

There is limited evidence that *SPRY2* expression, like that of *MIG-6*, influences ErbB-targeted therapy in human cancers [101, 120]. Breast cancer patients with low *SPRY2* expression show poorer response to trastuzumab treatment, and have a significant lower survival rate relative to those with high *SPRY2* expression [101]. Low *SPRY2* expression is usually associated with high ErbB2/HER2 in breast cancer, while reconstituting SPRY2 may enhance the sensitivity of breast cancer cells *in vitro* to trastuzumab treatment [101]. In colon cancer cells, low *SPRY2* expression is associated with less sensitivity to gefitinib, whereas its ectopic overexpression can enhance gefitinib responsiveness [120].

6. Conclusion and perspective

The negative feedback loops to receptor tyrosine kinases of MIG-6 and of SPRY2 provide crucial intersecting points for tumor suppressor genes and oncogenes, placing them in the same signaling networks for regulating physiologic and oncogenic activity. This sophisticated regulatory mechanism allows timely attenuation of RTK signaling by those TSGs, and ensures a proper cellular response following growth factor stimulation. MIG-6 and SPRY2 are no more than the representatives for the class of TSGs involved in RTK signaling regulation, and we believe there are more such TSGs in the human genome either remain to be discovered or have already been revealed (such as other SPRY family members). To date, most of the studies on MIG-6 and SPRY2 have focused on their activity in regulating selected RTKs such as EGFR, MET and FGFR. Their roles in regulating most other RTKs and the clinical relevance of such regulation remain largely unknown. Also, there are conflicting results on how MIG-6 and SPRY2 may regulate the RTK signaling, on both the receptor and the downstream signaling levels, and further studies are required to address those issues. Although EGFR and other RTKs like MET appear to be regulated slightly differently by MIG-6 and SPRY2, it remains to be determined to what extent these TSGs may provide signaling specificity to different RTKs. Beyond all aforementioned issues, a broader question might be why an RTK network needs multiple negative feedback regulators to fine-tune its signaling; might the regulators function differently for each RTK in a temporal and spatial manner i.e. at right place, on right time and for right target?

While conventional mechanisms such as mutation or promoter methylation may contribute to the inactivation of MIG-6 or SPRY2 tumor suppressor roles, their activities are also likely silenced by unconventional means in cancer. For example, in most cancer types investigated so far, *MIG-6* expression is not down-regulated by direct methylation or histone deacetylation in its promoter, but rather by an indirect mechanism involving other unidentified transcriptional factor(s) or transcriptional co-regulator(s). It is also striking to see that different promoter methylation status in *MIG-6* or *SPRY2* gene is observed in different cancer types: methylated in some cancers, but unmethylated in others. The cause of such difference is a curiosity, but if only for TSG down-modulation, genome instability in the cancer cell environment could provide many mechanisms.

Clinically, it is important to understand how tumor suppressor genes may affect the therapeutic outcome of RTK-targeted therapies, which can be effective in treating certain human cancers. In a limited number of studies, low expression of *MIG-6* or *SPRY2* is associated with poorer patient responses to ErbB-targeted therapies (i.e., the EGFR inhibitor gefitinib and the ErbB2/HER2 inhibitor Herceptin). The question is, can *MIG-6* or *SPRY2* expression be used in conjunction with RTK status to select patients for RTK-targeted "personalized" cancer therapy? The approach sounds plausible, given that those TSGs negatively regulate the RTK signaling activities, but extensive studies will certainly be required before implementing such measures.

Acknowledgements

Thanks to David Nadziejka for critical reading and technical editing, and to the Van Andel Foundation for the financial support.

Author details

Yu-Wen Zhang* and George F. Vande Woude

*Address all correspondence to: YuWen.Zhang@vai.org

Van Andel Research Institute, Grand Rapids, Michigan, USA

References

[1] Sherr CJ. Principles of tumor suppression. Cell. 2004 Jan 23;116(2):235-46.

[2] Payne SR, Kemp CJ. Tumor suppressor genetics. Carcinogenesis. 2005 Dec;26(12): 2031-45.

[3] Jones PA, Baylin SB. The fundamental role of epigenetic events in cancer. Nat Rev Genet. 2002 Jun;3(6):415-28.

[4] Lujambio A, Lowe SW. The microcosmos of cancer. Nature. 2012 Feb 16;482(7385): 347-55.

[5] Blume-Jensen P, Hunter T. Oncogenic kinase signalling. Nature. 2001 May 17;411(6835):355-65.

[6] Schlessinger J. Cell signaling by receptor tyrosine kinases. Cell. 2000 Oct 13;103(2): 211-25.

[7] Hynes NE, Lane HA. ERBB receptors and cancer: the complexity of targeted inhibitors. Nat Rev Cancer. 2005 May;5(5):341-54.

[8] Birchmeier C, Birchmeier W, Gherardi E, Vande Woude GF. Met, metastasis, motility and more. Nat Rev Mol Cell Biol. 2003 Dec;4(12):915-25.

[9] Zhang YW, Vande Woude GF. Mig-6, signal transduction, stress response and cancer. Cell Cycle. 2007 Mar 1;6(5):507-13.

[10] Kim HJ, Bar-Sagi D. Modulation of signalling by Sprouty: a developing story. Nat Rev Mol Cell Biol. 2004 Jun;5(6):441-50.

[11] Pirone DM, Carter DE, Burbelo PD. Evolutionary expansion of CRIB-containing Cdc42 effector proteins. Trends Genet. 2001 Jul;17(7):370-3.

[12] Makkinje A, Quinn DA, Chen A, Cadilla CL, Force T, Bonventre JV, et al. Gene 33/ Mig-6, a transcriptionally inducible adapter protein that binds GTP-Cdc42 and activates SAPK/JNK. A potential marker transcript for chronic pathologic conditions, such as diabetic nephropathy. Possible role in the response to persistent stress. J Biol Chem. 2000 Jun 9;275(23):17838-47.

[13] Burbelo PD, Drechsel D, Hall A. A conserved binding motif defines numerous candidate target proteins for both Cdc42 and Rac GTPases. J Biol Chem. 1995 Dec 8;270(49):29071-4.

[14] Pante G, Thompson J, Lamballe F, Iwata T, Ferby I, Barr FA, et al. Mitogen-inducible gene 6 is an endogenous inhibitor of HGF/Met-induced cell migration and neurite growth. J Cell Biol. 2005 Oct 24;171(2):337-48.

[15] Tsunoda T, Inokuchi J, Baba I, Okumura K, Naito S, Sasazuki T, et al. A novel mechanism of nuclear factor kappaB activation through the binding between inhibitor of nuclear factor-kappaBalpha and the processed NH(2)-terminal region of Mig-6. Cancer Res. 2002 Oct 15;62(20):5668-71.

[16] Mabuchi R, Sasazuki T, Shirasawa S. Mapping of the critical region of mitogene-inducible gene-6 for NF-kappaB activation. Oncol Rep. 2005 Mar;13(3):473-6.

[17] Fiorentino L, Pertica C, Fiorini M, Talora C, Crescenzi M, Castellani L, et al. Inhibition of ErbB-2 mitogenic and transforming activity by RALT, a mitogen-induced signal transducer which binds to the ErbB-2 kinase domain. Mol Cell Biol. 2000 Oct; 20(20):7735-50.

[18] Fiorini M, Ballaro C, Sala G, Falcone G, Alema S, Segatto O. Expression of RALT, a feedback inhibitor of ErbB receptors, is subjected to an integrated transcriptional and post-translational control. Oncogene. 2002 Sep 19;21(42):6530-9.

[19] Hackel PO, Gishizky M, Ullrich A. Mig-6 is a negative regulator of the epidermal growth factor receptor signal. Biol Chem. 2001 Dec;382(12):1649-62.

[20] Anastasi S, Fiorentino L, Fiorini M, Fraioli R, Sala G, Castellani L, et al. Feedback inhibition by RALT controls signal output by the ErbB network. Oncogene. 2003 Jul 3;22(27):4221-34.

[21] Shen F, Lin Q, Gu Y, Childress C, Yang W. Activated Cdc42-associated kinase 1 is a component of EGF receptor signaling complex and regulates EGF receptor degradation. Mol Biol Cell. 2007 Mar;18(3):732-42.

[22] van Laar T, Schouten T, van der Eb AJ, Terleth C. Induction of the SAPK activator MIG-6 by the alkylating agent methyl methanesulfonate. Mol Carcinog. 2001 Jun; 31(2):63-7.

[23] Messina JL, Hamlin J, Larner J. Effects of insulin alone on the accumulation of a specific mRNA in rat hepatoma cells. J Biol Chem. 1985 Dec 25;260(30):16418-23.

[24] Descot A, Hoffmann R, Shaposhnikov D, Reschke M, Ullrich A, Posern G. Negative regulation of the EGFR-MAPK cascade by actin-MAL-mediated Mig6/Errfi-1 induction. Mol Cell. 2009 Aug 14;35(3):291-304.

[25] Xu D, Patten RD, Force T, Kyriakis JM. Gene 33/RALT is induced by hypoxia in cardiomyocytes, where it promotes cell death by suppressing phosphatidylinositol 3-

kinase and extracellular signal-regulated kinase survival signaling. Mol Cell Biol. 2006 Jul;26(13):5043-54.

[26] Mohn KL, Laz TM, Melby AE, Taub R. Immediate-early gene expression differs between regenerating liver, insulin-stimulated H-35 cells, and mitogen-stimulated Balb/c 3T3 cells. Liver-specific induction patterns of gene 33, phosphoenolpyruvate carboxykinase, and the jun, fos, and egr families. J Biol Chem. 1990 Dec 15;265(35): 21914-21.

[27] Haber BA, Mohn KL, Diamond RH, Taub R. Induction patterns of 70 genes during nine days after hepatectomy define the temporal course of liver regeneration. J Clin Invest. 1993 Apr;91(4):1319-26.

[28] Reschke M, Ferby I, Stepniak E, Seitzer N, Horst D, Wagner EF, et al. Mitogen-inducible gene-6 is a negative regulator of epidermal growth factor receptor signaling in hepatocytes and human hepatocellular carcinoma. Hepatology. Apr;51(4):1383-90.

[29] Mateescu RG, Todhunter RJ, Lust G, Burton-Wurster N. Increased MIG-6 mRNA transcripts in osteoarthritic cartilage. Biochem Biophys Res Commun. 2005 Jul 1;332(2):482-6.

[30] Ferby I, Reschke M, Kudlacek O, Knyazev P, Pante G, Amann K, et al. Mig6 is a negative regulator of EGF receptor-mediated skin morphogenesis and tumor formation. Nat Med. 2006 May;12(5):568-73.

[31] Ballaro C, Ceccarelli S, Tiveron C, Tatangelo L, Salvatore AM, Segatto O, et al. Targeted expression of RALT in mouse skin inhibits epidermal growth factor receptor signalling and generates a Waved-like phenotype. EMBO Rep. 2005 Aug;6(8):755-61.

[32] Jin N, Cho SN, Raso MG, Wistuba I, Smith Y, Yang Y, et al. Mig-6 is required for appropriate lung development and to ensure normal adult lung homeostasis. Development. 2009 Oct;136(19):3347-56.

[33] Zhang YW, Su Y, Lanning N, Swiatek PJ, Bronson RT, Sigler R, et al. Targeted disruption of Mig-6 in the mouse genome leads to early onset degenerative joint disease. Proc Natl Acad Sci U S A. 2005 Aug 16;102(33):11740-5.

[34] Jeong JW, Lee HS, Lee KY, White LD, Broaddus RR, Zhang YW, et al. Mig-6 modulates uterine steroid hormone responsiveness and exhibits altered expression in endometrial disease. Proc Natl Acad Sci U S A. 2009 May 26;106(21):8677-82.

[35] Zhang YW, Staal B, Su Y, Swiatek P, Zhao P, Cao B, et al. Evidence that MIG-6 is a tumor-suppressor gene. Oncogene. 2007 Jan 11;26(2):269-76.

[36] Guy GR, Wong ES, Yusoff P, Chandramouli S, Lo TL, Lim J, et al. Sprouty: how does the branch manager work? J Cell Sci. 2003 Aug 1;116(Pt 15):3061-8.

[37] Christofori G. Split personalities: the agonistic antagonist Sprouty. Nat Cell Biol. 2003 May;5(5):377-9.

[38] Wakioka T, Sasaki A, Kato R, Shouda T, Matsumoto A, Miyoshi K, et al. Spred is a Sprouty-related suppressor of Ras signalling. Nature. 2001 Aug 9;412(6847):647-51.

[39] Guy GR, Jackson RA, Yusoff P, Chow SY. Sprouty proteins: modified modulators, matchmakers or missing links? J Endocrinol. 2009 Nov;203(2):191-202.

[40] Lao DH, Chandramouli S, Yusoff P, Fong CW, Saw TY, Tai LP, et al. A Src homology 3-binding sequence on the C terminus of Sprouty2 is necessary for inhibition of the Ras/ERK pathway downstream of fibroblast growth factor receptor stimulation. J Biol Chem. 2006 Oct 6;281(40):29993-30000.

[41] Lao DH, Yusoff P, Chandramouli S, Philp RJ, Fong CW, Jackson RA, et al. Direct binding of PP2A to Sprouty2 and phosphorylation changes are a prerequisite for ERK inhibition downstream of fibroblast growth factor receptor stimulation. J Biol Chem. 2007 Mar 23;282(12):9117-26.

[42] Aranda S, Alvarez M, Turro S, Laguna A, de la Luna S. Sprouty2-mediated inhibition of fibroblast growth factor signaling is modulated by the protein kinase DYRK1A. Mol Cell Biol. 2008 Oct;28(19):5899-911.

[43] Hall AB, Jura N, DaSilva J, Jang YJ, Gong D, Bar-Sagi D. hSpry2 is targeted to the ubiquitin-dependent proteasome pathway by c-Cbl. Curr Biol. 2003 Feb 18;13(4): 308-14.

[44] Nadeau RJ, Toher JL, Yang X, Kovalenko D, Friesel R. Regulation of Sprouty2 stability by mammalian Seven-in-Absentia homolog 2. J Cell Biochem. 2007 Jan 1;100(1): 151-60.

[45] DaSilva J, Xu L, Kim HJ, Miller WT, Bar-Sagi D. Regulation of sprouty stability by Mnk1-dependent phosphorylation. Mol Cell Biol. 2006 Mar;26(5):1898-907.

[46] Hacohen N, Kramer S, Sutherland D, Hiromi Y, Krasnow MA. sprouty encodes a novel antagonist of FGF signaling that patterns apical branching of the Drosophila airways. Cell. 1998 Jan 23;92(2):253-63.

[47] Nutt SL, Dingwell KS, Holt CE, Amaya E. Xenopus Sprouty2 inhibits FGF-mediated gastrulation movements but does not affect mesoderm induction and patterning. Genes Dev. 2001 May 1;15(9):1152-66.

[48] Tefft JD, Lee M, Smith S, Leinwand M, Zhao J, Bringas P, Jr., et al. Conserved function of mSpry-2, a murine homolog of Drosophila sprouty, which negatively modulates respiratory organogenesis. Curr Biol. 1999 Feb 25;9(4):219-22.

[49] Chambers D, Mason I. Expression of sprouty2 during early development of the chick embryo is coincident with known sites of FGF signalling. Mech Dev. 2000 Mar 1;91(1-2):361-4.

[50] Mailleux AA, Tefft D, Ndiaye D, Itoh N, Thiery JP, Warburton D, et al. Evidence that SPROUTY2 functions as an inhibitor of mouse embryonic lung growth and morphogenesis. Mech Dev. 2001 Apr;102(1-2):81-94.

[51] Chi L, Zhang S, Lin Y, Prunskaite-Hyyrylainen R, Vuolteenaho R, Itaranta P, et al. Sprouty proteins regulate ureteric branching by coordinating reciprocal epithelial Wnt11, mesenchymal Gdnf and stromal Fgf7 signalling during kidney development. Development. 2004 Jul;131(14):3345-56.

[52] Shim K, Minowada G, Coling DE, Martin GR. Sprouty2, a mouse deafness gene, regulates cell fate decisions in the auditory sensory epithelium by antagonizing FGF signaling. Dev Cell. 2005 Apr;8(4):553-64.

[53] Taketomi T, Yoshiga D, Taniguchi K, Kobayashi T, Nonami A, Kato R, et al. Loss of mammalian Sprouty2 leads to enteric neuronal hyperplasia and esophageal achalasia. Nat Neurosci. 2005 Jul;8(7):855-7.

[54] Zhang S, Lin Y, Itaranta P, Yagi A, Vainio S. Expression of Sprouty genes 1, 2 and 4 during mouse organogenesis. Mech Dev. 2001 Dec;109(2):367-70.

[55] Lo TL, Yusoff P, Fong CW, Guo K, McCaw BJ, Phillips WA, et al. The ras/mitogen-activated protein kinase pathway inhibitor and likely tumor suppressor proteins, sprouty 1 and sprouty 2 are deregulated in breast cancer. Cancer Res. 2004 Sep 1;64(17):6127-36.

[56] Casci T, Vinos J, Freeman M. Sprouty, an intracellular inhibitor of Ras signaling. Cell. 1999 Mar 5;96(5):655-65.

[57] Kramer S, Okabe M, Hacohen N, Krasnow MA, Hiromi Y. Sprouty: a common antagonist of FGF and EGF signaling pathways in Drosophila. Development. 1999 Jun; 126(11):2515-25.

[58] Anastasi S, Baietti MF, Frosi Y, Alema S, Segatto O. The evolutionarily conserved EBR module of RALT/MIG6 mediates suppression of the EGFR catalytic activity. Oncogene. 2007 Dec 13;26(57):7833-46.

[59] Zhang X, Pickin KA, Bose R, Jura N, Cole PA, Kuriyan J. Inhibition of the EGF receptor by binding of MIG6 to an activating kinase domain interface. Nature. 2007 Nov 29;450(7170):741-4.

[60] Zhang X, Gureasko J, Shen K, Cole PA, Kuriyan J. An allosteric mechanism for activation of the kinase domain of epidermal growth factor receptor. Cell. 2006 Jun 16;125(6):1137-49.

[61] Frosi Y, Anastasi S, Ballaro C, Varsano G, Castellani L, Maspero E, et al. A two-tiered mechanism of EGFR inhibition by RALT/MIG6 via kinase suppression and receptor degradation. J Cell Biol. 2010 May 3;189(3):557-71.

[62] Avraham R, Yarden Y. Feedback regulation of EGFR signalling: decision making by early and delayed loops. Nat Rev Mol Cell Biol. 2011 Feb;12(2):104-17.

[63] Madshus IH, Stang E. Internalization and intracellular sorting of the EGF receptor: a model for understanding the mechanisms of receptor trafficking. J Cell Sci. 2009 Oct 1;122(Pt 19):3433-9.

[64] Teo M, Tan L, Lim L, Manser E. The tyrosine kinase ACK1 associates with clathrin-coated vesicles through a binding motif shared by arrestin and other adaptors. J Biol Chem. 2001 May 25;276(21):18392-8.

[65] Ying H, Zheng H, Scott K, Wiedemeyer R, Yan H, Lim C, et al. Mig-6 controls EGFR trafficking and suppresses gliomagenesis. Proc Natl Acad Sci U S A. 2010 Apr 13;107(15):6912-7.

[66] Tari AM, Lopez-Berestein G. GRB2: a pivotal protein in signal transduction. Semin Oncol. 2001 Oct;28(5 Suppl 16):142-7.

[67] Liu N, Matsumoto M, Kitagawa K, Kotake Y, Suzuki S, Shirasawa S, et al. Chk1 phosphorylates the tumour suppressor Mig-6, regulating the activation of EGF signalling. EMBO J. 2012 May 16;31(10):2365-77.

[68] van Hemert MJ, Steensma HY, van Heusden GP. 14-3-3 proteins: key regulators of cell division, signalling and apoptosis. Bioessays. 2001 Oct;23(10):936-46.

[69] Tong J, Taylor P, Jovceva E, St-Germain JR, Jin LL, Nikolic A, et al. Tandem immunoprecipitation of phosphotyrosine-mass spectrometry (TIPY-MS) indicates C19ORF19 becomes tyrosine-phosphorylated and associated with activated epidermal growth factor receptor. J Proteome Res. 2008 Mar;7(3):1067-77.

[70] Guha U, Chaerkady R, Marimuthu A, Patterson AS, Kashyap MK, Harsha HC, et al. Comparisons of tyrosine phosphorylated proteins in cells expressing lung cancer-specific alleles of EGFR and KRAS. Proc Natl Acad Sci U S A. 2008 Sep 16;105(37): 14112-7.

[71] Rikova K, Guo A, Zeng Q, Possemato A, Yu J, Haack H, et al. Global survey of phosphotyrosine signaling identifies oncogenic kinases in lung cancer. Cell. 2007 Dec 14;131(6):1190-203.

[72] Lee CC, Putnam AJ, Miranti CK, Gustafson M, Wang LM, Vande Woude GF, et al. Overexpression of sprouty 2 inhibits HGF/SF-mediated cell growth, invasion, migration, and cytokinesis. Oncogene. 2004 Jul 1;23(30):5193-202.

[73] Lim J, Wong ES, Ong SH, Yusoff P, Low BC, Guy GR. Sprouty proteins are targeted to membrane ruffles upon growth factor receptor tyrosine kinase activation. Identification of a novel translocation domain. J Biol Chem. 2000 Oct 20;275(42):32837-45.

[74] Impagnatiello MA, Weitzer S, Gannon G, Compagni A, Cotten M, Christofori G. Mammalian sprouty-1 and -2 are membrane-anchored phosphoprotein inhibitors of growth factor signaling in endothelial cells. J Cell Biol. 2001 Mar 5;152(5):1087-98.

[75] Tefft D, Lee M, Smith S, Crowe DL, Bellusci S, Warburton D. mSprouty2 inhibits FGF10-activated MAP kinase by differentially binding to upstream target proteins. Am J Physiol Lung Cell Mol Physiol. 2002 Oct;283(4):L700-6.

[76] Lim J, Yusoff P, Wong ES, Chandramouli S, Lao DH, Fong CW, et al. The cysteine-rich sprouty translocation domain targets mitogen-activated protein kinase inhibito-

ry proteins to phosphatidylinositol 4,5-bisphosphate in plasma membranes. Mol Cell Biol. 2002 Nov;22(22):7953-66.

[77] Li X, Brunton VG, Burgar HR, Wheldon LM, Heath JK. FRS2-dependent SRC activation is required for fibroblast growth factor receptor-induced phosphorylation of Sprouty and suppression of ERK activity. J Cell Sci. 2004 Dec 1;117(Pt 25):6007-17.

[78] Hanafusa H, Torii S, Yasunaga T, Nishida E. Sprouty1 and Sprouty2 provide a control mechanism for the Ras/MAPK signalling pathway. Nat Cell Biol. 2002 Nov;4(11): 850-8.

[79] Sasaki A, Taketomi T, Kato R, Saeki K, Nonami A, Sasaki M, et al. Mammalian Sprouty4 suppresses Ras-independent ERK activation by binding to Raf1. Nat Cell Biol. 2003 May;5(5):427-32.

[80] Yusoff P, Lao DH, Ong SH, Wong ES, Lim J, Lo TL, et al. Sprouty2 inhibits the Ras/MAP kinase pathway by inhibiting the activation of Raf. J Biol Chem. 2002 Feb 1;277(5):3195-201.

[81] Chandramouli S, Yu CY, Yusoff P, Lao DH, Leong HF, Mizuno K, et al. Tesk1 interacts with Spry2 to abrogate its inhibition of ERK phosphorylation downstream of receptor tyrosine kinase signaling. J Biol Chem. 2008 Jan 18;283(3):1679-91.

[82] Wong ES, Fong CW, Lim J, Yusoff P, Low BC, Langdon WY, et al. Sprouty2 attenuates epidermal growth factor receptor ubiquitylation and endocytosis, and consequently enhances Ras/ERK signalling. EMBO J. 2002 Sep 16;21(18):4796-808.

[83] Rubin C, Litvak V, Medvedovsky H, Zwang Y, Lev S, Yarden Y. Sprouty fine-tunes EGF signaling through interlinked positive and negative feedback loops. Curr Biol. 2003 Feb 18;13(4):297-307.

[84] Fong CW, Leong HF, Wong ES, Lim J, Yusoff P, Guy GR. Tyrosine phosphorylation of Sprouty2 enhances its interaction with c-Cbl and is crucial for its function. J Biol Chem. 2003 Aug 29;278(35):33456-64.

[85] Tsygankov AY, Teckchandani AM, Feshchenko EA, Swaminathan G. Beyond the RING: CBL proteins as multivalent adapters. Oncogene. 2001 Oct 1;20(44):6382-402.

[86] Lee SA, Ladu S, Evert M, Dombrowski F, De Murtas V, Chen X, et al. Synergistic role of Sprouty2 inactivation and c-Met up-regulation in mouse and human hepatocarcinogenesis. Hepatology. 2010 Aug;52(2):506-17.

[87] Girard L, Zochbauer-Muller S, Virmani AK, Gazdar AF, Minna JD. Genome-wide allelotyping of lung cancer identifies new regions of allelic loss, differences between small cell lung cancer and non-small cell lung cancer, and loci clustering. Cancer Res. 2000 Sep 1;60(17):4894-906.

[88] Nomoto S, Haruki N, Tatematsu Y, Konishi H, Mitsudomi T, Takahashi T. Frequent allelic imbalance suggests involvement of a tumor suppressor gene at 1p36 in the

pathogenesis of human lung cancers. Genes Chromosomes Cancer. 2000 Jul;28(3): 342-6.

[89] Fujii T, Dracheva T, Player A, Chacko S, Clifford R, Strausberg RL, et al. A preliminary transcriptome map of non-small cell lung cancer. Cancer Res. 2002 Jun 15;62(12): 3340-6.

[90] Caren H, Fransson S, Ejeskar K, Kogner P, Martinsson T. Genetic and epigenetic changes in the common 1p36 deletion in neuroblastoma tumours. Br J Cancer. 2007 Nov 19;97(10):1416-24.

[91] Duncan CG, Killela PJ, Payne CA, Lampson B, Chen WC, Liu J, et al. Integrated genomic analyses identify ERRFI1 and TACC3 as glioblastoma-targeted genes. Oncotarget. Aug;1(4):265-77.

[92] Poetsch M, Woenckhaus C, Dittberner T, Pambor M, Lorenz G, Herrmann FH. Significance of the small subtelomeric area of chromosome 1 (1p36.3) in the progression of malignant melanoma: FISH deletion screening with YAC DNA probes. Virchows Arch. 1999 Aug;435(2):105-11.

[93] Tseng RC, Chang JW, Hsien FJ, Chang YH, Hsiao CF, Chen JT, et al. Genomewide loss of heterozygosity and its clinical associations in non small cell lung cancer. Int J Cancer. 2005 Nov 1;117(2):241-7.

[94] Zhang YW, Staal B, Dykema KJ, Furge KA, Vande Woude GF. Cancer-Type Regulation of MIG-6 Expression by Inhibitors of Methylation and Histone Deacetylation. PLoS One. 2012;7(6):e38955.

[95] Li Z, Dong Q, Wang Y, Qu L, Qiu X, Wang E. Downregulation of Mig-6 in nonsmall-cell lung cancer is associated with EGFR signaling. Mol Carcinog. Jul 7.

[96] Anastasi S, Sala G, Huiping C, Caprini E, Russo G, Iacovelli S, et al. Loss of RALT/MIG-6 expression in ERBB2-amplified breast carcinomas enhances ErbB-2 oncogenic potency and favors resistance to Herceptin. Oncogene. 2005 Jun 30;24(28):4540-8.

[97] Ruan DT, Warren RS, Moalem J, Chung KW, Griffin AC, Shen W, et al. Mitogen-inducible gene-6 expression correlates with survival and is an independent predictor of recurrence in BRAF(V600E) positive papillary thyroid cancers. Surgery. 2008 Dec; 144(6):908-13; discussion 13-4.

[98] Lin CI, Du J, Shen WT, Whang EE, Donner DB, Griff N, et al. Mitogen-inducible gene-6 is a multifunctional adaptor protein with tumor suppressor-like activity in papillary thyroid cancer. J Clin Endocrinol Metab. Mar;96(3):E554-65.

[99] Lo TL, Fong CW, Yusoff P, McKie AB, Chua MS, Leung HY, et al. Sprouty and cancer: the first terms report. Cancer Lett. 2006 Oct 28;242(2):141-50.

[100] McKie AB, Douglas DA, Olijslagers S, Graham J, Omar MM, Heer R, et al. Epigenetic inactivation of the human sprouty2 (hSPRY2) homologue in prostate cancer. Oncogene. 2005 Mar 24;24(13):2166-74.

[101] Faratian D, Sims AH, Mullen P, Kay C, Um I, Langdon SP, et al. Sprouty 2 is an independent prognostic factor in breast cancer and may be useful in stratifying patients for trastuzumab therapy. PLoS One. 2011;6(8):e23772.

[102] Fong CW, Chua MS, McKie AB, Ling SH, Mason V, Li R, et al. Sprouty 2, an inhibitor of mitogen-activated protein kinase signaling, is down-regulated in hepatocellular carcinoma. Cancer Res. 2006 Feb 15;66(4):2048-58.

[103] Song K, Gao Q, Zhou J, Qiu SJ, Huang XW, Wang XY, et al. Prognostic significance and clinical relevance of Sprouty 2 protein expression in human hepatocellular carcinoma. Hepatobiliary Pancreat Dis Int. 2012 Apr;11(2):177-84.

[104] Sutterluty H, Mayer CE, Setinek U, Attems J, Ovtcharov S, Mikula M, et al. Down-regulation of Sprouty2 in non-small cell lung cancer contributes to tumor malignancy via extracellular signal-regulated kinase pathway-dependent and -independent mechanisms. Mol Cancer Res. 2007 May;5(5):509-20.

[105] Gao M, Patel R, Ahmad I, Fleming J, Edwards J, McCracken S, et al. SPRY2 loss enhances ErbB trafficking and PI3K/AKT signalling to drive human and mouse prostate carcinogenesis. EMBO Mol Med. 2012 May 31.

[106] Fritzsche S, Kenzelmann M, Hoffmann MJ, Muller M, Engers R, Grone HJ, et al. Concomitant down-regulation of SPRY1 and SPRY2 in prostate carcinoma. Endocr Relat Cancer. 2006 Sep;13(3):839-49.

[107] Velasco A, Pallares J, Santacana M, Gatius S, Fernandez M, Domingo M, et al. Promoter hypermethylation and expression of sprouty 2 in endometrial carcinoma. Hum Pathol. 2011 Feb;42(2):185-93.

[108] Kwak HJ, Kim YJ, Chun KR, Woo YM, Park SJ, Jeong JA, et al. Downregulation of Spry2 by miR-21 triggers malignancy in human gliomas. Oncogene. 2011 May 26;30(21):2433-42.

[109] Sanchez A, Setien F, Martinez N, Oliva JL, Herranz M, Fraga MF, et al. Epigenetic inactivation of the ERK inhibitor Spry2 in B-cell diffuse lymphomas. Oncogene. 2008 Aug 21;27(36):4969-72.

[110] Frank MJ, Dawson DW, Bensinger SJ, Hong JS, Knosp WM, Xu L, et al. Expression of sprouty2 inhibits B-cell proliferation and is epigenetically silenced in mouse and human B-cell lymphomas. Blood. 2009 Mar 12;113(11):2478-87.

[111] Feng YH, Wu CL, Tsao CJ, Chang JG, Lu PJ, Yeh KT, et al. Deregulated expression of sprouty2 and microRNA-21 in human colon cancer: Correlation with the clinical stage of the disease. Cancer Biol Ther. 2011 Jan 1;11(1):111-21.

[112] Holgren C, Dougherty U, Edwin F, Cerasi D, Taylor I, Fichera A, et al. Sprouty-2 controls c-Met expression and metastatic potential of colon cancer cells: sprouty/c-Met upregulation in human colonic adenocarcinomas. Oncogene. 2010 Sep 23;29(38): 5241-53.

[113] Hanahan D, Weinberg RA. Hallmarks of cancer: the next generation. Cell. 2011 Mar 4;144(5):646-74.

[114] Xu D, Makkinje A, Kyriakis JM. Gene 33 is an endogenous inhibitor of epidermal growth factor (EGF) receptor signaling and mediates dexamethasone-induced suppression of EGF function. J Biol Chem. 2005 Jan 28;280(4):2924-33.

[115] Xu J, Keeton AB, Wu L, Franklin JL, Cao X, Messina JL. Gene 33 inhibits apoptosis of breast cancer cells and increases poly(ADP-ribose) polymerase expression. Breast Cancer Res Treat. 2005 Jun;91(3):207-15.

[116] Nagashima T, Ushikoshi-Nakayama R, Suenaga A, Ide K, Yumoto N, Naruo Y, et al. Mutation of epidermal growth factor receptor is associated with MIG6 expression. FEBS J. 2009 Sep;276(18):5239-51.

[117] Naruo Y, Nagashima T, Ushikoshi-Nakayama R, Saeki Y, Nakakuki T, Naka T, et al. Epidermal growth factor receptor mutation in combination with expression of MIG6 alters gefitinib sensitivity. BMC Syst Biol.5:29.

[118] Tsavachidou D, Coleman ML, Athanasiadis G, Li S, Licht JD, Olson MF, et al. SPRY2 is an inhibitor of the ras/extracellular signal-regulated kinase pathway in melanocytes and melanoma cells with wild-type BRAF but not with the V599E mutant. Cancer Res. 2004 Aug 15;64(16):5556-9.

[119] Miyoshi K, Wakioka T, Nishinakamura H, Kamio M, Yang L, Inoue M, et al. The Sprouty-related protein, Spred, inhibits cell motility, metastasis, and Rho-mediated actin reorganization. Oncogene. 2004 Jul 22;23(33):5567-76.

[120] Feng YH, Tsao CJ, Wu CL, Chang JG, Lu PJ, Yeh KT, et al. Sprouty2 protein enhances the response to gefitinib through epidermal growth factor receptor in colon cancer cells. Cancer Sci. 2010 Sep;101(9):2033-8.